AF324867

Stem Cell Engineering

Science Policy Reports

The series Science Policy Reports presents the endorsed results of important studies in basic and applied areas of science and technology. They include, to give just a few examples: panel reports exploring the practical and economic feasibility of a new technology; R&D studies of development opportunities for particular materials, devices or other inventions; reports by responsible bodies on technology standardization in developing branches of industry.

Sponsored typically by large organizations – government agencies, watchdogs, funding bodies, standards institutes, international consortia – the studies selected for Science Policy Reports will disseminate carefully compiled information, detailed data and in-depth analysis to a wide audience. They will bring out implications of scientific discoveries and technologies in societal, cultural, environmental, political and/or commercial contexts and will enable interested parties to take advantage of new opportunities and exploit on-going development processes to the full.

For further volumes:
http://www.springer.com/series/8882

Robert M. Nerem • Jeanne Loring
Todd C. McDevitt • Sean P. Palecek
David V. Schaffer • Peter W. Zandstra
Editors

Stem Cell Engineering

A WTEC Global Assessment

 Springer

Editors
Robert M. Nerem
Parker H. Petit Institute
  for Bioengineering and Bioscience
Georgia Institute of Technology
Atlanta, GA, USA

Todd C. McDevitt
Biomedical Engineering
Stem Cell Engineering
  Center Georgia Institute of Technology
Atlanta, GA, USA

David V. Schaffer
Berkeley Chemical and Biomolecular
  Engineering
University of California
Berkeley, CA, USA

Jeanne Loring
Chemical Physiology
The Centre for Regenerative Medicine
Scripps Research Institute
San Diego, CA, USA

Sean P. Palecek
Chemical and Biological Engineering
University of Wisconsin-Madison
Madison, WI, USA

Peter W. Zandstra
Terrence Donnelly Centre for Cellular
  and Biomolecular Research
University of Toronto
Toronto, ON, Canada

ISSN 2213-1965   ISSN 2213-1973 (electronic)
ISBN 978-3-319-05073-7   ISBN 978-3-319-05074-4 (eBook)
DOI 10.1007/978-3-319-05074-4
Springer Cham Heidelberg New York Dordrecht London

Library of Congress Control Number: 2014939959

Printed on acid-free paper

Springer is part of Springer Science+Business Media (www.springer.com)

WTEC Panel on Global Assessment of Stem Cell Engineering

Sponsored by the U.S. National Science Foundation (NSF), National Cancer Institute (NCI) of the National Institutes of Health (NIH), and National Institute of Standards and Technology (NIST).

Dr. Robert M. Nerem (Chair)
Parker H. Petit Institute for
    Bioengineering and Bioscience
Georgia Institute of Technology
315 Ferst Drive
Atlanta GA 30332-0363, USA

Dr. Jeanne Loring
Center for Regenerative Medicine
Scripps Research Institute,
    Department of Chemical Physiology
4122 Sorrento Valley Blvd., Ste. 107,
    SP30-3021
San Diego, CA 92121, USA

Dr. Todd C. McDevitt
Stem Cell Engineering Center
Parker H. Petit Institute for
    Bioengineering and Bioscience
Georgia Institute of Technology
315 Ferst Dr. NW Atlanta,
    GA 30332 Emory University/Georgia
    Tech, USA

Dr. Sean P. Palecek
Department of Chemical and Biological
    Engineering
University of Wisconsin-Madison
3637 Engineering Hall, 1415 Engineering Drive
Madison, WI 53706-1691, USA

Dr. David V. Schaffer
Department of Chemical Engineering
University of California-Berkeley
274 Stanley Hall, Mail Code 3220
Berkeley, CA 94720, USA

Dr. Peter W. Zandstra
Terrence Donnelly Centre for Cellular and
    Biomolecular Research (CCBR), Donnelly
    Building
160 College Street, Office #1116
University of Toronto
Toronto, Ontario M5S 3E1
Canada

Sponsor Representatives with WTEC Panel

Semahat S. Demir[1]
Program Director, Directorate for Engineering
Office of Emerging Frontiers in Research
  and Innovation
National Science Foundation
Arlington, VA 22230, USA

Kaiming Ye
Associate Program Director,
  Directorate of Engineering
Division of Chemical, Bioengineering, Environmental,
  and Transport Systems (ENG/CBET)
National Science Foundation
4201 Wilson Blvd., Rm. 565 S
Arlington, VA 22230, USA

Larry A. Nagahara
Director
Office of Physical Sciences-Oncology
National Cancer Institute, NIH
31 Center Drive, MSC 2580
Bethesda, MD 20892-2580, USA

Nicole Moore
Project Manager
Office of Physical Sciences-Oncology
National Cancer Institute, NIH
31 Center Drive, MSC 2580
Bethesda, MD 20892-2580, USA

Nastaran Zahir Kuhn
Project Manager
Office of Physical Sciences-Oncology
National Cancer Institute, NIH
31 Center Drive, MSC 2580
Bethesda, MD 20892-2580, USA

WTEC Participants in Site Visits

Hassan Ali
Project Manager
Hemant Sarin
Senior Policy Fellow

Frank Huband
Senior Vice President and General Counsel

---

[1] As of September 2012, President of Istanbul Kültür University, Istanbul, Turkey.

WTEC Panel Report on Global Assessment of Stem Cell Engineering

This document was sponsored by the National Science Foundation (NSF) and other agencies of the U.S. Government under an NSF cooperative agreement (ENG 0844639) with the World Technology Evaluation Center (WTEC). The Government has certain rights in this material. Any opinions, findings, and conclusions or recommendations expressed in this material are those of the authors and do not necessarily reflect the views of the United States Government, the authors' parent institutions, or WTEC.

December 2012

Robert M. Nerem (Chair)
Jeanne Loring
Todd C. McDevitt
Sean P. Palecek
David V. Schaffer
Peter W. Zandstra

World Technology Evaluation Center, Inc. (WTEC)

R.D. Shelton, President
Michael DeHaemer, Executive Vice President
Geoffrey M. Holdridge, Vice President for Government Services
Frank Huband, Senior Vice President and General Counsel
Patricia Foland, Vice President for Operations
Hassan Ali, Project Manager
Haydon Rochester, Jr., Report Editor

WTEC Mission

WTEC provides assessments of international research and development in selected technologies under awards from the National Science Foundation (NSF), the Office of Naval Research (ONR), and the National Institute of Standards and Technology (NIST). Formerly part of Loyola University Maryland, WTEC is now a separate nonprofit research institute. The Deputy Assistant Director for Engineering is NSF Program Director for WTEC. Sponsors interested in international technology assessments or related studies can provide support through NSF or directly through separate grants or GSA task orders to WTEC.

WTEC's mission is to inform U.S. scientists, engineers, and policymakers of global trends in science and technology. WTEC assessments cover basic research, advanced development, and applications. Panels of typically six technical experts conduct WTEC assessments. Panelists are leading authorities in their field, technically active, and knowledgeable about U.S. and foreign research programs. As part of the assessment process, panels visit and carry out extensive discussions with foreign scientists and engineers in their labs.

The WTEC staff helps select topics, recruits expert panelists, arranges study visits to foreign laboratories, organizes workshop presentations, and finally, edits and publishes the final reports. Dr. R.D. Shelton, President, is the WTEC point of contact: telephone 717 299 7130 or email Shelton@ScienceUS.org.

Executive Summary

In the last 15 years, our knowledge of stem cells has increased, seemingly at an exponential rate. The result is that there is an ever-increasing arsenal of stem cells. This arsenal includes embryonic stem cells, various types of adult stem cells, and what are called induced pluripotent stem cells; i.e., iPS cells, that are reprogrammed from fully differentiated cells such as a skin fibroblast. A few years ago, iPS cells were heralded as the "breakthrough of the year," and this year the key scientists whose work resulted in this technology shared the award for the Nobel Prize in physiology and medicine. There still are many questions to answer in regard to these cells; however, it is clear that they provide a unique tool for the stem cell field. Engineers have become increasingly involved in the area of stem cells, all the way from basic research to the variety of applications that are evolving. Concurrently there have been two other "streams of thinking" that have emerged.

One of these is that of interdisciplinary research. Here the 2009 National Academies report entitled "A New Biology for the 21st Century" makes the point that to achieve a deeper understanding of biology necessary to address the major problems that society is facing will require not just breaking down the "silos" within biology itself, but incorporating chemists, computational researchers, engineers, mathematicians, and physicists into basic biological research. Furthermore, only through such an integration of disciplines will it be possible to address major society problems. There is no area of biology where this might be more true than that of stem cells.

A second "stream of thinking" that has emerged is that of translational research. Not only are Federal agencies in the United States interested in fostering the translation of bench-top science into a variety of commercial/clinical applications, but this also has become a priority for many states. Furthermore, what is happening in this area in the United States simply "mirrors" what is taking place in the rest of the world. It is thus timely that a global assessment of stem cell engineering be conducted, and it is such an assessment that is reported here.

Study Objectives

What is stem cell engineering? As defined for the purposes of this study, it is not just tissue engineering and regenerative medicine, but the entire interface of engineering with the "world" of stem cells. It thus ranges from basic stem cell research to models and tools, to enabling and scalable technologies, to stem cell biomanufacturing and the development of stem cell-based applications and products. The objective of this global assessment of stem cell engineering has been to compare U.S. R&D with activities globally so as to identify gaps where engineers can make contributions, the major innovations emerging globally, barriers in the field, and opportunities for cooperation and collaboration. A major purpose is to provide information that can guide investments by funding agencies in the future. In this it again must be emphasized that stem cell engineering as addressed in this report ranges all the way from basic stem cell research to stem cell-based applications and products.

Panel Members

The panel was made up of the following individuals: Jeanne Loring, Ph.D., Scripps Research Institute; Todd C. McDevitt, Ph.D., Georgia Institute of Technology and Emory University; Robert M. Nerem, Ph.D., Georgia Institute of Technology (chair); Sean P. Palecek, Ph.D., University of Wisconsin, Madison; David V. Schaffer, Ph.D., University of California, Berkeley; and Peter W. Zandstra, Ph.D., University of Toronto.

Study Scope

The basic purpose of this global assessment of stem cell engineering has already been noted; however, to expand on what was earlier stated, the study will include not only relevant scientific/technical topics, but the role of engineering in the translation of our knowledge of stem cells to applications and, as much as possible, education and training in this area and government policies. The scientific and technical topics of interest included the following:

- Application of engineering and physical science principles in stem cell R&D
- Scalable expansion and differentiation of stem cells
- High-throughput screening and the application of microfluidics
- Real-time, non-destructive phenotyping
- Systems-based quantitative analysis
- Computational modeling approaches
- Bioprocessing and biomanufacturing
- Targeted delivery of stem cells

The above list is not meant to be totally inclusive, but to indicate the breadth of topics of interest in this study. All of the above as well as many others represent areas where engineers and the engineering approach either are or could be making contributions.

Study Process

There were many components to the process used in this global assessment of stem cell engineering. A foundation of course is provided by the knowledge of each of the panelists, knowledge not only about activities in the United States, but knowledge each of them had about activities in other parts of the world. To this foundation was added the following four components:

- Site visits in Asia and Europe
- Workshops in Atlanta and in Seoul, Korea
- Participation in the 3rd International Conference on Stem Cell Engineering
- Virtual site visits

For virtual site visits, information was gathered solely though the internet and/or by e-mail exchange. Examples of this are Australia and Iran. The term virtual site visit was also used for a site visit where only one panel member visited.

Although the WTEC panel was able to see much of the stem cell activities going on around the world, they certainly did not see everything. Even so, one can make the argument that the process outlined above in terms of the various components as well as the knowledge base each panelist had coming into this study provided for a global assessment of this field of stem cell engineering.

Principal Findings

From this global assessment of stem cell engineering as conducted by the WTEC panel, it is clear that engineers and the engineering approach with its quantitative, systems-based thinking can contribute much more to basic stem cell research than it has to date. As stated in the National Academies report on "A New Biology for the 21st Century," to achieve the deeper understanding of biology required in this century, there will need to be an integration of many disciplines into biological research, and this certainly includes engineering. Engineering analysis can be used to identify the components of highly complex stem cell systems and provide an understanding of how these components work together. Furthermore, computational models will be increasingly important in our efforts to achieve a better understanding of complex biological systems. In all of the above engineers are in a position to take a leadership role.

Engineers also can take the lead in developing new, innovative enabling technologies. This includes high-throughput screening techniques, improved culture and differentiation systems, and *in vitro* models engineered to be more physiologic. The last of these include organ-on-a-chip models and engineered *in vitro* tumor models that can lead to a better understanding of cancer.

Finally, for stem cell biomanufacturing there is a need for further advances in culture systems, techniques for real-time monitoring, and for process automation. Underpinning these specific application areas, computational modeling has an important role to play throughout the spectrum from discovery to translation.

One of the interesting things that came out of the site visits and this global assessment was observing some of interesting models that have been developed to translate bench-top stem cell science into clinical therapies and into commercial products. Four such models discussed in the report are as follows: the Berlin-Brandenburg Center for Regenerative Therapies, the Cell Therapy Catapult in the United Kingdom, the Centre for Commercialization of Regenerative Medicine in Canada, and the Tokyo Women's Medical University. There are of course other models for translation; however, it is these four that are discussed in the report.

It is clear to this WTEC panel that, for engineers to be accepted by biologists, they need to be viewed as understanding biological mechanisms and making a contribution to biology. Thus, for training programs to be successful, they need to include what might be called a "high level" of biology, and this is certainly what is done in the leading bioengineering programs in North America. Outside of North America an excellent example of a unique training program is that at Loughborough University. One of the outcomes of the Atlanta Workshop on Stem Cell Engineering was the agreement to establish an international school in the area of cell manufacturing. The initial offering of this school will take place at the end of April 2013 in Portugal.

Research today in general is very interdisciplinary, and this is certainly true of biology and the stem cell field. As part of this, collaborations almost become a necessity. These might be with an investigator at one's own institution, somewhere else in the city, or even at a longer distance. In today's world where research and the development of technology is done within the global community, collaborations can also exist between investigators in different countries. In fact, US investigators need to leverage the excellence of activities in other countries, and it thus was encouraging for the WTEC panel members to see the hosts of the different site visits being so open in the sharing of information and very interested in the possibility of collaborating.

Whatever the United States decides to do, there is urgency to it. This is because of the global competitiveness that exists in the area of stem cells. Not only are China, Japan and Korea making significant investments, but so are European countries such as The Netherlands and the United Kingdom. Furthermore, as important as basic bench-top stem cell research is and recognizing the major contributions that engineers and the engineering approach can make to basic research, ultimately the competition will be in terms of creating the enabling technologies, the new clinical therapies, and the innovative commercial products. It is here, i.e., in moving stem cells to the front of the bioeconomy being created for this twenty-first century, that engineering can play a critical role.

Conclusions

The assessment conducted by the WTEC panel confirms that there is a need for an increasing involvement of engineers in the field of stem cells and related technologies. Although one might argue that the United States today has a leadership role, to capitalize on this and build on the current existing momentum, and most importantly, to accelerate the translation of bench-top research into various applications including clinical therapies and commercial products, will require the United States to take bold steps. The panel thus offers the following conclusions.

1. The United States has a unique opportunity to maintain a leadership position in the stem cell field through the continued support of R&D that will provide a foundation for the generation of new markets and that will lead to economic growth.
2. Because of the contributions that engineers can make in all areas of the stem cell field, as elaborated in the global assessment reported here, this needs to include increased investment in engineering, applied research, and commercialization as it relates to stem cell research and related stem cell-based technologies.
3. A major component in this could be that the Federal agencies that support R&D establish a broad interagency program for stem cell engineering, one that provides grants to interdisciplinary teams that include engineers, computational researchers, and biologists as well as individuals from other disciplines.
4. Another component that would be beneficial is the establishment of new, innovative mechanisms that support academic-industry partnership and unique translational models that facilitate the translation of research into the private sector.
5. To address national workforce needs, the development of training programs at universities and advanced short courses should be encouraged and supported by Federal agencies.
6. Finally, in today's global economy and with the excellent activities taking place in other countries, the United States would benefit from forming strategic partnerships with other countries so as to leverage the existing and emerging strengths in institutions outside of the United States; to implement such partnerships will require binational grant programs with appropriate review mechanisms.

These conclusions align with the National Bioeconomy Blueprint released by the White House Office of Science and Technology Policy. It is up to the Federal agencies to implement a plan based on the conclusions from this assessment study. Without the implementation of the above, however, this unique opportunity could be lost. In this case, it might be possible that the United States in the future is relegated to the second tier of countries in this critical area of stem cell engineering, and ultimately the application of stem cell-based technologies for health and welfare. On the other hand, if implementation takes place in some form, and there is an urgency to do this, then the United States can expect to continue to be in a leadership position and at the forefront in advancing the sciences, developing new, innovative enabling technologies and platforms that lead to clinical therapies, to commercializing the results of stem cell research, and to the generation of new markets and economic growth based on advances in the stem cell field. Some of the results from this will be:

- The acceleration of the development of new drugs while at the same time reducing the costs of this development process
- The development of cell therapies that address diseases and conditions of injury for which today there are no real treatment options available for patients in need
- The growth of the twenty-first century bioeconomy in the United States based on advances in our knowledge of stem cells and the translation of this into applications and products

This has been the dream for at least 20 years; however, with the right strategy by the United States it can be realized and be the reality of tomorrow.

November 2012   Robert M. Nerem

Preface

Research on stem cells has been an important topic in medicine since the 1960s, and over the last decade has become more exciting, and even controversial, as an increasing number of therapies based on them become available. Development of advanced treatments based on stem cells has the potential to cure a wide variety of serious diseases and infirmities that are virtually incurable today, for example reversing paralysis due to spinal cord injuries. Supporters in the Federal government, a number of US states, and many foreign countries believe that increased funding for stem cell research will lead not only to great improvements in public health, but potentially large economic benefits from new biotechnology markets. Thus, the focus today is shifting from supporting purely basic research towards biotechnology and commercialization of stem cell related products and therapies.

By definition, stem cells are biological cells found in all multicellular organisms that can divide and differentiate into diverse specialized cell types, and can self-renew to produce more stem cells. In mammals, there are two broad types of stem cells: adult stem cells, which are found in various tissues and act as a repair system for the body, and embryonic stem cells, which can differentiate into all the different types of specialized cells found in the body. The highly plastic adult stem cells can be harvested from the body and are now routinely used in medical therapies, for example in bone marrow transplantation. Stem cells can also now be artificially grown and transformed into specialized cell types with characteristics consistent with the normal cells of various tissues such as muscles or nerves. Embryonic stem cells generated through therapeutic cloning have also been proposed as promising candidates for transformational new therapies. Most medical researchers anticipate that, in the future, technologies derived from such stem cell research will be available to treat a wide variety of diseases including cancer, Parkinson's disease, spinal cord injuries, amyotrophic lateral sclerosis, multiple sclerosis, and muscle damage, among a number of other serious impairments and conditions. Today, however, there still exists a great deal of social and scientific uncertainty surrounding stem cell research, and further research and public debate will be necessary to clarify the way forward.

The study we summarize here takes an even broader and more unique view of this area. That is, it is an assessment of "stem cell engineering," which we define as all activities from basic stem cell research to models and tools, to enabling and scalable technologies, to stem cell biomanufacturing and the development of stem cell-based applications and products, and therefore areas where engineering research is a key enabler. The objective of this global assessment of stem cell engineering has been to compare such US R&D with similar activities globally so as to identify gaps where engineers can make contributions, the major innovations emerging globally, barriers in the field, and opportunities for cooperation and collaboration. A major purpose is to provide information that can guide investments by funding agencies in the future.

These goals directly inform research being funded by our individual program offices. For example, the NSF Chemical, Bioengineering, Environmental and Transport Systems Division (CBET) generally supports research and education in the rapidly evolving fields of bioengineering and environmental engineering, and in particular expands the knowledge base of bioengineering at scales ranging from proteins and cells to organ systems, including mathematical models, devices, and instrumentation systems. Main themes include tissue engineering and the development of biological substitutes, biosensors, and devices that use a biological component, all of which are enabled by advances in stem cell engineering. Practical applications of stem cell engineering are potentially myriad, for example in research that will lead to the development of new technologies, devices, or software for persons with disabilities, such as the work supported by the CBET General & Age Related Disabilities Engineering (GARDE) program.

Similarly, stem cell engineering is directly relevant to the CBET Biomedical Engineering and Engineering Healthcare (BEEH) Cluster programs, which fund projects that integrate engineering and life science principles in solving biomedical problems, including deriving information from cells, tissues, organs, and organ systems, and new approaches to the design of structures and materials for eventual medical use, with applications towards the characterization, restoration, and/or substitution of normal functions in humans. Stem cell engineering is also a fundamental topic for the mission of the CBET Biomedical Engineering (BME) program, which seeks to develop novel ideas into discovery-level and transformative projects that integrate engineering and life science, including areas such as neural engineering and cellular biomechanics, with multiple goals which include development of technologies for tissue repair and regenerative medicine.

This study was conducted by a panel of experts of truly profound expertise and insight. It included Robert M. Nerem, Ph.D., Georgia Institute of Technology (chair); Jeanne Loring, Ph.D., Scripps Research Institute; Todd C. McDevitt, Ph.D., Georgia Institute of Technology and Emory University; Sean P. Palecek, Ph.D., University of Wisconsin, Madison; David V. Schaffer, Ph.D., University of California, Berkeley; and Peter W. Zandstra, Ph.D., University of Toronto. The breadth of their expertise, and the synergy with which they worked together, has created here a truly important product.

The study methodology itself was very well thought out and comprehensive. It involved physical site visits to over 40 research facilities in Europe and Asia. We would like to thank all of the hosts who welcomed us and shared information during our site visits. The study also included attendance by the panel members at relevant workshops in Atlanta and Seoul, participation in the 3rd International Conference on Stem Cell Engineering, and a number of "virtual" site visits which consisted of information gathering via the internet or by email exchange. In all, a very large body of high quality and topically diverse information was gathered, and subsequently synthesized into this report.

We would also like to acknowledge some of the WTEC staff who worked on the project. These include Duane Shelton, Frank Huband, Mike DeHaemer, Hassan Ali, Hemant Sarin, and Haydon Rochester, each of whom contributed significantly to the overall effort.

The report itself is structured so that each chapter provides a summary of a particular topic, written by one or more of the panel members. In summary:

Chapter "Introduction" is an introduction and overview of the study process and methodology, written by the panel chairman, Robert M. Nerem.

Chapter "Physical and Engineering Principles in Stem Cell Research" was written by David V. Schaffer, and discusses the physical and engineering principals involved in stem cell research.

Chapter "High-Throughput Screening, Microfluidics, Biosensors, and Real-Time Phenotyping" discusses technologies for high-throughput screening, microfluidics, bio-sensors, and real time phenotyping, and was authored by Sean P. Palecek

Chapter "Computational Modeling and Stem Cell Engineering" was contributed by Peter W. Zandstra and Geoff Clarke, and considers methods for computational modeling applied to stem cell engineering

Chapter "Stem Cell Bioprocessing and Biomanufacturing" covers the topic of stem cell bioprocessing and biomanufacturing, and was written by Todd C. McDevitt.

Each chapter of this report is supported by a comprehensive list of references, which in total cover many aspects of stem cell engineering activities assessed in the United States, Europe, and Asia. Other highlights of this study are to be found in the appendices, which present biographical information on the WTEC panel members and authors in Appendix A, the site reports of the panel's visits in Appendix B, and the reports of the virtual site visits in Appendix C, all of which contain a wealth of information. A glossary of abbreviations and acronyms is in Appendix D.

In addition to this report, the expert panel presented highlights of their findings at a public workshop at the National Science Foundation on May 24th, 2012. The meeting was webcast and video archived for convenient viewing by the public, which is available on the WTEC website, www.wtec.org.

Finally, we would like to thank other individuals who contributed resources for this study, including Anne Plant of the Biosystems and Biomaterials Division at NIST; Larry Nagahara, Nicole Moore, and Nastaran Kuhn, all of the Office of Physical Sciences-Oncology at the National Cancer Institute; Kesh Narayanan of the head office of the NSF Engineering Division; and Theresa Good of the CBET Biotechnology, Biochemical and Biomass Engineering Program. Their generous contributions provided the additional resources to make this important study even more substantive.

Arlington, VA, USA   Ted Conway, Ph.D.
Arlington, VA, USA   Kaiming Ye, Ph.D.
Arlington, VA, USA   Semahat S. Demir, Ph.D.[2]
November 2012

---

[2]Now President of İstanbul Kültür University, Istanbul, Turkey

Acknowledgments

We at WTEC wish to thank all the panelists for their valuable insights and their dedicated work in conducting this global assessment of stem cell engineering, and to thank all the site visit hosts for so generously sharing their time, expertise, and facilities with us. For their sponsorship of this important study, our sincere thanks goes to the National Science Foundation, National Cancer Institute, and National Institute of Standards and Technology.

R.D. Shelton
WTEC

Contents

List of Figures

List of Tables

Introduction

Robert M. Nerem

This report provides an assessment of the state of stem cell engineering (SCE) globally. This is based on a yearlong study that was conducted by six panel members and managed by the World Technology Evaluation Center (WTEC). This opening chapter provides background for this study, outlines the scope of the study, identifies the six panel members, describes the study process, provides an overview of the principal findings, and finally states conclusions that hopefully provide the basis for stem cell engineering moving forward and for the acceleration of the progress being made in the broad field of stem cells.

Background

In the last 15 years, our knowledge of stem cells has increased, seemingly at an exponential rate. The result is that there is an ever-increasing arsenal of stem cells. This arsenal includes embryonic stem cells, various types of adult stem cells, and what are called induced pluripotent stem cells, i.e., iPS cells, that are reprogrammed from fully differentiated cells such as a skin fibroblast. A few years ago iPS cells were heralded as "a significant breakthrough" and this year the key scientists whose work resulted in this technology shared the award for the Nobel Prize in physiology and medicine. There still are many questions to answer in regard to these cells; however, it is clear that they provide a unique tool for the stem cell field. In addition, engineers have become increasingly involved in the area of stem cells, all the way from basic research to the variety of applications that are evolving.

R.M. Nerem (✉)
Parker H. Petit Institute for Bioengineering and Bioscience, Georgia Institute of Technology, 315 Ferst Drive, 30332-0363 Atlanta, GA, USA
e-mail: robert.nerem@ibb.gatech.edu

R.M. Nerem et al. (eds.), *Stem Cell Engineering: A WTEC Global Assessment*,
Science Policy Reports, DOI 10.1007/978-3-319-05074-4_1,
© Springer International Publishing Switzerland 2014

Concurrently there have been two other "streams of thinking" that have emerged. One of these is that of interdisciplinary research. The 2009 National Academies report entitled "A New Biology for the 21st Century" makes the point that to achieve the deeper understanding of biology necessary to address the major problems that society is facing will require not just breaking down the "silos" within biology itself, but incorporating chemists, computational researchers, engineers, mathematicians, and physicists into basic biological research. Furthermore, only through such an integration of disciplines will it be possible to address major society problems. There is no area of biology where this might be more true than that of stem cells.

A second "stream of thinking" that has emerged is that of translational research. Not only are Federal agencies in the United States interested in fostering the translation of bench-top science into a variety of commercial/clinical applications, but this also has become a priority for many states. Furthermore, what is happening in this area in the United States simply "mirrors" what is taking place in the rest of the world. It is thus timely that a global assessment of stem cell engineering be conducted, and it is such an assessment that is reported here.

What is stem cell engineering? As defined for the purposes of this study, it is not just tissue engineering and regenerative medicine, but the entire interface of engineering with the "world" of stem cells. It thus ranges from basic stem cell research to models and tools, to enabling and scalable technologies, to stem cell biomanufacturing and the development of stem cell-based applications and products. It is in this context that this global assessment was conducted and that is reported here.

A preliminary workshop on Stem Cell Research for Regenerative Medicine (RM) and Tissue Engineering (TE) was held at National Science Foundation (NSF) on February 1–2, 2007. It was sponsored by NSF and also by the National Institutes of Health (NIH) and it was facilitated by WTEC (Shoichet and Caplan 2007). The workshop speakers presented an overview of the research activities in North America. The workshop confirmed the increasing convergence of these research areas in the drive toward clinical solutions that will address the deterioration of various human tissues and organs impacted by injury or disease. The workshop revealed that, although substantial research has been accomplished, there was much to be done to meet any expectations for improvement in human health and for commercial success. It was also clear that there was much to be learned abroad—other nations have been making rapid progress while the U.S. research community has been handicapped by Federal restrictions.

Bibliometric studies show the skyrocketing interest in the field—from 360 papers worldwide in 2005 to almost 1,000 just 4 years later. Using this simple filter, the United States leads the world in stem cell engineering, but not by much as the European Union countries as a whole are essentially equal to the United States, as also is a group of five top Asian countries. There thus are clearly valuable opportunities to be learned from research taking place overseas.

In May 2010, the NSF and others funded the Second International Conference on Stem Cell Engineering in Boston, MA. The conference emphasized how research in stem cell biology and engineering can combine to aid in the development of stem cell therapeutics and bioprocesses. The goal of the conference was to accelerate progress towards innovative solutions to basic and translation problems in regenerative medicine. Topics emphasized how quantitative approaches could yield an increased understanding of the biological mechanisms that underlie stem cell fate choices, cancer stem cells, iPS cells, technologies to study stem cell function, and the development of bioprocesses to culture stem cells for commercial applications. This conference not only provided background for this study, but was followed by the Third International Conference on Stem Cell Engineering held April 29–May 2, 2012 in Seattle, WA.

Scope of the Study

The purpose of this study, funded by NSF and also the National Cancer Institute at NIH and the National Institute for Standards and Technology (NIST), was to gather information on the worldwide status and trends in stem cell engineering, i.e., the interface of engineering with the world of stem cells. The study panelists gathered hands-on information on stem cell engineering activities abroad that will be used by the U.S. Government to modify its own programs. This report intends to critically analyze and compare the research in the United States with that being pursued in Asia and in Europe, to identify opportunities for collaboration, and to suggest ways to refine the thrust of U.S. research programs. To obtain the intended benefits, this study focused on a range of issues in which the R&D occurring abroad will best inform our own Government programs and the research community of the challenges, barriers, and opportunities in SCE. The study panel developed and refined the scope of the study, with the guidance of the sponsors. The scientific areas of focus for this study include:

- Understanding and controlling the signals for cellular response
- Formulating biomaterial scaffolds and the tissue matrix environment
- Scalable expansion and differentiation
- High-throughput screening and microfluidics
- Real-time, non-destructive phenotyping
- Systems-based quantitative analysis
- Computational modeling approaches
- Biomanufacturing and bioprocessing
- Targeted delivery of stem cells

Beyond the technical issues, this report also intends to address to the extent possible the following broader issues:

- Mechanisms for enhancing international and interdisciplinary cooperation in the field
- Opportunities for shortening the lead time for deployment of new SCE technologies emerging from the laboratory

- Long range research, educational, and infrastructure issues that need addressed to promote better progress in the field
- Current government R&D funding levels overseas compared to the United States, to the extent data are available

Prior Work at WTEC

With core funding and management from the NSF Directorate for Engineering, WTEC has conducted over 60 international R&D assessments. Other U.S. Federal agencies have also provided funding for various WTEC studies: Department of Defense, the Department of Energy, several institutes at NIH, NIST, and most other NSF directorates. Recently, panels of experts assembled by WTEC have assessed Asian and European R&D in nanotechnology, human-robot interaction, brain-computer interfaces, catalysis by nanostructured materials, and simulation-based engineering and science. Full text versions of the final reports are available free at http://wtec.org. WTEC also compiles cross-cutting findings to help evaluate national positions in science and technology; a recent example is Shelton and Leydesdorff (2012).

Panel Members

A panel of experts (Table 1) proposed by the chair and nominated by the sponsoring agencies, conducted this study, using the WTEC methodology of peer reviews of research abroad, visiting the sites of the research institutions and researchers who are noted for the most advanced work in Asia and Europe.

Some biographical information on each panel member is provided in Appendix A. It should be noted, however, that Dr. Jeanne Loring who was selected to be the resident stem cell biologist also has considerable experience in the biotech industry. Furthermore, the four other panelists have each co-chaired the International Conference on Stem Cell Engineering. This started with Dr. Schaffer co-chairing the inaugural meeting in 2008, Dr. Zandstra co-chairing the 2010 meeting, Dr. Palecek the 2012 meeting, and Dr. McDevitt who has been selected to co-chair the 2014 meeting. This is clear evidence that these four engineering members of the panel are recognized leaders in the field of stem cell engineering.

Panelist	Affiliation
Robert M. Nerem (Chair)	Georgia Institute of Technology
Jeanne Loring	The Scripps Research Institute
Todd McDevitt	Georgia Tech/Emory University
Sean Palecek	University of Wisconsin
David Schaffer	University of California at Berkeley
Peter Zandstra	University of Toronto

Table 1 Panelists and their affiliations

Study Process

There were many components to the process used in this global assessment of stem cell engineering. This starts with the knowledge provided by each of the panelists, knowledge not only about activities in the United States, but knowledge each of them had about activities in other parts of the world. To this foundation was added the following four components:

- Site visits in Asia and Europe
- Workshops in Atlanta and in Seoul, Korea
- Participation in the 3rd International Conference on Stem Cell Engineering
- Virtual site visits

The process actually began with a "kick-off" meeting at the National Science Foundation in Arlington, Virginia on June 22, 2011. This was followed by a series of conference calls that led to the Asia site visits, which occurred in November of 2011. The Asia site visits were carried out with the panel dividing into two teams, one that conducted site visits in China and the other site visits in Japan. At the end of the week the two teams came together for a meeting at Narita Airport outside Tokyo before traveling back to the United States. The countries visited are shown on the map in Fig. 1 and the institutions visited are listed in Table 2.

The next event or component in the study was the workshop held at Georgia Institute of Technology, December 15–16, 2011 on the topic of "Stem Cell Biomanufacturing." This workshop was financially supported by both Georgia Tech and Emory University Woodruff Health Sciences Center as well as WTEC and the British Consulate in Atlanta. The participants numbered approximately 40 with a mixture of academics and industry individuals. It included participants from the United States, Ireland, Japan, Korea, and the United Kingdom.

On January 17th, 2012 a workshop was held in Seoul, Korea on stem cell engineering. This workshop did not involve the panel, with the exception of the chair of the panel who was co-organizer of this meeting. The workshop was help at the Korean Institute of Science and Technology (KIST) with the host being Professor Soo Hyun Kim. More than 100 participants attended, and this meeting provided the opportunity to assess activities in stem cell engineering in Korea.

Next came the European site visits, which took place from February 26 through March 3, 2012. Here again the panel divided itself into two teams, and the countries visited are shown in Fig. 2 and the specific institutions visited are listed in Table 3.

The next component to this study was the participation of the panel in the 3rd International Conference on Stem Cell Engineering held in Seattle, Washington, April 29–May 2, 2012. At this conference the organizers provided the opportunity for what in effect was a town hall meeting with discussion taking place not only among the panelists seated up front who each were asked to make brief opening comments, but also with members of the audience.

Just three weeks later, a workshop was help at the National Science Foundation in Arlington, Virginia. At this one-day workshop on May 24, 2012, the WTEC panel

Fig. 1 Countries visited (*black star*) and virtual site visits (*white star*) in Asia

Table 2 Sites visited in Asia

China	Academy of Military Medical Sciences, Tissue Engineering Research Center
China	Chinese University of Hong Kong (CUHK)
China	Fudan University, Zhongsan Hospital
China	Institute of Biochemistry and Cell Biology, Shanghai Institutes for Biological Sciences
China	Institute of Biophysics, Chinese Academy of Sciences
China	Institute of Zoology, Chinese Academy of Sciences
China	National Natural Science Foundation of China (NSFC)
China	National Tissue Engineering Center, Shanghai Jiao Tong University School of Medicine
China	Peking University, The College of Life Sciences
China	Shanghai Jiao Tong University, School of Medicine
China	State Key Laboratory of Bioreactor Engineering
China	Tongji University School of Medicine
China	Tsinghua University, School of Medicine
Japan	Keio University, Yagami Campus
Japan	Kyoto University – CiRA (Center for iPS Cell Research and Application)
Japan	Okayama University, Graduate School of Medicine, Dentistry and Pharmaceutical Sciences
Japan	Osaka Univ. at TWMU (Kiro)
Japan	RIKEN Institute, Kobe
Japan	Tokyo Women's Medical University (Kano)
Japan	University of Tokyo, Hongo Campus, Department of Biomedical Engineering
Japan	University of Tokyo, Hongo Campus, Laboratory of Cell Growth and Differentiation
Japan	University of Tokyo, Komaba Campus, Research Center for Advanced Science and Technology
Japan	University of Tokyo, Komaba II Campus, Institute of Industrial Science
Japan	University of Tokyo, Shirokanedai Campus

Fig. 2 Countries visited (*black star*) and virtual site visits (*white star*) in Europe and Mideast

Table 3 Sites visited in Europe

France	Institute for Stem Cell Therapy and Exploration of Monogenic Diseases (I-STEM)
Germany	Berlin-Brandenburg Center for Regenerative Therapies
Germany	Fraunhofer Institute for Immunology and Cell Therapy
Germany	Institute for Medical Informatics and Biometry (IMB), Dresden University of Technology (TUD)
Germany	Life&Brain Center, Bonn
Germany	Lonza Cologne GmbH
Germany	Max Planck Institute for Molecular Biomedicine
Netherlands	Netherlands Initiative for Regenerative Medicine
Netherlands	Leiden University Medical Center
Sweden	Karolinska Institute and Karolinska University Hospitals
Sweden	Lund University Biomedical Centre (BMC)
Sweden	University of Uppsala
Switzerland	Basel Stem Cell Network (BSCN), University Hospital Basel and University of Basel
Switzerland	Laboratory of Stem Cell Bioengineering (LSCB), École Polytechnique Fédérale de Lausanne (EPFL)
Switzerland	Swiss Center for Regenerative Medicine (SCRM), University Hospital Zurich, and University of Zurich

Table 4 "Virtual" site visit reports

Australia	Stem Cells Australia
Iran	Royan Institute for Stem Cell Biology and Technology (RI-SCBT)
Korea	Workshop on Stem Cell Engineering
Korea	MEDIPOST, Co., Ltd.
Korea	Pharmicell Co., Ltd.
Portugal	Stem Cell Bioengineering Laboratory, Instituto Superior Técnico (IST)
Portugal	Instituto de Engenharia Biomédica (INEB)
Singapore	Bioprocessing Technology Institute
Singapore	National University of Singapore (NUS)

had the opportunity to provide through a series of oral presentations their assessment of activities globally. There were approximately 60 people in attendance; however, the audience was in fact much larger as the workshop was webcast, with 69 sites, an estimated 200 people around the world, looking at the presentations.

In addition to the above components, there were other mechanisms that are referred to by the panel as virtual site visits. This included site visits where information was gathered solely though the internet and/or by e-mail exchange. The term virtual site visit was also used for a site visit where only one panel member visited. The institutions/organizations that were assessed through virtual site visits are listed in Table 4.

There were some countries with active stem cell engineering activities that were not visited because the WTEC panelists believed that, through a variety of interactions, they had a reasonable idea of what was going on in that particular country.

One example is Ireland where, because of the close relationship between Georgia Tech and several universities in Ireland, considerable knowledge of stem cell activities already existed. Furthermore, Dr. Frank Barry from the National University of Ireland, Galway participated in the Atlanta workshop in December 2011. Another example is the United Kingdom. Here again there is a close relationship between Georgia Tech and Imperial College London. Furthermore, Dr. McDevitt has visited both Cambridge University and Loughborough University, and both Loughborough University and University College London were represented at the Atlanta workshop. A third and final example is Israel. The WTEC panel chair intended to visit this country at the end of March 2012; however, for personal reasons it was necessary for him to cancel the trip. Still because of the active participation of Israeli scientists and engineers in North American stem cell meetings, the WTEC panel believed that they had a reasonable idea of activities in Israel.

Finally, it must be noted that Canada has a particular concentration of stem cell engineering activity. Examples of groups include James Piret (University of British Columbia), Michael Kallos (Calgary), Eric Jeris (Waterloo), Peter Zandstra, Molly Shoichet, Julie Audet, Craig Simmons (Toronto) and Alain Grainer (Laval). A critical component in the establishment and growth of the Canadian stem cell engineering effort has been the availability of funding targeted specificity at bringing stem cell biologists and bioengineers together on both basic and translational research teams. Perhaps the best example of this funding strategy is the Canadian Stem Cell Network (www.stemcellnetowrk.ca), a federally funded National Center of Excellence (NCE) that has, over the last 13 years, invested over $42 million (not including partner cash and in-kind contributions) in interdisciplinary projects. These projects have in a number of cases been led by bioengineers and the work has benefited from this intimate interdisciplinary collaboration. The outcomes of the Stem Cell Network (SCN) are significantly greater than one would expect given the financial investment (962 peer-reviewed articles, of which 21 % appeared in high impact journals (impact factor >10), 399 patent applications, 60 issued patents, and 43 licenses granted). SCN-supported intellectual property has catalyzed the growth or launch of 11 start-up biotechnology companies and, critically, the SCN has also brought together teams around these basic discoveries and translational technologies to initiate nine phase I or II stem cell-based clinical trials. Globally across the SCN approximately 20 % of these activities have involved at least one engineer and one biologist/clinician. The Canadian government has continued to foster this interdisciplinary (and now multisectoral) activity with the recent funding of the Centre for the Commercialization of Regenerative Medicine (www.ccrm.ca), which is discussed later in this chapter.

Although the WTEC panel was able to see much of the stem cell activities going on around the world, they certainly did not see everything. They easily could have spent 2–3 weeks in both Asia and Europe, could have visited Australia, perhaps even India, and could have site visited activities in the Middle East. Even so, one can make the argument that the process outlined above in terms of the various components, as well as the knowledge base that each panelist had coming into this study, provided for a global assessment of this field of stem cell engineering. It is from this that the principal findings to be discussed next were derived.

Overview of the Report

In this section, the principal findings will be summarized. This section has been organized into four parts, representing the four chapters that follow, and then with some additional comments on translational models, education, opportunities for collaboration, and a brief summary of the status in the countries where an assessment of stem cell engineering was conducted, government policy, and conclusions. Appendix B has the detailed site visit reports from the 38 sites studied and Appendix C has "virtual" site visits from some additional organizations and laboratories. Also, a glossary of abbreviations and acronyms is given in Appendix D.

Engineering and Physical Sciences Principles in Stem Cell Research

The lead person for assessing this activity was Dr. David Schaffer, who has indicated that engineers be educated to do both analysis and synthesis. Through analysis one can identify key components of highly complex systems and then understand how these function collectively. From this one can also understand how the inputs from a cell's microenvironment can result in functional outputs.

Currently, it is well recognized by stem cell biologists that soluble components of the cellular microenvironment play important roles in regulating stem cell function including fate. Furthermore, many methods have been developed for controlling stem cells that involve serial or combinational application of a small number of factors, in many cases inspired by knowledge from developmental biology. What is not so widely recognized is the importance of biophysical cues in addition to those that are biochemical in nature. This is an area where engineers and the engineering approach can make a real contribution.

The engineering approach can also be used to develop novel systems that allow an investigator to pursue analysis by synthesis, i.e., the creation of new technologies and experimental systems that better enable basic investigations. An example is the development of innovative systems to control cell function, ones that can be used to monitor cells or that can be used in the separation of cells.

The microenvironments of stem cells are extremely complex in nature. The signals provided by a stem cell's microenvironment include complex networks of biochemical reactions as well as mechanical cues and electrostatic signals. Within this environment there are transport phenomena that must be taken into account. The engineering approach can contribute to the understanding of this complex microenvironment, ultimately to the engineering of synthetic niches for stem cells that provide the necessary input to result in the desired output. As an example, biomaterials can be engineered to be ideal platforms that can be used to elucidate principles by which biology controls stem cell function and fate. Furthermore, engineered biomaterials can be used to create culture systems for use in clinical translation.

High-Throughput Screening, Microfluidics, Biosensors, and Real-Time Phenotyping

The lead person for this part of our report is Dr. Sean Palecek from the University of Wisconsin, Madison. As he has pointed out, there are a number of challenges in engineering the stem cell microenvironment. These include identifying the factors that regulate stem cell fate and understanding the combined effects of different cues, constructing culture systems that apply the desired cues, and ultimately developing systems that allows one to direct the fate of stem cells.

High-throughput screening is an area where engineers and the engineering approach can make significant contributions. Challenges in this area include the designing of high-throughput cell analytic assays, systems that minimize false positives and negatives, and systems that allow for targeted screening. Such systems must be rapid, inexpensive, sensitive, specific and reproducible.

A technology that is taking on increasing importance in stem cell R&D is that of microfluidics. A variety of microfluidic systems have been developed, and such systems can facilitate the analysis of the dynamic response of stem cells to their microenvironmental cues, can be used to enable the isolation and analysis of clonal populations, and can provide proof-of-concept demonstration of the isolation of low abundance stem cells from a mixture of cells. Some of the challenges in the area of microfluidics include developing robust microfluidic culture platforms, incorporating integrated, real-time analysis, and the translation of microfluidic systems to commercial and/or clinical applications.

Another area is that of biosensors. Here stem cells can be a source for cells to be used in the creation of *in vitro* models of tissues. Using iPS cell technology, the cells could be either normal or represent a disease state. Such *in vitro* biosensors could be used in drug discovery for either drug screening or drug toxicity testing. Such systems are now called organ-on-a-chip, and these need to be designed to model tissue and organ level function in an *in vitro* microdevice. Important will be real-time analysis of the behavior of the system. If such systems can be engineered to provide in general a physiological environment, one that in many cases will need to be multicellular and have a three dimensional architecture, that may as a minimum supplement animal testing in the development of a drug and perhaps even replace it.

Computational Stem Cell Engineering

The lead person for this part of the report is Dr. Peter Zandstra from the University of Toronto. He has suggested that stem cell properties make these cells especially suitable for computational modeling approaches, with both fundamental and translational applications. An example is the rarity of stem cells in a population of cells. Because of this, signals may be diluted across many potential targets, the behavior of other cells may overwhelm that of the stem cells, and stochastic responses within a small population may be important determinants of behavior. Stem cells are rarely in equilibrium, and they are responsive to both local and global control. These characteristics are a challenge to experimental investigators, and the use of computational models together with experiments may help in unraveling the behavior of stem cells.

A variety of computational/mathematical approaches are being employed. These range from the use of differential equations, to Markov chains, to Boolean models, to Bayesian networks, to statistical mining. Questions that need to be answered include: does the model need to be deterministic or stochastic in nature? Is it a model of a single cell or of a population?

Computational modeling has the potential to emerge as a foundation to understanding complex systems such as stem cells. Computational models can provide tools for the design of experiments and novel technologies, can be used to predict system behavior and to devise methods for improving control, can provide novel insight into mechanisms, and can be utilized to both explore hypotheses regarding stem cell biology and to suggest the design of novel experiments. Computational models have already made a significant impact on the development of better ways to grow and control stem cells and have provided new fundamental insights into how stem cell fate decisions are made. The power of computational modeling in stem cell R&D will only increase with new and larger data sets and more sophisticated cell growth control strategies.

Stem Cell Bioprocessing and Biomanufacturing

The lead person from our WTEC panel in this fourth area is Dr. Todd McDevitt from Georgia Tech and Emory University. As defined by Dr. McDevitt, there is a slight difference between the term "bioprocessing" and the term "biomanufacturing." The former is the development of systems for the scalable growth and differentiation of stem cells while the latter is the implementation of bioprocessing for stem cell commercialization.

Current approaches have used formats and platforms optimized for biological engineering, i.e., where the cells are the vehicle for producing a product. They may involve systems engineered for optimal cell growth, but in general have a "hands on," manual processing of the various culture steps. The challenges to be addressed include the development of scalable culture systems, ones that scale "up," not "out," the incorporation of real-time monitoring, feedback control systems, and the development of robust, reproducible automated processes. All this is needed for the transformation of biomanufacturing from cells being the vehicle for the product to cells being themselves the product.

The issue of scale "out" versus scale "up" is an important one. If one has adherent cells and one wishes to increase the number of cells by a factor of 10, maybe even a 100, then one needs to increase the surface area by a factor of 10 or a 100. This at some point becomes impractical, and thus there is considerable need for suspension culture formats. Such formats include microcarrier beads with adherent cells on the outside of the bead, the use of cell aggregates either with or without materials, and the microencapsulation of stem cells. The use of suspension formats makes the significant increase in the number of cells much more doable.

Another issue is the multitude of parameters that are capable of affecting cell growth and phenotype. For process optimization there is a large experimental space that must be explored, and current high-throughput formats and screening platforms are inherently incapable of simulating bioprocess parameters.

Finally, stem cell manufacturing facilities need to be developed as closed culture systems with a miniaturized "foot print" and composed of modular elements. They need to incorporate real-time monitoring, feedback-based control, and learning-based algorithms. There thus are plenty of opportunities for engineers and the engineering approach to make a contribution to the further development of bioprocessing systems and to biomanufacturing.

Translational Models

One of the important aspects of this global assessment was identifying some of the interesting models that have been developed to translate bench-top stem cell science into clinical therapies and into commercialization. Four such models are listed below.

- Berlin-Brandenburg Center for Regenerative Therapies
- Cell Therapy Catapult in the United Kingdom
- Centre for Commercialization of Regenerative Medicine in Canada
- Tokyo Women's Medical University

The uniqueness of the Berlin-Brandenburg Center for Regenerative Therapies is that they do an opportunity analysis early in the development of a research project. There are three multidisciplinary platforms. These are basic science, biomaterials, and translation technology. Several of the groups within the Center are organized in a matrix structure, supporting the work of a particular host platform as well as those of other platforms by delivering basic technologies and principles. In addition, there is a Department of Clinical Development and Regulatory Affairs and a Department of Business Development. These support all projects within the center.

In the United Kingdom the Cell Therapy Catapult is one of seven such catapult initiatives established by the Technology Strategy Board of the U.K. government in order to create new industries. The Cell Therapy Catapult will support the development and commercialization of cell therapies and advanced therapeutics as well as the enabling technologies for manufacturing, quality control, and safety. It will be based in London, and it will be a center, independent of higher education institutions, but where academics, business, and clinicians can work together, focusing on the commercial development of innovative technologies.

In Canada, Dr. Peter Zandstra, a member of this WTEC panel, is the Chief Scientific Officer of the Centre for Commercialization of Regenerative Medicine. This is a federally incorporated, nonprofit organization supporting the development of technologies that accelerate the commercialization of stem cell- and biomaterials—based products and therapies. The business strategy is to enable unique translational platforms that address key barriers in regenerative medicine commercialization, integrate Canada's strength in stem cells and engage industry partners so as to make the Centre a global nexus for regenerative medicine commercialization.

Finally, at Tokyo Women's Medical University Dr. Teruo Okano heads the Institute of Advanced Biomedical Engineering and Science, and he has provided the leadership to create a unique activity. The focus has been on cell sheet tissue engineering, and the institute has partnered with the Waseda University's Graduate School of Bioscience and Medical Engineering. In 2008, the Tokyo Women's Medical University-Waseda University Joint Institution for Advanced Biomedical Sciences (TWIns) opened. There also is a partnership with Professor Masahiro Kino-Oka from Osaka University to develop a tissue factory for cell sheet manufacturing.

It is these four that are discussed in this report; however there are of course other models for translation. One of these is the Global Stem Cell and Regenerative Medicine Initiative recently established by the Korean Ministry of Health and Welfare as part of a national Korean strategy to exercise global leadership in the stem cell and regenerative medicine field. The operation and management of this initiative is being assisted by the Global Stem Cell and Regenerative Medicine Acceleration Center whose activities include strategic planning, project design, performance assessment, global networking, and many other supporting activities. This center has an international advisory board on which the author of this chapter has been asked to serve. The major focus of this initiative is on translational research to accelerate therapeutic development, clinical research aimed at the delivery of treatments, and infrastructure development to speed up commercialization. As this initiative is brand new, the exact details are still somewhat unclear; however, it will be interesting to see how this activity in Korea develops.

Education

It is clear to this WTEC panel that, for engineers to be accepted by biologists, they need to be viewed as understanding biological mechanisms and making a contribution to biology. Thus, for training programs to be successful, they need to include what might be called a "high level" of biology, and this is certainly what is done in the leading bioengineering programs in North America.

Outside of North America an excellent example of a unique training program is that at Loughborough University. The Doctoral Training Centre there was established with funding from the U.K.'s Engineering and Physical Sciences Research Council and in partnership with Keele University and with the University of Nottingham. There are more than 50 Ph.D. students in this program. The program introduces the students to the principles of bioprocessing and manufacturing and provides "hands on" research experiences with stem cells in existing and new, novel platforms. The intent is to train the future leaders of industry in the biomanufacturing area.

One of the outcomes of the Atlanta Workshop on Stem Cell Biomanufacturing was the agreement to establish an international school in the area of cell manufacturing. The initial offering of this school will be April 28–May 4, 2013 in Portugal. The school is being organized by faculty at Loughborough University in the U.K., Georgia Tech, and the Instituto Superior Técnico in Portugal; however, the participation of faculty, staff, and students from other universities also is anticipated.

Opportunities for Collaboration

Research today in general is very interdisciplinary in nature and this is true of biology and the stem cell field. This is certainly a theme for the National Academies report already referred to previously. As part of this, collaborations almost become a necessity. These might be with an investigator at one's own institution, somewhere else in the city, or even at a longer distance.

In today's world where research and the development of technology is done within the global community, collaborations can also exist between investigators in different countries. In fact, U.S. investigators need to leverage the excellence of activities in other countries, and it thus was encouraging for the WTEC panel members to see the hosts of the different sites visited being so open and very interested in the possibility of collaborating.

What is needed, however, if we are to encourage international collaborations are government programs that foster this. Included should be realistic levels of funding. Also, the review process needs to be one that uses a single review committee with membership from both of the countries sponsoring the program.

State of Stem Cell Engineering Outside of North America

Appendices B and C of this report contain the site visit reports for each institution. These site visit reports provide more detail than can be stated here; however, in the listing below for each country visited or in some other way assessed the state of stem cell engineering is briefly characterized.

Europe Sites

France: Observed some engineering involvement
Germany: Strong engineering involvement at the Berlin-Brandenberg Center for
   Regenerative Therapies and at the Fraunhofer Institutes

Ireland: A major stem cell center at NUI Galway with engineering involvement
Netherlands: A good integration of engineering with biology and medicine in NIRM
Portugal: Strong engineering involvement in bioprocessing
Sweden: Significant activity including translation into the clinic, some government
    funding, some engineering involvement of engineering and the physical sciences
Switzerland: Strong engineering and physical sciences involvement at EPFL and in
    Zurich and Basel (ETH)
United Kingdom: Major engineering activities, largely in bioprocessing and
    manufacturing

Asia Pacific Sites

Australia: A new stem cell initiative with the involvement of some engineers
China: Excellent young investigators, massive investments by the government, high
    impact biology, engineers involved in more traditional roles
Japan: A leader in iPS cells, engineering integrated with biology and medicine at Tokyo
    Women's Medical University, other engineering activities more independent
Korea: Significant activities with major government funding, a number of start-up
    companies, some engineering involvement
Singapore: Excellent Bioprocessing Technology Institute with engineering
    involvement

Other Countries

Iran: A major stem cell research institute but limited if any engineering involvement
Israel: Considerable activities involving both biologists and engineers, also some
    commercial activities

Government Policy

From the assessment conducted, it is clear to this panel that there are countries that
recognize the importance of investing in science and technology and are actually
doing it. A list of such countries includes China where the R&D budget continues to be
increased on an annual basis, Japan where it appears that the country has identified
as a priority making regenerative medicine a key component of their twenty-first
century economy, Korea where there is a new global regenerative medicine initiative,
and Singapore. This list also includes such European countries as Germany, The
Netherlands, Sweden, and the United Kingdom. Taking the United Kingdom as a
specific example, at the end of 2011 the British government launched a new strategy

for U.K. Life Sciences. This comprehensive strategy includes significant new investments in life sciences research and in the development and commercialization of research. The Cell Therapy Catapult initiative is a part of this strategy. The British government, in spite of the global economic recession and a very significant U.K. budget deficit, is doing this because its goal is for the U.K. to be the global "hub" for the life sciences in the future.

In contrast, in the United States the budgets of the Federal agencies that support R&D are "flat" with little indication that this situation will change soon. This is highlighted in a recent report entitled "Leadership in Decline" (Atkinson et al. 2012). This potential decline in leadership is true in the life sciences in general and certainly could happen in the stem cell area. This could threaten the U.S. leadership in the development of enabling technologies, new clinical therapies, and other innovative stem cell-based applications, areas where engineers and the engineering approach has a critical role to play. Having said that there is a potential for a decline in the historical leadership of the United States in the life sciences, the White House Office of Science and Technology Policy back on April 26, 2012 released a National Bioeconomy Blueprint. This document outlines what are called five strategic imperatives that potentially will result in the generation of new markets and economic growth. These are as follows:

1. Support R&D that will provide the foundation for the future bioeconomy;
2. Facilitate the translation of research to the market;
3. Develop and reform regulations so as to reduce barriers and increase the speed and predictability of regulatory processes and thus reduce costs;
4. Update training programs and provide institutional incentives for student training for national workforce needs; and
5. Identify and support opportunities for the development of public-private partnerships and precompetitive collaborations.

With the exception of the third recommendation above that deals with regulatory issues, the conclusions offered in the next section align with the above recommendations.

Conclusions

From this global assessment of stem cell engineering as conducted by the WTEC panel, it is clear that engineers and the engineering approach with its quantitative, systems-based thinking can contribute much more to basic stem cell research than it has to date. As stated in the National Academies report on "A New Biology for the 21st Century," to achieve the deeper understanding of biology required in this century there will need to be an integration of many disciplines into biological research and this certainly includes engineering. Engineering analysis can be used to identify the components of highly complex stem cell systems and provide an understanding of how these components work together. Furthermore, computational models will be increasingly important in our efforts to achieve a better understanding of complex biological systems. In all of the above engineers are in a position to take a leadership role.

Engineers also can take the lead in developing new, innovative enabling technologies. This includes high-throughput screening techniques, improved culture and differentiation systems, and *in vitro* models engineered to be more physiologic. The last of these include organ-on-a-chip models and also engineered *in vitro* tumor models that can lead to a better understanding of cancer.

Finally, for stem cell biomanufacturing there is a need for further advances in culture systems, techniques for real-time monitoring, and for process automation. Underpinning these specific application areas, computational modeling has an important role to play throughout the spectrum from discovery to translation.

In summary, from the assessment conducted there is a need for an increasing involvement of engineers in the field of stem cells and related technologies. Although one might argue that the United States today has a leadership role, to capitalize on this and to build on the current existing momentum, and most importantly to accelerate the translation of bench-top research into various applications including clinical therapies and into commercialization, will require the United States taking bold steps. The panel thus offers the following conclusions.

- The United States has a unique opportunity to maintain a leadership position in the stem cell field through the continued support of R&D that will provide a foundation for the generation of new markets and that will lead to economic growth.
- Because of the contributions that engineers can make in all areas of the stem cell field, as elaborated in the global assessment reported here, this needs to include increased investment in engineering, applied research, and commercialization as it relates to stem cell research and related stem cell-based technologies.
- A major component in this could be that the Federal agencies that support R&D should establish a broad interagency program for stem cell engineering, one that provides grants to interdisciplinary teams that include engineers, computational researchers, and biologists as well as individuals from other disciplines.
- Another component that would be beneficial is the establishment of new, innovative mechanisms that support academic-industry partnerships and unique translational models that facilitate the translation of research into the private sector.
- To address national workforce needs, the development of training programs at universities and advanced short courses should be encouraged and supported by Federal agencies.
- Finally, in today's global economy and with the excellent activities taking place in other countries, the United States would benefit from forming strategic partnerships with other countries so as to leverage the existing and emerging strengths in institutions outside of the United States; to implement such partnerships will require binational grant programs with appropriate review mechanisms.

As noted in the previous section, these conclusions align with the National Bioeconomy Blueprint released by the White House Office of Science and

Technology Policy. It is up to the Federal agencies to implement a plan based on the conclusions from this assessment study. Without the implementation of the above, however, this unique opportunity could be lost. In this case, it might be possible that the United States in the future is relegated to the second tier of countries in this critical area of stem cell engineering. On the other hand, if implementation takes place in some form, and there is an urgency to do this, then the United States can expect to continue to be in a leadership position and at the forefront in advancing the sciences, developing new, innovative enabling technologies and platforms that lead to clinical therapies, to commercializing the results of stem cell research, and to the generation of new markets and economic growth based on advances in the stem cell field. Some of the results from this will be:

- The acceleration of the development of new drugs while at the same time reducing the costs of this development process
- The development of cell therapies that address diseases and conditions of injury for which today there are no real treatment options available for patients in need
- The growth of the twenty-first century bioeconomy in the United States based on advances in our knowledge of stem cells and the translation of this into applications and products

This has been the dream for at least 20 years; however, with the right strategy by the United States it can be realized and be the reality of tomorrow.

Acknowledgements This study would not have been possible without sustained support from several U.S. Government program officers: Michael Reischman, Kesh Narayanan, Semahat Demir, Ted Conway, and Theresa Good of the National Science Foundation, Anne Plant of the National Institute of Standards and Technology, and Larry Nagahara of the National Cancer Institute at NIH. We also wish to thank Joe Gold, Ron McKay, and Jon Rowley who served as consultants to this study. Most importantly we extend our sincere thanks to the foreign hosts for their hospitality and generosity in sharing with us their research results, their facilities, and their insights. And most of all, I especially appreciate the hard work and the long days endured by the delegation making the site visits. This all contributed to making this study a success.

References

Atkinson, R.D., S.J. Ezell, I.V. Giddings, L.A. Stewart, and S.M. Andes. 2012. *Leadership in decline: Assessing U.S. international competitiveness in biomedical research*. Washington, DC: The Information Technology and Innovation Foundation and United for Medical Research. 23 pp.

National Research Council. 2009. A new biology for the 21st century. Washington, DC: The National Academies Press. 112 pp.

Shoichet, M., and A. Caplan. 2007. Stem Cell Research for Regenerative Medicine and Tissue Engineering, Workshop Report. http://wtec.org. Accessed March 3, 2012.

Shelton, R.D., and L. Leydesdorff. 2012. Publish or patent: Bibliometric evidence for empirical tradeoffs in national funding strategies. Proceedings of the 13th International Conference on Scientometrics and Informetrics, pp. 763–774. July, 2011, Durban.

Physical and Engineering Principles in Stem Cell Research

David V. Schaffer

Introduction

Biological research in general is becoming increasingly interdisciplinary, and stem cell research in particular has strong potential to become progressively more so. In this field, there has for example been a growing recognition that, while biochemical signals play critical roles in regulating the behavior and fate decisions of stem cells, biology presents regulatory information to cells not only in the binary absence or presence of a given molecule, but also numerous biophysical aspects of these regulatory cues. These include mechanics, topographical features at multiple size scales, electrostatics, spatiotemporal variation in the presentation of biochemical cues, transport phenomena, and biochemical reaction kinetics. As a result, there are considerable opportunities for physical scientists and engineers to become increasingly involved in stem cell research, not only to gain basic insights into new mechanisms in stem cell biology but to create new technologies to advance this field. Within this report, chapter "High-throughput Screening, Microfluidics, Biosensors, and Real-time Phenotyping" discusses the development of technologies to discover novel signals that regulate stem cell behavior, and chapter "Computational Modeling and Stem Cell Engineering" reviews progress in the development of mathematical models that quantitatively investigate the underlying regulatory mechanisms. The present chapter will review research into now biophysical features of the microenvironment or niche regulate the behavior of a stem cell.

D.V. Schaffer (✉)
Berkeley Chemical and Biomolecular Engineering, University of California,
274 Stanley Hall, Mail Code 3220, Berkeley, CA 94720, USA
e-mail: schaffer@berkeley.edu

R.M. Nerem et al. (eds.), *Stem Cell Engineering: A WTEC Global Assessment*,
Science Policy Reports, DOI 10.1007/978-3-319-05074-4_2,
© Springer International Publishing Switzerland 2014

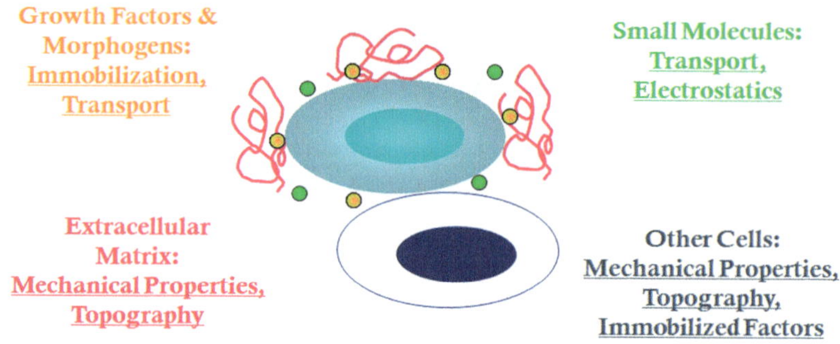

Fig. 1 Schematic of the stem cell niche (Courtesy of the author)

Biochemical and Biophysical Information in the Stem Cell Niche

During development and throughout adulthood, stem cells reside within specialized regions of tissue that continuously present them with regulatory cues, and this repertoire of signals is collectively referred to as the stem cell niche (Schofield 1978; Watt et al. 2000; Scadden 2006). The niche presents a stem cell with considerable molecular information, in the form of soluble molecules; extracellular matrix (ECM) proteins, glycosaminoglycans, and proteoglycans; growth factors and morphogens that may be soluble or immobilized to the ECM; and cues presented from the surface of neighboring cells (Fig. 1). Soluble small molecules, soluble and immobilized proteins, ECM components, and intercellular components collaborate to regulate stem cell behavior. In addition, there are numerous physical and engineering principles that modulate the manner in which these components present information, including mechanical properties, spatial organization, and temporal variation in the presentation of cues, topographical features of the niche on the nanoscale and microscale, mass transport properties, and electrostatics.

Due to efforts in genetics, developmental biology, and cell biology, it is well recognized that biochemical cues within the niche play critical roles in regulating stem cell function. As a prominent example, forward genetics approaches in model organisms are a classical approach to identify novel factors that play roles in organismal development, in many cases via regulating stem cells in developing tissues. For instance, the segment polarity gene hedgehog was originally discovered in a random mutagenesis screen in *Drosophila melanogaster* for embryonic lethal phenotypes (Nusslein-Volhard and Wieschaus 1980). The Jessell lab later found that the vertebrate homolog Sonic hedgehog (Shh) played a critical role in the differentiation of motor neurons in the developing spinal cord (Roelink et al. 1994), and the same lab subsequently demonstrated that Shh—in combination with other developmentally important factors—could help guide or instruct the differentiation of mouse embryonic stem cells (mESCs) into motoneurons in culture (Wichterle et al. 2002). In an analogous approach, collections of factors previously discovered via forward genetics and other approaches can be screened for their potential effects on particular stem cell populations. As one such example of a candidate approach, the founding member of the Wnt family of proteins was originally discovered first in *Drosophila* as a gene whose mutation led to the absence of wings (and was thus named wingless (Sharma and Chopra 1976)) and in vertebrates as a gene whose transcriptional activation promotes mammary tumorigenesis (Nusse et al. 1984). Given the importance of Wnt family members subsequently demonstrated for numerous tissues (van Amerongen and Nusse 2009), they have been considered as prominent candidate regulators of stem cell function. In one such important study, Wnt3 was demonstrated to regulate the neuronal differentiation of neural stem cells in the adult brain (Lie et al. 2005).

In addition to clearly demonstrated role of many biochemical cues in regulating stem cell function, as exemplified in the forward genetics and candidate molecule studies cited above, there are numerous biophysical features of the niche that may offer additional regulatory control over stem cells. However, the biophysical properties of a tissue are not monogenic, i.e., they depend on the properties of many molecules and genes. For example, the mechanical properties of a tissue are determined by its constituent materials, including cells and ECM. Analogously, the topography of a tissue depends on the identities of its ECM and cells as well as the history of their assembly, and mass transport properties vary with the tissue interstitial space and potential fluid flow. As a result, these properties do not arise in a straightforward manner in genetic screens, which perhaps contributes to the fact that they have not been as broadly studied in stem cell research as biochemical factors.

However, an emerging theme in the nascent field of stem cell engineering is to use *in vitro* engineered systems—ranging from synthetic materials to microfluidic devices—to systematically vary these biophysical properties, i.e., to provide them with an "x-axis" in a manner that is not currently possible using genetic approaches. While there are inherent challenges with this approach—including demonstrating the *in vivo* relevance of findings, as well as integrating engineering and biology approaches to explore underlying mechanisms—these engineering approaches have broadened the field's view of the stem cell niche (Saha et al. 2007; Discher et al. 2009; Guilak et al. 2009; Lutolf et al. 2009; Keung et al. 2010).

This chapter discusses the application of engineered microenvironments—or systems that emulate the niche—to vary and thereby investigate the effects of biophysical properties of tissues on stem cell behavior. In addition, while a number of these studies discover new phenomena in the biophysical regulation of stem cell function, there has been increasing progress in understanding mechanisms by which cells respond to these cues. Finally, the application of physical and engineering approaches to create additional technologies to study stem cell function will be discussed, as well as future opportunities for engineers and biologists in the stem cell field.

Mechanoregulation of Stem Cell Function

There are many mechanical properties of tissues that could potentially regulate cell function. The elastic modulus is the linear proportionality constant between the stress applied to a material and its strain or deformation. Though elastic modulus is sometimes used interchangeably with stiffness, the former is an intensive property of the material, whereas the latter extensive property depends on material geometry. In addition, elastic stress-strain relationships can be nonlinear. Furthermore, many tissues are viscoelastic, or have both elasticity and viscosity, i.e., a fluid property describing resistance to deformation by either shear or tensile stress. Like elasticity, viscosity can also be linear or nonlinear, and in all cases these material properties of a tissue can vary in space and time (Humphrey 2003). In principle, a resident cell may be able to sense and respond to any or all of these material properties.

Static Mechanical Properties

Given the complexities of a material's mechanical properties, the field has made strong progress by initially focusing on linear properties, with a strong focus on elastic modulus. In 1997, to study the effects of material stiffness on cell migration, Pelham and Wang developed a linearly elastic, bioactive material—specifically a polyacrylamide (PA) hydrogel coated with the ECM protein collagen—in which the modulus could be varied by changing the proportion of crosslinker during polymerization. Aided by this system, in landmark work Engler and Discher (Engler et al. 2006) demonstrated that the lineage choice of differentiating mesenchymal stem cells (MSCs) is strongly influenced by substrate stiffness, such that cells developed into neuron-like cells on soft PA gels, myoblasts on intermediate stiff-nesses, and osteocytes on harder substrates (Fig. 2). They thus proposed that MSCs on different stiffnesses differentiate into cell fates associated with tissues that cor-respond to those stiffnesses. In work that extended mechanoregulation to another stem cell type, Saha et al. demonstrated that neural stem cells (NSCs) preferentially differentiate into neurons when cultured on soft materials and astrocytes on hard materials (Saha et al. 2008a, b). Additionally, Banerjee et al. found that the effects of stiffness on NSC differentiation extended to cells embedded in three-dimensional (3D) materials (Banerjee et al. 2009). Moreover, the extent of maturation of neurons differentiated from NSCs was enhanced on soft vs. stiff gels (Teixeira et al. 2009).

In addition to differentiation, modulus can influence stem cell self-renewal. For example, it was shown that substrate stiffness strongly impacts the ability of muscle stem cells (also termed satellite cells) to undergo self-renewal in culture. Muscle stem cells were isolated from muscle and grown on soft of stiff substrates composed of a polyethylene glycol hydrogel. Cells grown on the former but not the latter stiffer material were able to expand and, upon implantation into adult muscle,

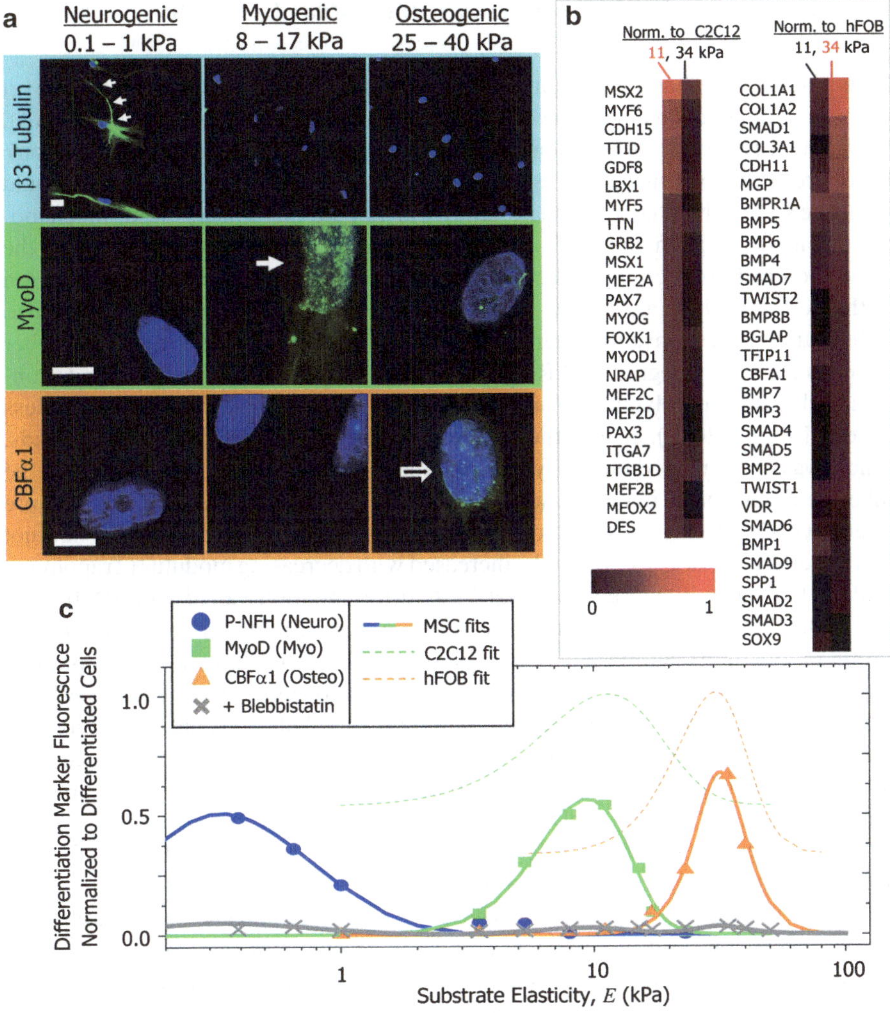

Fig. 2 Substrate stiffness directs mesenchymal stem cell differentiation (From Engler et al. 2006). (**a**) The neuronal marker β-tubulin III is expressed in MSCs differentiated on soft gels, the muscle transcription factor MyoD1 is expressed on substrates of intermediate elasticity, and the osteoblast factor CBFα1 is expressed on stiff substrates. (**b**) Microarray profiles are shown for MSCs differentiated on 11 vs. 34 kPa matrices, showing upregulation of markers indicative of muscle or osteogenic differentiation. (**c**) Differentiation marker expression as a function of substrate stiffness reveals optimal differentiation into a given lineage at the stiffness characteristic of that lineage

contribute to the tissue (Gilbert et al. 2010). Furthermore, mouse embryonic stem cell self-renewal is promoted on soft substrates, accompanied by downregulation of cell-matrix tractions (Chowdhury et al. 2010). Finally, the development of an innovative high-throughput system for analyzing the effects of stiffness, and other microenvironmental properties, on stem cell function promises to accelerate progress in this area (Gobaa et al. 2011).

Mechanisms for Stiffness Regulation of Stem Cell Fate

At its essence, mechanoregulation of stem cell function requires the cell to convert an extracellular mechanical cue into an intracellular biochemical response (i.e., activation or repression of genes involved in stem cell self-renewal or differentiation). There are numerous mechanisms by which material mechanical properties could influence stem cell behavior. One possibility is that the ECM itself is the "mechanosensor," as it has been shown that forces involved in cell adhesion can unfold the ECM protein fibronectin and thereby expose additional biochemical information to the cell (Smith et al. 2007). Alternatively, in work showing that MSCs also differentiate in response to material stiffness in 3D, it was shown that the number of bonds between integrins and RGD peptides (i.e., synthetic adhesive ligands containing the arginine-glycine-aspartic acid motif) varied biphasically with stiffness (Huebsch et al. 2010). It was thus proposed that the number of adhesive bonds, which was in turn modulated by cellular reorganization of matrix to cluster the adhesive ligands near integrins, was correlated with downstream cell fate. In more recent work, investigators found that the porosity of polyacrylamide, but not polydimethylsiloxane (PDMS) gels, increased with decreasing modulus (Trappmann et al. 2012). They likewise found that MSC differentiation varied with PA but not with PDMS stiffness, leading them to propose that differences in the number of points or positions within ECM proteins that were crosslinked or anchored to materials of different stiffness was responsible for apparent mechanosensitive cell differentiation. However, this intriguing finding should also be analyzed in light of mechanosensitive NSC and MSC differentiation on hydrogels functionalized with either RGD peptides (Saha et al. 2007; Huebsch et al. 2010) or large ECM proteins (Engler et al. 2006; Keung et al. 2011).

The next question is how cells sense and respond to adhesive bonds. Tension between adhesion receptors and the extracellular matrix is transmitted across the membrane and into the cytoskeleton. These forces are accompanied by the assembly of focal adhesions and biochemical activation of enzymes (e.g., kinases) within this structure. It is clear that the former, specifically actin-myosin contractility, is necessary for stem cell mechanosensitivity. In their original work, Engler and Discher showed that addition of blebbistatin, a myosin II inhibitor, both blocks both cell cortical stiffening and myogenic differentiation as a function of substrate stiffness (Engler et al. 2006). Fu and colleagues, as an innovative alternative method for varying substrate stiffness, generated molded arrays of elastomeric PDMS posts (Fu et al. 2010). By culturing MSCs on posts of variable high height, which thereby require variable magnitudes of cellular force to bend the posts, they again showed that substrate compliance regulates cell fate. Furthermore, they demonstrated that early, transient addition of an inhibitor of Rho kinase, an enzyme that both stabilizes actin filaments and promotes myosin contractility, decreased the extent of osteogenic differentiation 7 days later. Recently, Keung and others showed that NSCs elevate RhoA and Cdc42 (but not Rac1) activity on stiffer substrates and that RhoA and Cdc42 inhibition block the ability of cells to stiffen as a function of substrate stiffness

(Keung et al. 2011). Furthermore, RhoA/Cdc42 activation or inhibition promoted astrocytic or neuronal differentiation respectively *in vitro*, in a manner depending on myosin contractility, and upregulation of RhoA activity in NSCs within the adult brain blocked neuronal differentiation.

These results firmly establish a role for actin-myosin contractility in stem cell mechanoregulation. However, it is unclear how such changes in the cellular cytoskeleton translate into changes in gene expression that drive cell differentiation. An important link was discovered when Dupont and colleagues implicated YAP and TAZ—transcriptional coactivators typically associated with the Hippo signaling pathway—in MSC mechanoregulation. Specifically, they found that these molecules localize to the nucleus in MSCs on stiff but not soft substrates, in a manner dependent on Rac1 and RhoA activity but not Hippo signaling. Furthermore, their knockdown ablates the effects of stiff substrates on MSC differentiation (Dupont et al. 2011). Future efforts will likely elucidate the link between the cytoskeleton and YAP/TAZ, the mechanisms of action for YAP/TAZ in the nucleus, and whether these mechanisms are general to other stem cell types.

Shear Flow

In addition to mechanical interactions with solid phase components of the niche, in some cases stem cells may be exposed to fluid flow and therefore shear forces, particularly within the cardiovascular system. As an early example in this area, Yamamoto et al. found that endothelial progenitor cells, isolated from human blood, responded to laminar flow in adherent culture by increasing their proliferation, enhancing their expression of endothelial markers such as vascular endothelial growth factor (VEGF) receptors, and showed increased tube formation, an *in vitro* readout indicative of vascular formation activity (Yamamoto et al. 2003). In subsequent work, they sorted a fraction of mouse embryonic stem cells (mESCs) that express VEGF receptor 2 (otherwise known as Flk-1) and found that exposing these cells to laminar flow for 24–96 h induced the upregulation of a number of endothelial cell markers in a Flk-1-dependent manner (Yamamoto et al. 2005).

The effects of shear flow on stem cells have also been investigated mechanistically. For example, analogous to the work of Yamamoto, Zeng and others found that shear flow induced the endothelial differentiation of mES cells, and they further explored the mechanisms underlying this process (Zeng et al. 2006). Specifically, shear increased the deacetylation of p53 by histone deacetylase 3 (HDAC3), the activated p53 upregulated expression of the p21, and this cell cycle inhibitor contributed to mESC differentiation. Furthermore, they found that Flk-1 activation by shear, and subsequent activation of the kinase Akt, were required for the HDAC3 activity. Illi and colleagues found, under somewhat similar culture conditions, that shear rapidly altered patterns of histone acetylation in mESCs, and after 24 h of shear exposure, not only endothelial but also cardiovascular markers were upregulated (Illi et al. 2005).

In additional work that is both very creative and practical, investigators explored a potential relationship between mechanical forces and early embryonic development. Naruse and colleagues noted that during passage through the oviduct, embryos are exposed to shear flow, compression, and stretching. In effort to address problems in fertility, they hypothesized that a mechanically active culture system could improve the quality and viability of embryos (Matsuura et al. 2010). The resulting device, a tilting embryonic culture system (TECS) that exposes cells to cyclic shear flow by oscillatory tilting, has enhanced embryo quality for usage in *in vitro* fertilization in recent clinical studies.

Cyclic Strain

In addition to the static mechanical properties of organs and tissues, due to the action of the heart and lungs, many tissues experience cyclic strain with a ~1 Hz frequency. Furthermore, organismal motion made possible by the musculoskeletal system also exposes these and other tissues to strain. Based on these considerations, the effects of cyclic strain on stem cells have been investigated.

In important work, Saha and others applied biaxial cyclic strain to human embryonic stem cells (hESCs) cultured on a deformable elastic substrate and found that strains >10 % inhibited cell differentiation, independent of strain frequency (Saha et al. 2006). They subsequently found that this strain promoted the secretion of transforming growth factor beta (TGF-β), activin, and Nodal into the medium, which in turn contributed to the inhibition of differentiation and promotion of self-renewal via autocrine/paracrine signaling (Saha et al. 2008a, b). In conceptually related work, Shimizu and colleagues applied uniaxial cyclic strain to mESCs cultured on silicone membranes, with 4–12 % strain at 1 Hz for 24 h. Cells aligned perpendicular to the direction of strain, and importantly differentiated into a vascular smooth muscle cell fate in a mechanism dependent on the upregulation of platelet derived growth factor receptor beta (PDGFRβ) (Shimizu et al. 2008).

In work that blended cyclic strain with topographical cues (an area discussed greater detail below), Kurpinski and others plated MSCs on elastic substrates patterned with microgrooves to align cell polarity (Kurpinski et al. 2006). When uniaxial strain was applied parallel but not perpendicular to the direction of cell alignment, MSC proliferation and expression of smooth muscle cell markers increased substantively. This work indicates that not only the magnitude and frequency of strain, but its orientation relative to that of the cell, are important.

Dynamic Mechanical Forces During Tissue Morphogenesis

During the process of organismal development, it is appreciated that cells and multicellular structures exert forces on one another in a manner critical for morphogenesis (Ray et al. 2008). Recent innovative investigations have utilized embryonic stem cells as model systems to address fundamental questions about the role of forces in tissue development. Specifically, the development of the optic cup (a structure that contains the retina and underlying retinal pigment epithelium) initially involves the evagination of the nascent structure to create a "bud," followed by invagination from the surface of the bud to create a double walled cup (Fig. 3n) where the outer wall becomes the retinal pigment epithelium and the inner wall the neuroretina (Eiraku et al. 2011, 2012). There has been a longstanding debate in the field about the mechanism that enables the invagination to occur, and specifically whether the physical forces necessary for this process are intrinsic to the retinal tissue or require the external action of another structure such as the lens. Sasai and colleagues showed that under certain conditions in mESC aggregates, the evagination and subsequently the invagination occur, implicating an intrinsic or autonomous mechanism (Fig. 3). They proposed that at the onset of the invagination, the ring of cells that would eventually form the rim of the cup exert cytoskeleton-dependent forces to form a wedge shape and by extension create a flattened outer side of the bud. Subsequent rounds of cell division on that flattened side lead to increased surface area and thereby generate compressive forces that bend the layer inward to yield the cup. This highly innovative blend of stem cell and developmental biology yielded insights not only into fundamental mechanisms of tissue development but also offers future translational promise for the ability of stem cells to generate complex tissue structure in culture.

Topographical and Shape Features of the Stem Cell Niche

In addition to providing resident stem cells with a mechanical milieu, niches offer features that can alter the shape of a cell. On the microscale, ECM, neighboring cells, and in some cases mineralized tissue can modulate and even constrain the surface area or volume available to, and therefore the shape of, a cell in a manner important for its function (Paluch and Heisenberg 2009). Likewise, on the nanoscale, ECM proteins often assemble into fibers and other structural features that modulate the topographical features that an adherent cell experiences. Advances in lithography and in materials science have enabled investigators to investigate the effects of these features on stem cell behavior (Kolind et al. 2012).

Cell Shape

In seminal work, which predated investigation of mechanical properties on stem cell differentiation, McBeath et al. used microcontact printing to pattern adhesive islands of different sizes onto a surface (Fig. 4) (McBeath et al. 2004). When MSCs were seeded onto these substrates, it was found that large 10,000 μm^2 islands permitted cell spreading and promoted osteogenic differentiation, whereas small 1,024 μm^2

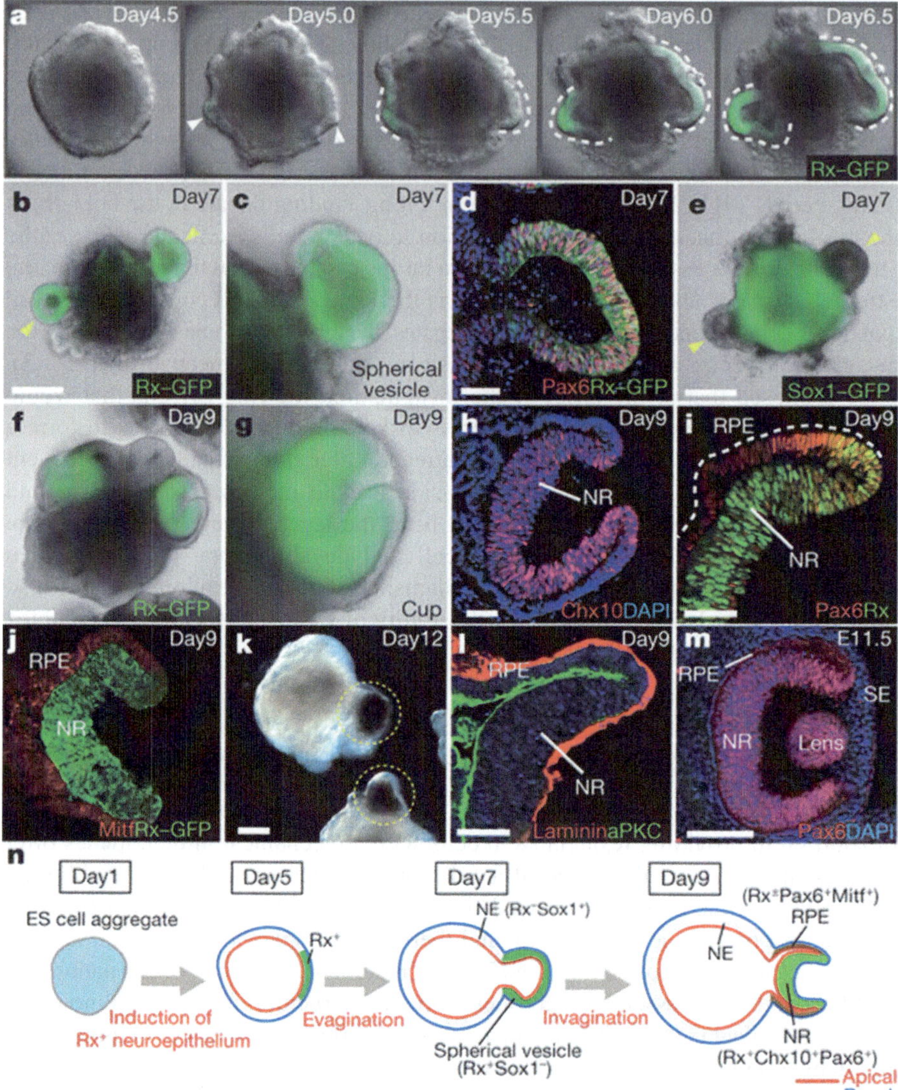

Fig. 3 Development of an optic cup in embryonic stem cell cultures (From Eiraku et al. 2011). (**a**) Mouse embryonic stem cells expressing GFP under the control of the retina and anterior neural fold homeobox promoter were grown in aggregates. Over 6.5 days, evaginated structures reminiscent of the nascent optic cup were progressively formed. (**b, c, f, g**) The resulting structure fully budded to form a vesicle, which then invaginated to form a cup. (**d, e, h, i**) The resulting structures were initially positive for the retinal marker Pax6 but negative for the immature neuroectodermal marker Sox1, and subsequently expressed neuroretinal marker Chx10. (**j**) The outer shell of the cup, corresponding to retinal pigment epithelium, expressed the marker MITF and (**k**) subsequently generated pigment. (**l**) The polarized cell layers expressed the apical marker aPKC and laminin in a surrounding basement membrane. (**m**) The corresponding structure of the E11.5 mouse eye. (**n**) A schematic for optic cup self-formation. *NE* neuroectoderm, *NR* neuroretina, *RPE* retinal pigment epithelium, *SE* surface ectoderm

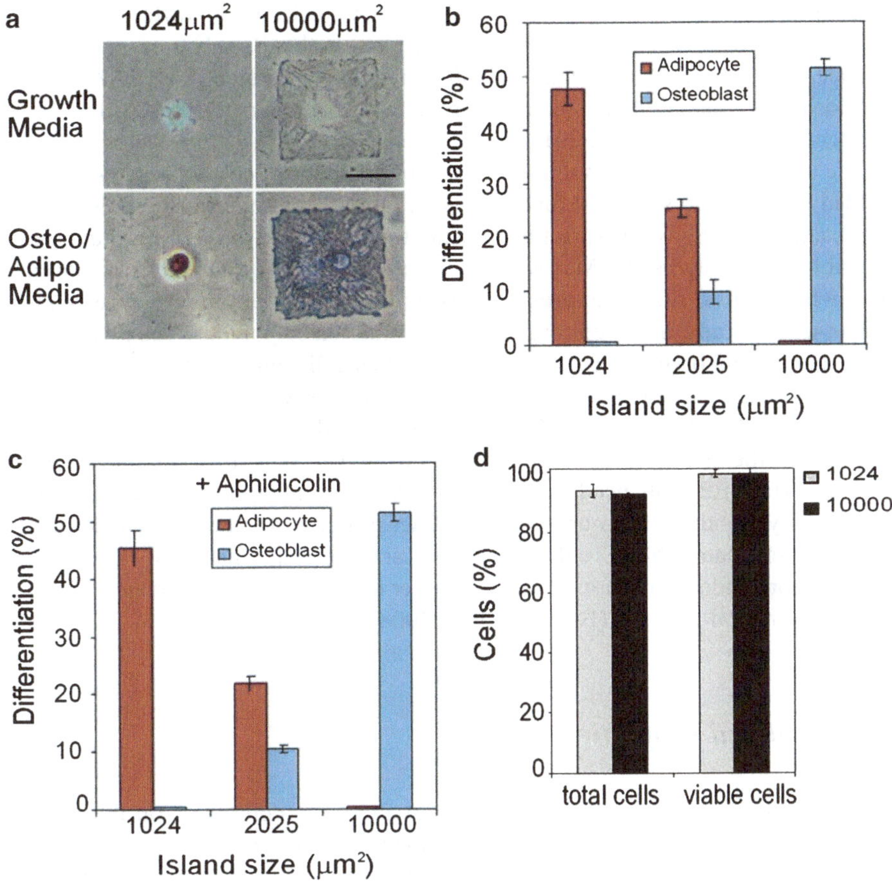

Fig. 4 Cell shape directs mesenchymal stem cell differentiation (From McBeath et al. 2004). (**a**) Human MSCs were plated onto small (1,024 μm²) or large (10,000 μm²) adhesive patterns coated with fibronectin after 1 week in growth or mixed medium. Under mixed differentiation conditions, cells differentiated into adipocytes (*red*) or osteocytes (*blue*). (**b** and **c**) Proportion of MSC differentiation into adipocytes or osteoblasts after 1 week of culture on 1,024, 2,025, or 10,000 μm² islands, either without (**b**) or with (**c**) the mitotic inhibitor aphidicolin. (**d**) Total cells and viable cells within the cultures

islands did not enable substantial cell spreading promoted adipogenic differentiation. Furthermore, cell spreading led to activation of RhoA, and its inhibition promoted to adipocyte differentiation. In subsequent work from this group, which extended the principle to another MSC fate decision, Gao and colleagues showed that upon exposure to TGF-β3, well spread MSCs underwent smooth muscle cell (SMC) differentiation. In contrast, rounded MSCs differentiated into chondrocytes (Gao et al. 2010). Furthermore, the SMC fate was dependent on Rac1 activation and subsequent N-cadherin expression. These results thus indicate that controlling cell shape, which has an impact on the cytoskeleton, invokes cellular mechanical mechanisms analogous to those observed for MSCs cultured on different stiffness substrates. In fact, Dupont and others showed that both stiff substrates and micro-contact printed islands that enable MSC spreading promoted YAP/TAZ nuclear translocation, which as discussed above plays a key role in mechanosensitive MSC differentiation (Dupont et al. 2011).

The use of materials technologies to control the shape of cell aggregates has also yielded insights into stem cell function. For example, microfabricated polydimeth-ylsiloxane (PDMS) wells surrounded with functionalized protein-resistant, self-assembled monolayers (SAMs) were used to create hESC aggregates of uniform size, which led to higher expression levels of the pluripotency marker Oct-4. In another key study, hESC culture inside microfabricated polyethylene glycol wells yielded embryoid bodies (EBs, an embryonic stem cell aggregate that differentiates in a manner bearing some similarities to that of an early embryo) of various sizes (Karp et al. 2007; Hwang et al. 2009). Higher endothelial cell differentiation was observed in the smaller (150 µm) EBs, due to higher Wnt5a expression, whereas larger (450 µm) EBs enhanced cardiogenesis, as a result of higher Wnt11 expression. Interestingly, another group used microcontact printing of adhesive islands on 2D substrates to control hESC colony size and showed that smaller hESC colonies became more endoderm-biased, whereas larger colonies exhibited greater differen-tiation into neural lineages (Bauwens et al. 2008).

Topographical Properties

In addition to microenvironmental properties that alter cell shape on the micron scale, topographical cues—such as the organization of the ECM into fibers (Singh et al. 2010)—offer a cell with features that can modulate its shape at the nanometer scale. Such topographical cues are considered to provide features intermediate between a 2D and a 3D microenvironment, and they can be generated synthetically by several techniques, including electrospinning, self-assembly of materials, and lithography based methods.

In early work in this area, coculture of adult NSCs with astrocytes on microgrooves patterned into polystyrene led to significantly higher extents of neuronal differentiation compared to flat surfaces (Recknor et al. 2006). Another study explored the effects of electrospun fibers of polyethersulfone with different dimensions on the behavior of adult NSCs, and they found that fibers of small dimension (283 nm) promoted oligodendrocyte specification, whereas larger fibers (749 nm) increased neuronal differentiation (Christopherson et al. 2009; Fig. 5).

Leong and colleagues investigated the behavior of human MSCs on a 350 nm grating topography and found that cells exhibited smaller and more dynamic focal adhesions, which apparently underlay a faster cell migration along the direction of the grating (Kulangara et al. 2012). They interestingly found that cells specifically on this grating size had lower expression levels of the focal adhesion protein zyxin, which likely contributed to the focal adhesion behavior. Furthermore,

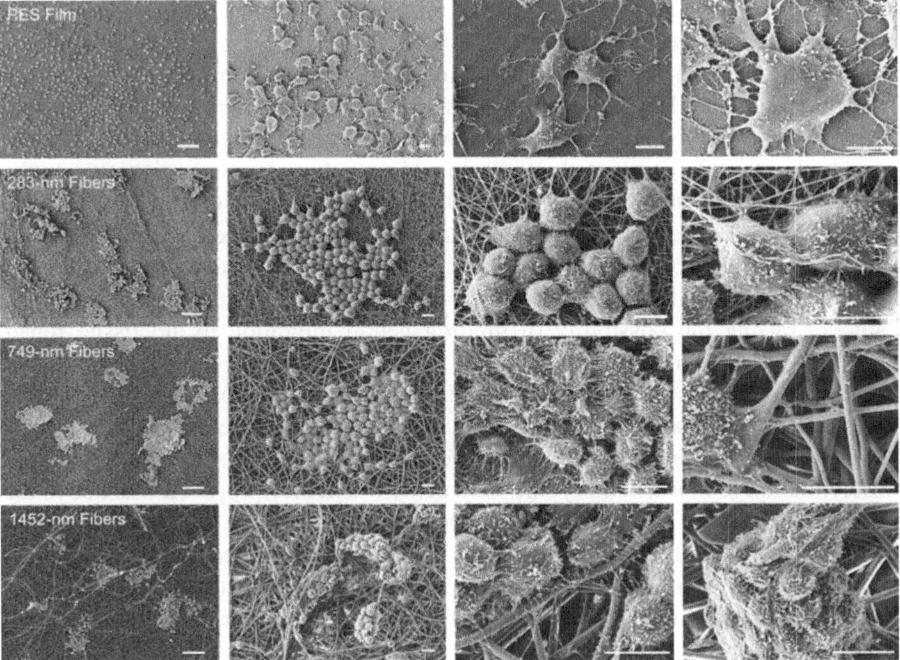

Fig. 5 Substrate topography modulates neural stem cell differentiation (From Christopherson et al. 2009). Adult rat neural stem cells were cultured on flat polyethersulfone (PES) films, or electrospun PES fibers of various dimensions. Scanning electron microscopy images of cultured in serum free medium and FGF-2 for 5 days show strong engagement with the surface topography. Under differentiating conditions, and relative to cultures on tissue culture polystyrene, cells showed a 40 % increase in oligodendrocyte specification on the 283 nm fibers and a 20 % increase in neuronal differentiation on the 749 nm PES fibers

addition of the neuronal inducing factor retinoic acid led to higher extents of neuronal marker expression from MSCs on the 350 nm spaced grating (Yim et al. 2007).

Electric Fields

The role of electrophysiology in the cardiovascular and nervous systems is well appreciated, and several investigators have explored the possibility that electric fields could play a role in regulating the function of stem cells in these tissues. In seminal work in this area, Radisic et al. (2004) seeded neonatal cardiomyocytes onto a collagen sponge and subjected them to a square wave electrical field with 1 Hz frequency, to emulate the natural electrophysiological environment of the heart. Cells became aligned with the direction of the field, exhibited a substantial increase in contractile amplitude, and expressed higher levels of various cardiac protein markers compared to cells that were not electrically stimulated. In another key study, electric fields designed to mimic neuronal activity enhanced the differentiation of muscle precursor cells (Serena et al. 2008).

The role of electric fields in the maturation and neurite outgrowth of neurons has long been studied (Hinkle et al. 1981), and new materials, for example an electrically conducting polymer, have been shown to enhance neurite outgrowth and neuronal maturation (Schmidt et al. 1997). In addition, the role of electric fields on the behavior of neural stem cells has recently been studied. For example, oscillating electric fields were found to favor NSC survival at a 1 Hz frequency, and to promote astrocytic over neuronal differentiation at 1 Hz (Matos and Cicerone 2010). Under a direct current, adult subependymal neural precursors were found to migrate to the cathode, raising the possibility of guiding neural migration to aid tissue repair (Babona-Pilipos et al. 2011). Also under direct current, adult hippocampal neural progenitor cells experienced lower viability yet a higher proportion of neuronal differentiation (Ariza et al. 2010). Finally, Kabiri and colleagues showed that incorporation of carbon nanotubes into an aligned nanofiber scaffold enhanced the neuronal differentiation of murine embryonic stem cells, which they proposed was due to the resulting increase in the conductivity of the material (Kabiri et al. 2012).

In addition, the effects of electric fields have been investigated on stem cells other than those of the heart and nervous system. For instance, application of direct current to mESC-derived embryoid bodies apparently enhanced endothelial differentiation in a manner dependent on the formation of reactive oxygen species (Sauer et al. 2005). In another example, pulsed electromagnetic field stimulation of human MSC cultures promoted the expression of osteogenic markers over a 28 day period (Tsai et al. 2009).

In summary, the role of electric fields in regulating the function of cardiac, neural, and other cells is being increasingly studied, though not yet to the level of activity of mechanical properties. There may thus be additional opportunities to diversity investigation in this area.

Mass Transfer Influences on Stem Cell Behavior

The transport of mass through tissues can exert strong effects on the local composition of those tissues, including the convection and diffusion of nutrients, oxygen, and signaling molecules. Furthermore, the mass transfer properties of the tissue can modulate the spatial distribution of locally produced molecules. As a result, mass transport phenomena play a role in modulating the behavior of stem cells within their niche, and on a larger scale aid in establishing useful or functional heterogeneity within a tissue.

Gradient Formation and Morphogenesis

As originally proposed by Turing 60 years ago, gradients of signaling factors secreted from signaling centers can aid in tissue patterning and differentiation during development (Turing 1990). The cells undergoing patterning in developing tissues are often stem cells, and a number of studies have modeled the dynamic effects of morphogen gradients on cell lineage commitment and morphogenesis (Reeves et al. 2006; Saha and Schaffer 2006; Torii 2012).

Such transport limitations can actually pose challenges in stem cell culture systems. In many situations it would be desirable to uniformly differentiate stem cells into a specific cell type for therapeutic application; however, transport limitations of both components from the medium as well as factors produced by the cells themselves, particularly within cellular aggregates such as EBs, can yield highly heterogeneous cultures of differentiated cells. One creative approach to overcome the problem of the diffusion of culture medium components into an embryoid body is to embed the factors, or specifically microspheres for controlled release of such factors, into the EBs (Carpenedo et al. 2009; Bratt-Leal et al. 2011). This approach offers considerable promise for reducing heterogeneity. Additionally, the optimized assembly of stem cells into aggregates can access mechanisms of pattern formation utilized in organismal development, which can thereby yield considerable insights into developmental mechanisms as well as create more complex structures of potential future utility for tissue engineering (Eiraku et al. 2011; Suga et al. 2011; Eiraku and Sasai 2012).

Oxygen

The atmospheric oxygen concentration (20 %) is higher than that in most organs of the body, despite the close proximity (~100 µm) of cells to capillaries in vascularized tissues (Chow et al. 2001). This consideration raises the possibility that stem cells could behave differently in atmospheric vs. niche oxygen levels.

In seminal work in this area, Koller and colleagues found that mononuclear cells isolated from human cord blood and bone marrow proliferated more rapidly and maintained higher frequencies of several colony forming cells when cultured in reduced vs. atmospheric oxygen (Koller et al. 1992). As a key example within the nervous system, Studer and others demonstrated that reduced oxygen (3 %) enhanced the survival and proliferation of neural precursors, as well as enhanced their differentiation into dopaminergic neurons from 18 to 56 % (Studer et al. 2000). Reduced oxygen upregulated the expression of several proteins, including erythropoietin (EPO), and EPO addition to the medium partially recapitulated the effects of reduced oxygen. In another study, culture in 5 % oxygen enhanced clonogenic neural crest stem cell differentiation into a sympathoadrenal lineage (Morrison et al. 2000). The development of small-scale culture systems in which oxygen can be easily and readily controlled would enable additional study.

Development of Novel Technologies to Study Stem Cells

In addition to offering principles to guide the discovery of novel mechanisms for stem cell regulation, the physical sciences and engineering offer technologies that enable new experimental investigations of stem cells. These include the various innovative material systems discussed above, as well as novel materials technologies developed to manipulate and apply mechanical loads to cells. For example, Ikuta and colleagues at the University of Tokyo have used photo-patterned 3D polymerization of materials to create various devices that can subsequently be interfaced with optical trapping for actuation. With the use of a robotic arm for operator control, they generated a pincer device that can apply defined mechanical loads to cells, for either measurement or mechanical perturbation.

In addition, the ability to genetically manipulate a stem cell is useful both for exploring molecular mechanisms involved in cellular function, as well as in the future for enhancing the therapeutic potential of those cells. Ma and colleagues conjugated magnetic particles to gene delivery vehicles, which enabled magnetically guided gene delivery to cells *in vitro* or *in vivo* (Li et al. 2008). While this technology has not yet been applied to stem cells, it will be promising to do so. Recent developments in protein engineering to create better gene delivery vehicles to stem cells (Asuri et al. 2012), as well as site-specific nucleases to aid in homologous recombination in stem cells (Hockemeyer et al. 2011), are also promising biological approaches for genetically manipulating stem cells.

Another capability that is important for stem cell research, and in particular future translational efforts, is cell separations. Investigators at Lund University have generated a microfluidic device that uses sound waves to separate blood cells based on differences in density, a process termed acoustophoresis (Dykes et al. 2011). In addition, in a dielectrophoretic separation, Miyata and colleagues used alternating electric fields in conjunction with patterned surfaces to achieve cell separations. This process has promise both in separating, for example, differentiated from undifferentiated stem cells, but also in patterning cell deposition on a surface. As a final example, Scadden and colleagues have applied pulsed electric fields to cell mixtures containing hematopoietic stem cells (HSCs) (Eppich et al. 2000). The fields selectively introduce pores into the membranes of larger cells, and in certain parameter ranges the resulting toxicity provides a means to ablate differentiated cells while preserving the smaller HSCs.

Therefore, principles and practices from materials science, physics, electrical engineering, and protein engineering will continue to make new technologies that benefit stem cell research.

Future Directions

Biomaterials Development

As discussed in the Introduction, a number of biophysical properties of the cellular microenvironment are difficult to manipulate and vary genetically, as these properties depend on contributions from more than a single gene. As a result, the systems described in this chapter have been not only interesting but also critical for assessing the effects of these properties on the function of stem cells. There are considerable opportunities for further innovation, especially in materials development. For example, stem cell lineage commitment is an inherently dynamic process, and elucidating the roles of biophysical cues in regulating the process would benefit from the ability to temporally vary cues. While it is straightforward to dynamically change soluble biochemical cues, and even mechanical inputs such as shear and strain, it is challenging to vary static mechanical properties, topographical cues, and immobilized biochemical cues. However, there has been considerable progress in using, for example, photoresponsive materials where light exposure can alter crosslinking and therefore vary mechanical properties in both space and time (Kloxin et al. 2010; Guvendiren and Burdick 2012). Furthermore, using materials with shape memory enables the dynamic variation of topographical inputs (Le et al. 2011). Finally, the capability to temporally vary local biochemical cues within a material has recently been developed (Kloxin et al. 2009). The future application of such systems to problems in stem cell biology promises novel insights.

In addition, a number of the systems described above enable systematic investigation of the role of an individual microenvironmental property on cell function, yet the niche of course simultaneously exposes cells to many of these properties. The development of advanced systems for multiparameter control of the cellular microenvironment promises to yield insights into how these inputs combinatorially modulate cell function (Gobaa et al. 2011). While there have thus been a number of promising advances in materials engineering, one area that could benefit from additional advances is the study of cell-cell interactions within the niche. Specifically, the ratios and relative positions of various cells within the niche are likely tightly regulated, and investigations of such systems will benefit from new technologies to precisely control the positioning of multiple cell types in culture.

Mechanistic Elucidation

Investigating the mechanisms by which physical and engineering properties of the cellular microenvironment modulate stem cell behavior requires expertise not only in the creation and fabrication of systems to vary these properties but also in the molecular elucidation of cell responses. That it is, these studies require expertise in both the physical sciences and engineering and in biology. As described above, in a number of studies that have melded these fields, there has been progress in elucidating mechanisms by which static mechanical properties, shear flow, and topography modulate stem cell function. That said, there are many unknowns in this field that will benefit from collaborations among investigators with complementary expertise. Furthermore, in general the interface between the physical sciences, engineering, and biology represents a major opportunity to train a new generation of scientists capable of highly interdisciplinary research.

Additional Opportunities

In addition to elucidating fundamental roles of biophysical properties on stem cell function, as described above, there are many opportunities to develop novel technologies to benefit both fundamental and translational stem cell research. One critical capability for mechanistic study is the ability to genetically manipulate a stem cell—both to add genetic material and to conduct gene targeting or genome editing. Furthermore, the ability to isolate and investigate specific cell types will benefit from both new affinity agents and novel cell separation modalities. As with basic investigations, each of these areas requires expertise in physical sciences, engineering, and biology.

Global Assessment and Conclusions

The United States is currently a leader in studying the roles of physical sciences and engineering principles in stem cell research. This position to date has benefitted from a broad and deep community of engineers, physical scientists, and materials scientists who have increasingly investigated not only applied but also fundamental questions in the biological sciences. Other countries with pronounced strengths in this area include Japan, Germany, and Switzerland, which have also invested in stem cell biology and engineering research in a manner that is progressively converging.

While the leadership role of the United States in materials science and engineering has clearly benefitted the stem cell engineering field, other countries have recently played an increasing role in the application of physical sciences and engineering to develop new technologies to study stem cells. These include imaging technologies, cell separation technologies, and especially high-throughput "microenvironmental screening" methodologies. Furthermore, both strong government (e.g., Switzerland and Sweden) and private foundation (e.g., Fraunhofer Institutes in Germany and I-STEM in France) support for technology development, application, and commercialization are models that merit deeper study.

In the future, knowledge of biophysical and biochemical regulation of stem cell properties will increasingly be integrated, which will progressively increase our appreciation of the complex means by which the niche orchestrates the various behavioral choices available to resident stem cells. These studies will also provide increasing levels of quantitative data that can be integrated into computational and modeling efforts. Furthermore, the resulting knowledge will aid in the development of cell culture systems for the reproducible, scalable, and economical expansion and differentiation of stem cells for therapeutic application.

References

Ariza, C. A., A.T. Fleury, C.J. Tormos, V. Petruk, S. Chawla, J. Oh, D.S. Sakaguchi, and S.K. Mallapragada. 2010. The influence of electric fields on hippocampal neural progenitor cells. *Stem Cell Rev.* 6:585–600.

Asuri, P., M.A. Bartel, T. Vazin, J.-H. Jang, T.B. Wong, and D.V. Schaffer. 2012. Directed evolution of adeno-associated virus for enhanced gene delivery and gene targeting in human pluripotent stem cells. *Mol. Ther.* 20:329–338.

Babona-Pilipos, R., I.A. Droujinine, M.R. Popovic, and C.M. Morshead. 2011. Adult subependymal neural precursors, but not differentiated cells, undergo rapid cathodal migration in the presence of direct current electric fields. *PLoS One* 6:e23808.

Banerjee, A., M. Arha, S. Choudhary, R.S. Ashton, S.R. Bhatia, D.V. Schaffer, and R.S. Kane. 2009. The influence of hydrogel modulus on the proliferation and differentiation of encapsulated neural stem cells. *Biomaterials* 30:4695–4699.

Bauwens, C.L., R. Peerani, S. Niebruegge, K.A. Woodhouse, E. Kumacheva, M. Husain, and P.W. Zandstra. 2008. Control of human embryonic stem cell colony and aggregate size heterogeneity influences differentiation trajectories. *Stem Cells* 26:2300–10.

Bratt-Leal, A.M., R.L. Carpenedo, M.D. Ungrin, P.W. Zandstra, and T.C. McDevitt. 2011. Incorporation of biomaterials in multicellular aggregates modulates pluripotent stem cell differentiation. *Biomaterials* 32:48–56.

Carpenedo, R.L., A.M. Bratt-Leal, R.A. Marklein, S.A. Seaman, N.J. Bowen, J.F. McDonald, and T.C. McDevitt. 2009. Homogeneous and organized differentiation within embryoid bodies induced by microsphere-mediated delivery of small molecules. *Biomaterials* 30:2507–2515.

Chow, D.C., L.A. Wenning, W.M. Miller, and E.T. Papoutsakis. 2001. Modeling pO2 distributions in the bone marrow hematopoietic compartment. I. Krogh's model. *Biophysical Journal* 81:675–684.

Chowdhury, F., Y. Li, Y.C. Poh, T. Yokohama-Tamaki, N. Wang, and T.S. Tanaka. 2010. Soft substrates promote homogeneous self-renewal of embryonic stem cells via downregulating cell-matrix tractions. *PLoS One* 5:e15655.

Christopherson, G.T., H. Song, and H.-Q. Mao. 2009. The influence of fiber diameter of electrospun substrates on neural stem cell differentiation and proliferation. *Biomaterials* 30:556–564.

Discher, D.E., D.J. Mooney, and P.W. Zandstra. 2009. Growth factors, matrices, and forces combine and control stem cells. *Science* 324:1673–1677.

Dupont, S., L. Morsut, M. Aragona, E. Enzo, S. Giulitti, M. Cordenonsi, F. Zanconato, J. Le Digabel, M. Forcato, S. Bicciato, N. Elvassore, and S. Piccolo. 2011. Role of YAP/TAZ in mechanotransduction. *Nature* 474:179–183.

Dykes, J., A. Lenshof, I.-B. Åstrand-Grundström, T. Laurell, and S. Scheding. 2011. Efficient removal of platelets from peripheral blood progenitor cell products using a novel micro-chip based acoustophoretic platform. *PLoS One* 6:e23074.

Eiraku, M., T. Adachi, and Y. Sasai. 2012. Relaxation-expansion model for self-driven retinal morphogenesis. *BioEssays* 34:17–25.

Eiraku, M. and Y. Sasai. 2012. Self-formation of layered neural structures in three-dimensional culture of ES cells. *Curr. Opin. Neurobiol.*, March 9, 2012.

Eiraku, M., N. Takata, H. Ishibashi, M. Kawada, E. Sakakura, S. Okuda, K. Sekiguchi, T. Adachi, and Y. Sasai. 2011. Self-organizing optic-cup morphogenesis in three-dimensional culture. *Nature* 472:51–56.

Engler, A.J., S. Sen, H.L. Sweeney, and D.E. Discher. 2006. Matrix elasticity directs stem cell lineage specification. *Cell* 126:677–689.

Eppich, H.M., R. Foxall, K. Gaynor, D. Dombkowski, N. Miura, T. Cheng, S. Silva-Arrieta, R.H. Evans, J.A. Mangano, F.I. Preffer, and D.T. Scadden. 2000. Pulsed electric fields for selection of hematopoietic cells and depletion of tumor cell contaminants. *Nat. Biotech.* 18:882–887.

Fu, J., Y.-K. Wang, M.T. Yang, R.A. Desai, X. Yu, Z. Liu, and C.S. Chen. 2010. Mechanical regulation of cell function with geometrically modulated elastomeric substrates. *Nat. Meth.* 7:733–736.

Gao, L., R. McBeath, and C.S. Chen. 2010. Stem cell shape regulates a chondrogenic versus myogenic fate through Rac1 and N-cadherin. *Stem Cells* 28:564–572.

Gilbert, P.M., K.L. Havenstrite, K.E.G. Magnusson, A. Sacco, N.A. Leonardi, P. Kraft, N.K. Nguyen, S. Thrun, M.P. Lutolf, and H.M. Blau. 2010. Substrate elasticity regulates skeletal muscle stem cell self-renewal in culture. *Science* 329:1078–1081.

Gobaa, S., S. Hoehnel, M. Roccio, A. Negro, S. Kobel, and M.P. Lutolf. 2011. Artificial niche microarrays for probing single stem cell fate in high throughput. *Nat. Meth.* 8:949–955.

Guilak, F., D.M. Cohen, B.T. Estes, J.M. Gimble, W. Liedtke, and C.S. Chen. 2009. Control of stem cell fate by physical interactions with the extracellular matrix. *Cell Stem Cell* 5:17–26.

Guvendiren, M. and J.A. Burdick. 2012. Stiffening hydrogels to probe short- and long-term cellular responses to dynamic mechanics. *Nat. Commun.* 3:792.

Hinkle, L., C.D. McCaig, and K.R. Robinson. 1981. The direction of growth of differentiating neurones and myoblasts from frog embryos in an applied electric field. *J. Physiol.* 314:121–35.

Hockemeyer, D., H. Wang, S. Kiani, C.S. Lai, Q. Gao, J.P. Cassady, G.J. Cost, L. Zhang, Y. Santiago, J.C. Miller, B. Zeitler, J.M. Cherone, X. Meng, S.J. Hinkley, E.J. Rebar, P.D. Gregory, F.D. Urnov, and R. Jaenisch. 2011. Genetic engineering of human pluripotent cells using TALE nucleases. *Nat. Biotech.* 29:731–734.

Huebsch, N., P.R. Arany, A.S. Mao, D. Shvartsman, O.A. Ali, S.A. Bencherif, J. Rivera-Feliciano, and D.J. Mooney. 2010. Harnessing traction-mediated manipulation of the cell/matrix interface to control stem-cell fate. *Nat. Mater.* 9:518–526.

Humphrey, J.D. 2003. Review Paper: Continuum biomechanics of soft biological tissues. *Proc. Roy. Soc. Lond. Ser. A: Math., Phys. Eng. Sci.* 459:3–46.

Hwang, Y.S., B.G. Chung, D. Ortmann, N. Hattori, H.C. Moeller, and A. Khademhosseini. 2009. Microwell-mediated control of embryoid body size regulates embryonic stem cell fate via differential expression of WNT5a and WNT11. *Proc. Natl. Acad. Sci. USA* 106:16978–16983.

Illi, B., A. Scopece, S. Nanni, A. Farsetti, L. Morgante, P. Biglioli, M.C. Capogrossi, and C. Gaetano. 2005. Epigenetic histone modification and cardiovascular lineage programming in mouse embryonic stem cells exposed to laminar shear stress. *Circ. Res.* 96:501–8.

Kabiri, M., M. Soleimani, I. Shabani, K. Futrega, N. Ghaemi, H.H. Ahvaz, E. Elahi, and M.R. Doran. 2012. Neural differentiation of mouse embryonic stem cells on conductive nanofiber scaffolds. *Biotechnol. Lett.* 34:1357–1365.

Karp, J.M., J. Yeh, G. Eng, J. Fukuda, J. Blumling, K.Y. Suh, J. Cheng, A. Mahdavi, J. Borenstein, R. Langer, and A. Khademhosseini. 2007. Controlling size, shape and homogeneity of embryoid bodies using poly(ethylene glycol) microwells. *Lab Chip* 7:786–94.

Keung, A.J., E.M. de Juan-Pardo, D.V. Schaffer, and S. Kumar. 2011. Rho GTPases mediate the mechanosensitive lineage commitment of neural stem cells. *Stem Cells* 29:1886–97.

Keung, A.J., S. Kumar, and D.V. Schaffer. 2010. Presentation counts: microenvironmental regulation of stem cells by biophysical and material cues. *Ann. Rev. Cell. Dev. Biol.* 26:533–56.

Kloxin, A.M., J.A. Benton, and K.S. Anseth. 2010. *In situ* elasticity modulation with dynamic substrates to direct cell phenotype. *Biomaterials* 31:1–8.

Kloxin, A.M., A.M. Kasko, C.N. Salinas, and K.S. Anseth. 2009. Photodegradable hydrogels for dynamic tuning of physical and chemical properties. *Science* 324:59–63.

Kolind, K., K.W. Leong, F. Besenbacher, and M. Foss. 2012. Guidance of stem cell fate on 2D patterned surfaces. *Biomaterials* 33:6626–6633.

Koller, M.R., J.G. Bender, W.M. Miller, and E.T. Papoutsakis. 1992. Reduced oxygen tension increases hematopoiesis in long-term culture of human stem and progenitor cells from cord blood and bone marrow. *Exp. Hematol.* 20:264–70.

Kulangara, K., Y. Yang, J. Yang, and K.W. Leong. 2012. Nanotopography as modulator of human mesenchymal stem cell function. *Biomaterials* 33:4998–5003.

Kurpinski, K., J. Chu, C. Hashi, and S. Li. 2006. Anisotropic mechanosensing by mesenchymal stem cells. *Proc. Natl. Acad. Sci. USA* 103:16095–100.

Le, D.M., K. Kulangara, A.F. Adler, K.W. Leong, and V.S. Ashby. 2011. Dynamic topographical control of mesenchymal stem cells by culture on responsive poly($\square$-caprolactone) surfaces. *Adv. Mat.* 23:3278–3283.

Li, W., N. Ma, L.-L. Ong, A. Kaminski, C. Skrabal, M. Ugurlucan, P. Lorenz, H.-H. Gatzen, K. Lützow, A. Lendlein, B.M. Pützer, R.-K. Li, and G. Steinhoff. 2008. Enhanced thoracic gene delivery by magnetic nanobead-mediated vector. *J. Gene Med.* 10:897–909.

Lie, D.C., S.A. Colamarino, H.J. Song, L. Desire, H. Mira, A. Consiglio, E.S. Lein, S. Jessberger, H. Lansford, A.R. Dearie, and F.H. Gage. 2005. Wnt signalling regulates adult hippocampal neurogenesis. *Nature* 437:1370–5.

Lutolf, M.P., P.M. Gilbert, and H.M. Blau. 2009. Designing materials to direct stem-cell fate. *Nature* 462:433–441.

Matos, M.A. and M.T. Cicerone. 2010. Alternating current electric field effects on neural stem cell viability and differentiation. *Biotech. Progr.* 26:664–670.

Matsuura, K., N. Hayashi, Y. Kuroda, C. Takiue, R. Hirata, M. Takenami, Y. Aoi, N. Yoshioka, T. Habara, T. Mukaida, and K. Naruse. 2010. Improved development of mouse and human embryos using a tilting embryo culture system. *Reprod. Biomed. Online* 20:358–64.

McBeath, R., D.M. Pirone, C.M. Nelson, K. Bhadriraju, and C.S. Chen. 2004. Cell shape, cytoskeletal tension, and RhoA regulate stem cell lineage commitment. *Dev. Cell* 6:483–495.

Morrison, S.J., M. Csete, A.K. Groves, W. Melega, B. Wold, and D.J. Anderson. 2000. Culture in reduced levels of oxygen promotes clonogenic sympathoadrenal differentiation by isolated neural crest stem cells. *J. Neuroscience* 20:7370–7376.

Nusse, R., A. van Ooyen, D. Cox, Y.K. Fung, and H. Varmus. 1984. Mode of proviral activation of a putative mammary oncogene (int-1) on mouse chromosome 15. *Nature* 307:131–136.

Nusslein-Volhard, C., and E. Wieschaus. 1980. Mutations affecting segment number and polarity in *Drosophila*. *Nature* 287:795–801.

Paluch, E., and C.-P. Heisenberg. 2009. Biology and Physics of Cell Shape Changes in Development. *Curr. Biol.* 19:R790-R799.

Radisic, M., H. Park, H. Shing, T. Consi, F.J. Schoen, R. Langer, L.E. Freed, and G. Vunjak-Novakovic. 2004. Functional assembly of engineered myocardium by electrical stimulation of cardiac myocytes cultured on scaffolds. *Proc. Nat. Acad. Sci. USA* 101(52):18129–18134, doi:10.1073/pnas.0407817101.

Ray, K., S. David, and S. Paul. 2008. The forces that shape embryos: physical aspects of convergent extension by cell intercalation. *Phys. Biol.* 5:015007.

Recknor, J.B., D.S. Sakaguchi, and S.K. Mallapragada. 2006. Directed growth and selective differentiation of neural progenitor cells on micropatterned polymer substrates. *Biomaterials* 27:4098–4108.

Reeves, G.T., C.B. Muratov, T. Schüpbach, and S.Y. Shvartsman. 2006. Quantitative models of developmental pattern formation. *Dev. Cell* 11:289–300.

Roelink, H., A. Augsburger, J. Heemskerk, V. Korzh, S. Norlin, A. Ruiz i Altaba, Y. Tanabe, M. Placzek, T. Edlund, T.M. Jessell, and J. Dodd. 1994. Floor plate and motor neuron induction by vhh-1, a vertebrate homolog of hedgehog expressed by the notochord. *Cell* 76:761–775.

Saha, K., A.J. Keung, E.F. Irwin, Y. Li, L. Little, D.V. Schaffer, and K.E. Healy. 2008a. Substrate modulus directs neural stem cell behavior. *Biophys. J.* 95:4426–4438.

Saha, K., J.F. Pollock, D.V. Schaffer, and K.E. Healy. 2007. Designing synthetic materials to control stem cell phenotype. *Curr. Opin. Chem. Biol.* 11:381–387.

Saha, K., and D.V. Schaffer. 2006. Signal dynamics in Sonic hedgehog tissue patterning. *Development* 133:889–900.

Saha, S., L. Ji, J.J. de Pablo, and S.P. Palecek. 2006. Inhibition of human embryonic stem cell differentiation by mechanical strain. *J. Cell. Physiol.* 206:126–137.

Saha, S., L. Ji, J.J. de Pablo, and S.P. Palecek. 2008b. TGF^2/Activin/Nodal pathway in inhibition of human embryonic stem cell differentiation by mechanical strain. *Biophys. J.* 94:4123–4133, doi:10.1529/biophysj.107.119891.

Sauer, H., M.M. Bekhite, J. Hescheler, and M. Wartenberg. 2005. Redox control of angiogenic factors and CD31-positive vessel-like structures in mouse embryonic stem cells after direct current electrical field stimulation. *Exp. Cell Res.* 304:380–90.

Scadden, D.T. 2006. The stem-cell niche as an entity of action. *Nature* 441:1075–1079.

Schmidt, C.E., V.R. Shastri, J.P. Vacanti, and R. Langer. 1997. Stimulation of neurite outgrowth using an electrically conducting polymer. *Proc. Natl. Acad. Sci. USA* 94:8948–8953.

Schofield, R. 1978. Relationship between spleen colony-forming cell and hematopoietic stem-cell — hypothesis. *Blood Cells* 4:7–25.

Serena, E., M. Flaibani, S. Carnio, L. Boldrin, L. Vitiello, P. De Coppi, and N. Elvassore. 2008. Electrophysiologic stimulation improves myogenic potential of muscle precursor cells grown in a 3D collagen scaffold. *Neurol. Res.* 30:207–14.

Sharma, R.P. and V.L. Chopra. 1976. Effect of the Wingless (wg1) mutation on wing and haltere development in *Drosophila melanogaster. Dev. Biol.* 48:461–465.

Shimizu, N., K. Yamamoto, S. Obi, S. Kumagaya, T. Masumura, Y. Shimano, K. Naruse, J.K. Yamashita, T. Igarashi, and J. Ando. 2008. Cyclic strain induces mouse embryonic stem cell differentiation into vascular smooth muscle cells by activating PDGF receptor β. *J. Appl. Physiol.* 104:766–772.

Singh, P., C. Carraher, and J.E. Schwarzbauer. 2010. Assembly of fibronectin extracellular matrix. *Ann. Rev. Cell. Dev. Biol.* 26:397–419.

Smith, M.L., D. Gourdon, W.C. Little, K.E. Kubow, R.A. Eguiluz, S. Luna-Morris, and V. Vogel. 2007. Force-induced unfolding of fibronectin in the extracellular matrix of living cells. *PLoS Biol.* 5:e268.

Studer, L., M. Csete, S.-H. Lee, N. Kabbani, J. Walikonis, B. Wold, and R. McKay. 2000. Enhanced proliferation, survival, and dopaminergic differentiation of CNS precursors in lowered oxygen. *J. Neuroscience* 20:7377–7383.

Suga, H., T. Kadoshima, M. Minaguchi, M. Ohgushi, M. Soen, T. Nakano, N. Takata, T. Wataya, K. Muguruma, H. Miyoshi, S. Yonemura, Y. Oiso, and Y. Sasai. 2011. Self-formation of functional adenohypophysis in three-dimensional culture. *Nature* 480:57–62.

Teixeira, A.I., S. Ilkhanizadeh, J.A. Wigenius, J.K. Duckworth, O. Inganäs, and O. Hermanson. 2009. The promotion of neuronal maturation on soft substrates. *Biomaterials* 30:4567–4572.

Torii, K.U. 2012. Two-dimensional spatial patterning in developmental systems. *Trends Cell Biol.* 22:438–446.

Trappmann, B., J.E. Gautrot, J.T. Connelly, D.G.T. Strange, Y. Li, M.L. Oyen, M.A. Cohen Stuart, H. Boehm, B. Li, V. Vogel, J.P. Spatz, F.M. Watt, and W.T.S. Huck. 2012. Extracellular-matrix tethering regulates stem-cell fate. *Nat. Mater.* 11:642–649.

Tsai, M.T., W.J. Li, R.S. Tuan, and W.H. Chang. 2009. Modulation of osteogenesis in human mesenchymal stem cells by specific pulsed electromagnetic field stimulation. *J. Orthop. Res.* 27:1169–1174.

Turing, A.M. 1990. The chemical basis of morphogenesis. 1953. *Bull. Math. Biol.* 52(1–2):153–197; discussion 119–152.

van Amerongen, R., and R. Nusse. 2009. Towards an integrated view of Wnt signaling in development. *Development* 136:3205–3214.

Watt, F.M., Hogan, and B.L.M. Hogan. 2000. Out of Eden: stem cells and their niches. *Science* 287:1427–1430.

Wichterle, H., I. Lieberam, J.A. Porter, and T.M. Jessell. 2002. Directed differentiation of embryonic stem cells into motor neurons. *Cell* 110:385–97.

Yamamoto, K., T. Sokabe, T. Watabe, K. Miyazono, J.K. Yamashita, S. Obi, N. Ohura, A. Matsushita, A. Kamiya, and J. Ando. 2005. Fluid shear stress induces differentiation of Flk-1-positive embryonic stem cells into vascular endothelial cells *in vitro*. *Am. J. Physiol. Heart Circ. Physiol.* 288:H1915–24.

Yamamoto, K., T. Takahashi, T. Asahara, N. Ohura, T. Sokabe, A. Kamiya, and J. Ando. 2003. Proliferation, differentiation, and tube formation by endothelial progenitor cells in response to shear stress. *J. Appl. Physiol.* 95:2081–8.

Yim, E.K.F., S.W. Pang, and K.W. Leong. 2007. Synthetic nanostructures inducing differentiation of human mesenchymal stem cells into neuronal lineage. *Exp. Cell Res.* 313:1820–1829.

Zeng, L., Q. Xiao, A. Margariti, Z. Zhang, A. Zampetaki, S. Patel, M.C. Capogrossi, Y. Hu, and Q. Xu. 2006. HDAC3 is crucial in shear- and VEGF-induced stem cell differentiation toward endothelial cells. *J. Cell Biol.* 174:1059–1069.

High-Throughput Screening, Microfluidics, Biosensors, and Real-Time Phenotyping

Sean P. Palecek

Introduction

Key opportunities for the field of stem cell engineering involve identification of cues that regulate stem cell fate, constructing a systems level understanding of how cells sense and process information provided by the microenvironment, and designing environments to elicit the desired cell fate. Meeting these opportunities will be facilitated by collaborative, interdisciplinary interactions among engineers, scientists, and clinicians. Chapters "Physical and Engineering Principles in Stem Cell Research" and "Computational Modeling and Stem Cell Engineering" in this report address the principles by which physical cues can affect stem cells and how mathematical modeling can provide insight into mechanisms of stem cell regulation. For these efforts to be successful, spatial and dynamic control over the microenvironment is needed. This chapter will focus on how recent advances in cell culture platform design and manufacture permit systematic application of regulatory cues to stem cells, and the insight these systems have provided in stem cell biology and engineering.

Stem Cell Microenvironment

The stem cell microenvironment is defined as the set of cues provided to the cell *in vitro* or *in vivo*. The microenvironment is an expansion of the stem cell niche concept, the physiologic environment that regulates the quiescence and maintenance of

S.P. Palecek (✉)
Department of Chemical and Biological Engineering, University of Wisconsin-Madison, 3637 Engineering Hall, 1415 Engineering Drive, Madison, WI 53706-1691, USA
e-mail: palecek@engr.wisc.edu

R.M. Nerem et al. (eds.), *Stem Cell Engineering: A WTEC Global Assessment*, Science Policy Reports, DOI 10.1007/978-3-319-05074-4_3,

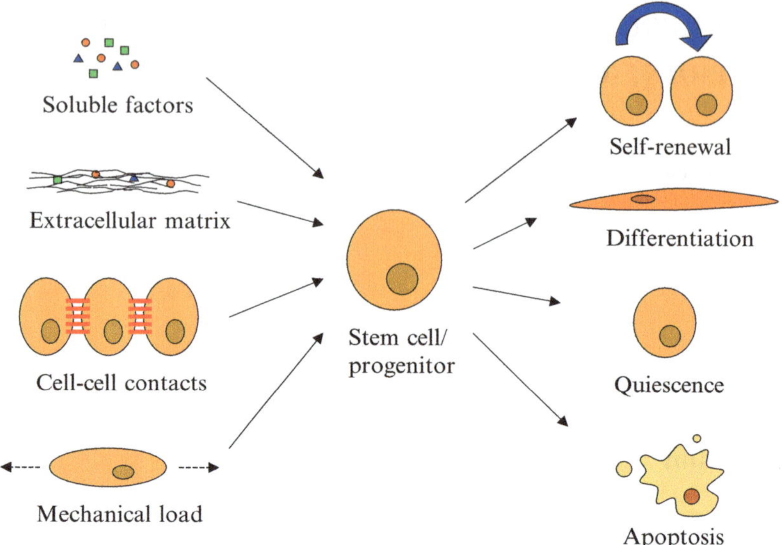

Fig. 1 Schematic of the stem cell microenvironment. Soluble factors, extracellular matrix components, intercellular interactions, and biomechanical cues synergize to regulate cell fate (From Metallo et al. 2007)

stem cells under normal conditions and the mobilization of stem cells for tissue generation in response to damage or disease (Scadden 2006). Chemical and mechanical cues presented in the microenvironment include short and long-range soluble factors, extracellular matrix, cell-cell contact, and mechanical forces (Fig. 1). These cues are sensed by receptors on or in the stem cell, and initiate chemical signaling cascades that affect gene transcription and protein translation, thereby governing a stem cell's status. Signals that guide stem cell fates are not only encoded in the composition of the microenvironment, but also the spatial organization and temporal presentation of the cues.

Stem cell microenvironment engineering often attempts to recreate the native stem cell niche *in vitro* to maintain stem cell potency and facilitate stem cell expansion and differentiation. Alternatively, design of *in vitro* microenvironments permits systematic analysis of how a particular cue or set of cues regulates stem cell fate. Libraries of microenvironments also can be used as tools to screen for conditions that produce desired fates.

High-Throughput Screening

High-throughput screening (HTS) is a method for discovery of active compounds that elicit a desired biological response. HTS is often used to generate lead compounds for drug development and can also be used to identify unknown biochemical interactions. Often, robotic liquid handling systems apply large chemical libraries to biological samples in microwell plates and automated analytic techniques collect data on the biological effects of the compounds. Hits identified in the primary screen are typically validated in a secondary screen and the mechanism of action of confirmed hits can be assessed by standard biochemical techniques.

High-throughput screens are a powerful approach to identify chemical compounds that regulate a cellular phenotype, and require no prior knowledge about mechanisms that regulate the phenotype. Appropriate design of the screening process is critical for achieving a successful outcome. The phenotypic readout should be rapid, inexpensive, and ideally quantitative. In addition to direct phenotypic assessment, reporter cell lines or fluorescent probes are commonly used to analyze molecular changes in cells subjected to HTS.

Application of HTS approaches to stem cells have identified small molecule compounds that enhance stem cell survival, facilitate stem cell self-renewal, direct stem cell differentiation, and improve the efficiency of stem cell reprogramming. Immobilized arrays of matrix proteins, peptides, and synthetic materials have been used to identify defined substrates for stem cell culture. Combinatorial screening platforms have uncovered synergies between different cues and provided insight into how pathways combine environmental signals to regulate stem cell fate. Examples of these advances will be discussed in this chapter.

Microfluidics

Microfluidics involves precise manipulation of small volumes of fluids. Initial applications of microfluidics were primarily in the fields of chemistry and physics, but recent integration of microfluidics with biological cells has produced new insight into microenvironmental regulation of cellular behavior. At the microscale surface tension and fluid energy dissipation are low. Reynolds numbers are often very small so that molecular transport at fluid interfaces is dominated by diffusion, leading to low mixing rates and establishment of predictable molecular gradients.

Microfluidic devices have facilitated analysis of the dynamic response of the stem cell to the chemical and mechanical microenvironment by enabling precise application of cues to stem cells. A recent review by Young and Beebe discusses concepts related to microfluidic control of the cellular microenvironment including state-of-the-art technologies and remaining challenges (Young and Beebe 2010). Stem cell research studies using microfluidic devices have provided insight into mechanisms of autocrine and paracrine signaling and mechanotransduction in different types of stem cells. The ability to culture and analyze individual cells in parallel in microfluidic devices has allowed researchers to assess stem cell population heterogeneity that exists over the stem cell donor source, culture conditions, passage number, and other environmental selections applied to the cells. Microfluidic devices have also been used to separate low abundance stem cells from larger cell populations.

Stem Cell Biosensors

A biosensor is a tool that integrates a biological sensor with an analytic method to detect and report upon components in the microenvironment. In contrast to molecular biosensors, cell-based biosensors can provide a direct readout of the effects of the environment on cell phenotype. Cells used in biosensors are often engineered to have desired properties, including a real-time readout such as expression of a fluorescent or enzymatic reporter, or sensitivity and specificity of detection. Stem cells are a promising source of cells for whole cell biosensors because of their ability to generate a large number of normal cells from a single clonal source. Induced pluripotent stem cells derived from patients can also produce cells that possess disease phenotypes for integration into biosensors. Many cell types important in sensing applications (e.g., neural, cardiac, hepatic) are difficult to obtain and expand from primary sources, and animal models and cell lines fail to adequately recapitulate human responses. Thus, stem cell-based biosensors have the potential to revolutionize toxicity testing, drug discovery and evaluation, environmental monitoring, and clinical sample analysis.

High-Throughput Screening of Factors Regulating Stem Cell Fates

One strategy to regulate stem cell fates involves construction of *in vitro* microenvironments inspired by the *in vivo* niche. However, this rational design strategy is limited by knowledge of developmental biology and physiology, a relatively poor understanding of how different cues synergize to control signaling and cell fate, and the possibility that developmental pathways other than those present *in vivo* may regulate cell fate *in vitro*. HTS platforms offer the potential to identify factors that regulate stem cell fates, to improve upon a lead compound, to optimize presentation of cues, and to identify combinatorial interactions of known regulatory factors. Recent advances in spatially patterning cells, development of compound and materials libraries, and screening methodologies have enhanced our understanding of factors that control stem cell fates, leading to improved stem cell culture and differentiation systems.

Screening Chemical Libraries

Standard HTS methodologies, involving the application of large natural products or combinatorial chemical libraries to stem cells cultured in microwell plates, have been used to identify compounds that regulate stem cell self-renewal and differentiation. In one example, Alves et al. at the University of Twente screened a 1,280 compound library of compounds with known pharmacological activity for the

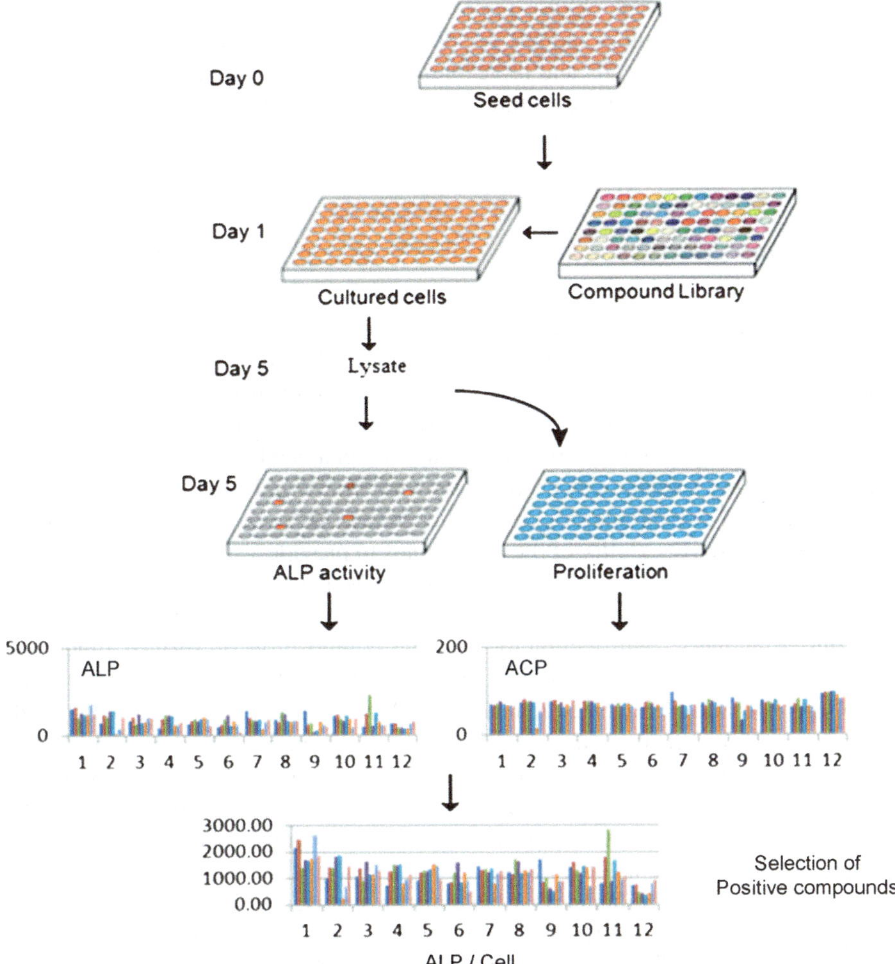

Fig. 2 Schematic of a high-throughput screen of osteogenic enhancers in human mesenchymal stem cells (hMSCs) (From Alves et al. 2011). hMSCs were plated into 96 well plates in osteogenic medium. After 4 days, alkaline phosphatase (*ALP*) and acid phosphatase (*ACP*) activity were measured in a fluorescent plate reader to assess osteogenesis and proliferation, respectively. Positive compounds were confirmed by flow cytometry and subjected to further evaluation

ability to increase osteogenesis in human mesenchymal stem cells (hMSCs) (Alves et al. 2011). High-throughput fluorimetric assays of alkaline phosphatase and acidic phosphatase were used to assess osteogenesis and cell proliferation, respectively (Fig. 2). This study identified novel lead compounds, which would have been difficult to predict based on their mechanism of action, that induced osteogenesis with greater activity than previously known chemical compounds. Another study used 384-well plate screening of a 2,880 compound library to identify novel compounds that promote short-term hESC self-renewal and direct differentiation toward specific lineages (Desbordes et al. 2008). Other high-throughput screens have identified small molecule compounds and siRNAs that promote the survival and differentiation of pluripotent stem cells and molecules involved in regulating reprogramming of somatic cells to a pluripotent state (Andrews et al. 2010; Barbaric et al. 2010; Outten et al. 2011; Xu et al. 2010).

Biomaterials Arrays

The standard chemical screening platforms developed and refined by the pharmaceutical industry work well for identifying soluble chemical factors that regulate stem cell fate, but are not effective at identifying other components of the microenvironment, including matrices and scaffolds, intercellular interactions, and mechanical forces. The development of biomaterial arrays has enabled the identification of how the chemical and mechanical properties of polymers and other materials affect stem cell fates. For example, Mei and colleagues at MIT screened a combinatorial polymer library for the ability of the compounds to provide a substrate that supports human embryonic stem cell (hESC) self-renewal (Mei et al. 2010). The investigators related the biological effects of the polymers to properties including elastic modulus, roughness, hydrophobicity, and the composition of surface functional groups. Another recent study investigated the effects of material topography on hMSC proliferation and osteogenesis (Unadkat et al. 2011). Poly(lactic acid) substrates with identical chemical compositions were printed in arrays with 2,176 random surface topographies. Machine learning algorithms related material topography to stem cell fate and identified features that instruct MSC proliferation and differentiation. Such efforts may prove valuable in designing materials that provide defined matrices for stem cell culture.

Patterned arrays of biomaterials have also been used to identify peptides that support hESC self-renewal when immobilized to a substrate. Laura Kiessling's lab at the University of Wisconsin–Madison used phage display to identify a library of peptides that bind to embryonic carcinoma cells, then created photopatterned arrays of these cells using self-assembled monolayers of alkanethiols on a gold surface (Derda et al. 2010). This array was screened for the ability to maintain self-renewal in hESCs, and two peptides that support hESC culture in defined conditions were identified.

Combinatorial arrays coupled with high-throughput analysis can be used to identify synergistic and antagonistic interactions between factors that regulate signaling pathways controlling stem cell development. For example, an array that contained mixtures of extracellular matrix proteins and growth factors known to be involved in neural development was screened for ability induce differentiation in human neural progenitors (Soen et al. 2006; Fig. 3). Dose and time-dependent analysis of cell differentiation revealed novel insights into interactions between pathways regulating cell fate. In another study, Bhatia and colleagues constructed a microwell-based system to assess the combinatorial interactions of growth

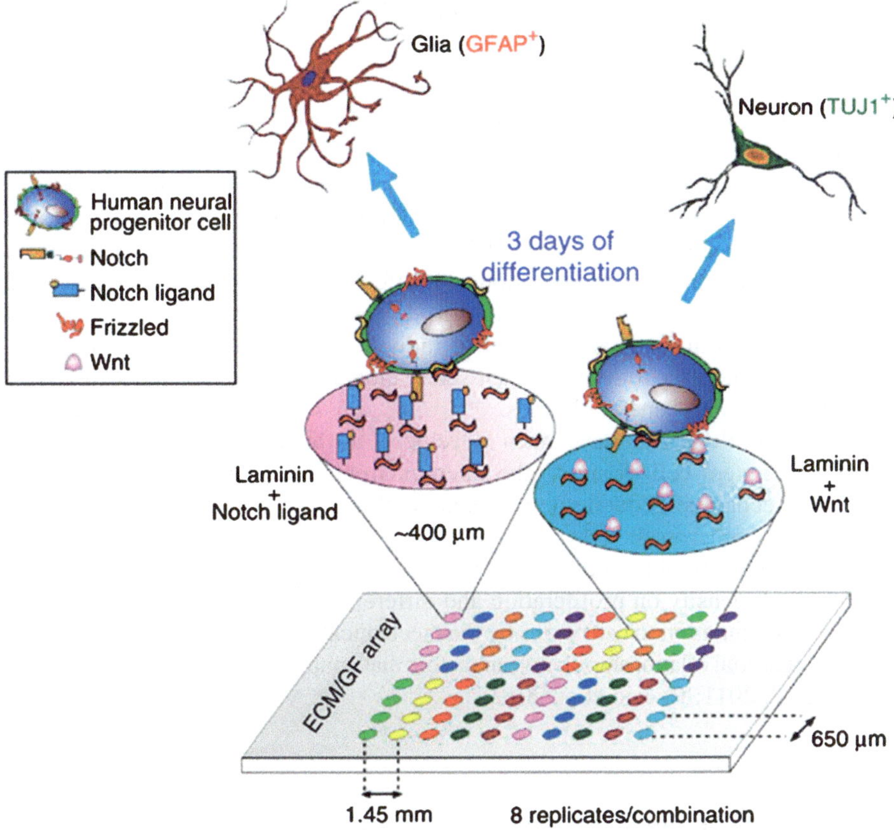

Fig. 3 Schematic of combinatorial screening of extracellular matrix (*ECM*) and growth factor (*GF*) effects on stem cells (From Soen et al. 2006). Arrays of pre-mixed combinations of proteins were printed using a noncontact piezoelectric arrayer. Human neural progenitor cells were seeded onto the arrays and cultured under differentiation-promoting conditions for 3 days. Proliferation and differentiation responses were analyzed by immunostaining for Tuj1, GFAP, and BrdU

factors and extracellular matrix proteins on embryonic stem cell fates, including differentiation to cardiac lineages (Flaim et al. 2008). A microwell chip was developed by Rosenthal and others to provide defined microenvironments consisting of soluble factors, matrices, and cell-cell contact to single cells or small clusters of stem cells (Rosenthal et al. 2007). This culture system was used to determine that intercellular interactions repress mESC colony initiation.

Array-based screening methods have the potential to elucidate how intercellular interactions regulate stem cell fate, including multicellular tissue organization. Patterns of identical composition, but different sizes and shapes, have been constructed to determine that mechanical interactions between cells affect multicellular organization of MSCs, which in turn regulates osteogenic and adipogenic fates (Ruiz and Chen 2008).

Arrays for Clonal Analysis of Stem Cells

Arrays of materials can be used to screen for how chemical and physical properties of the substrate regulates stem cell fate. However, heterogeneity in stem cell populations can lead to stem cell clones responding to the same environment in different manners. Arrays consisting of identical features can be used to probe the extent of heterogeneity that exists in a stem cell population. Ashton and colleagues created micropatterned substrates that enabled formation of microspheres from neural stem cell clones and monitored proliferation and differentiation of neurospheres derived from individual clones (Ashton et al. 2007). The spatial registry in this system permits isolation and characterization of individual neurospheres. This platform, which allows the user to monitor then select clonal-derived neurospheres, is also useful for screening genetic libraries.

3D microwell arrays have been developed as another platform to analyze responses of stem cell clones to the microenvironment. Matthias Lutolf's lab at École Polytechnique Fédérale de Lausanne (EPFL) constructed 3D poly(ethylene glycol) (PEG) microwell arrays that contained gradients in concentrations of two different proteins (Gobaa et al. 2011). These arrays were used to identify effects of MSC seeding density on proliferation and differentiation to adipocytes, regulation of MSC differentiation to osteocytes by the mechanical properties of the PEG matrix, and combinatorial relationships between factors that control neural stem cell expansion (Gobaa et al. 2011; Roccio et al. 2012).

Microfluidic Devices for Stem Cell Culture and Characterization

Microfluidic devices provide precise, dynamic control of the fluid environment surrounding a stem cell. Because of the small sample size required in microfluidics, these devices can be incorporated into high-throughput screens, assessing effects of spatial gradients and patterns of soluble cues and via dynamic changes in fluid composition. Microfluidic devices are amenable to automation, which can provide reproducible and precisely controlled pumping, mixing, temperature regulation, and integrated sensing.

Microfluidic devices enable the analysis of stem cell function at the microscale, including autocrine and paracrine signaling. Flow can be used to provide or remove factors in the environment of precisely-arranged stem cells to identify mechanisms of stem cell regulation. In addition, microfluidic devices have been applied to stem cell processing by enabling cell separations.

Microfluidics for Understanding Mechanisms of Stem Cell Regulation

The greatest impact of microfluidics to date on the field of stem cell engineering has been as a tool to elucidate mechanisms of stem cell fate regulation. For example, Peter Zandstra's lab at the University of Toronto used perfusion flow in a microfluidic channel to regulate accumulation of cell-secreted paracrine factors in the cell environment (Moledina et al. 2012). Using a modeling approach that accounted for fluid flow and cell position in the microchannel, the investigators identified that paracrine signaling is able to stimulate murine embryonic stem cell (mESC) self-renewal in the absence of leukemia inhibitory factor (LIF). Ellison and colleagues also used an integrated computational modeling and microfluidic culture approach to identify the presence of distinct cell-secreted factors that regulate mESC viability in serum-containing and defined culture media (Ellison et al. 2009).

Microfluidic devices are also useful in studying the application of mechanical forces, including shear stress, to stem cells and their derivatives. Toh and Voldman (2011) fabricated a microfluidic device to apply a known shear stress to ESCs cultured on a substrate and found that shear stress negatively impacts ESC colony growth. Using inhibitors and proteases in conjunction with shear stress, the investigators found that ESCs sense shear stress through the extracellular matrix, and that shear affects cell fate via Fgf5 signaling. An optimization of medium perfusion rates in a microchannel containing hESCs found that, at low flow rates, nutrient depletion and waste accumulation impair cell expansion while high flow rates can detach cells (Titmarsh et al. 2011).

Because of their small size, microfluidic devices are particularly valuable for analyzing the behavior of single cells in response to a dynamic microenvironment, including clonal behavior of stem cells. In one such study, Lecault and colleagues at the University of British Columbia constructed arrays of nanoliter chambers connected by microchannels (Lecault et al. 2011). Pumps and valves regulated flow between the chambers, allowing programmed dynamic application of soluble factors to hematopoietic stem cell clones seeded into the chambers. When coupled with live cell image analysis, this platform was used to identify the precise time point at which Steel factor stimulation stimulated the exit of adult hematopoietic stem cells from quiescence. This culture system combines the dynamic advantages of microfluidics with the spatial segregation of microwells to provide the opportunity to perform HTS at the clonal scale. Such studies have the potential to provide insight into regulation of stem cells and stem cell population heterogeneity.

In another example, Gomez-Sjorberg and colleagues developed a high-throughput microfluidic screening platform for proliferation, differentiation, and motility of hMSCs (Gomez-Sjoberg et al. 2007). This platform uses a multiplexing scheme to create 96 distinct experimental conditions in which cells can be cultured for several weeks.

Spatial Patterning in Microfluidic Devices

The small length scales in microfluidic chambers typically result in laminar flow profiles, even at relatively high fluid velocities. By introducing different reagents to a microfluidic channel by spatially-segregated entrance ports, concentration gradients across the channel can be constructed. The nature of these gradients depends on convective flow through the channel and diffusion across the channel. Such devices can be used to identify how a population of stem cells responds to signaling gradients, which are important regulators of tissue development *in vivo*. In addition, these devices can present gradients to stem cells that elicit distinct cell fates and create spatially-patterned cell populations. Zhang and colleagues at Columbia University established a gradient of doxycycline in a microfluidic device containing MSCs engineered to express doxycycline-inducible bone morphogenetic protein 2 (BMP2) (Zhang et al. 2011). This created a gradient in BMP2 signaling which produced spatially-patterned osteogenesis.

Philippe Renaud's lab at EPFL used a microfluidic device to construct gradients of neural growth factor (NGF) and B27 in a 3D hydrogel (Kunze et al. 2011). Cortical neurons were seeded in the hydrogels, and synapse formation observed in response to the NGF/B27 gradients. The investigators observed synergies in NGF and B27 in inducing synapses and demonstrated the ability to control synapse density using gradients of these neurotrophic factors. A similar approach could be used to regulate stem cell expansion, differentiation, or organization in a 3D microenvironment. Lii and others constructed a microfluidic chip that uses pneumatically actuated valves to perfuse reagents into 3D extracellular matrix gels containing mESCs (Lii et al. 2008). Channels above the gel permitted rapid flow and gel-embedded cells were protected from shear forces.

Microfluidic Separation and Characterization of Stem Cells

The precise spatial control of fluid flow in a microfluidic device offers the potential to improve stem cell separations. Low throughput of microfluidic devices may limit applications in larger scale bioprocessing operations, but microfluidic devices have the ability to isolate low abundance cells for subsequent expansion or characterization. Mehmet Toner and colleagues constructed a platform that can separate low abundance circulating tumor cells (CTCs) from peripheral blood samples based on cellular interactions with anti-body coated microposts in microfluidic channels (Nagrath et al. 2007; Stott et al. 2010). A microfluidic strategy to capture circulating endothelial progenitor cells from peripheral blood based on adhesion to antibody-coated substrate was also reported (Plouffe et al. 2009). Similar strategies to isolate other rare stem and progenitor cell populations from tissues may be promising if specific surface markers are available.

Stem cell properties other than surface affinity may also be the basis for microfluidic separations. Investigators at Lund University used forces generated by an acoustic standing wave to separate platelets from peripheral blood progenitor cells in apheresis products (Dykes et al. 2011). Microfluidic channels at the outlet of the device allowed the researchers to collect purified cell populations. A spiral microfluidic system that separates cells based on cell diameter was used to isolate bone marrow hMSC populations in different stages of the cell cycle (Lee et al. 2011).

Microfluidic capture of CTCs has been integrated with downstream culture, including clonal expansion as 3D spheroids by injection of a hydrogel matrix to encapsulate captured CTCs (Bichsel et al. 2012). Also, automated imaging platforms have been integrated with microfluidic culture to enable single cell-based characterization and high-throughput screening of stem cells based on marker or reporter expression (Kamei et al. 2009, 2010). Microfluidic flow can be used to deliver compounds to specific regions of a culture. For example, localized delivery of an enzyme can remove specific colonies for collection and analysis, and different cell stains can be provided to distinct regions of the culture channel (Villa-Diaz et al. 2009).

Stem Cell Biosensors

Whole Cell Biosensors

Stem cell biosensors have been constructed to report on the state of the stem cell itself as well as the microenvironment. The most common type of stem cell biosensor employs a genetic promoter-reporter construct to produce a fluorescent protein or an enzyme when the stem cell is in a particular differentiation state. For example, human embryonic stem cells expressing GFP under control of the OCT4 promoter can be noninvasively monitored for the loss of the pluripotency marker while somatic cells containing this construct will gain GFP expression following reprogramming to a pluripotent state (Gerrard et al. 2005; Huangfu et al. 2008). Numerous cell lines expressing GFP or other reporters under lineage-specific promoters have been constructed to study differentiation or purify desired cell populations. For example, NKX2-5 eGFP cells were used to purify cardiac-committed hESCs and identify specific surface markers on these cells (Elliott et al. 2011). Similarly, a nestin-eGFP reporter hESC line has been used to identify neural progenitors during differentiation (Noisa et al. 2010). Stem cells have also been engineered to express reporters when certain developmental signaling pathways, such as canonical Wnt signaling or Notch, have been activated (Davidson et al. 2012; Fre et al. 2011).

Biosensors can be constructed to assess cell differentiation state based on expression of surface markers. Surface plasmon resonance (SPR) has been used for real-time analysis of osteogenic differentiation in mesenchymal stem cells based on upregulation of OB-cadherin expression in cells cultured on an SPR substrate (Kuo et al. 2011).

Stem cell biosensors have also been developed to probe the composition of microenvironments *in vitro* and *in vivo*. Fluorescently-labeled aptamers that bind platelet-derived growth factor (PDGF) have been coupled to the surface of mesenchymal stem cells and used to spatially map PDGF concentrations near the stem cell niche *in vivo* (Zhao et al. 2011). Other biosensors report on stem cell phenotypes. For example, Ali Khademhosseini's lab developed a cardiotoxicity biosensor that integrated mESC-derived cardiomyocytes with an automated imaging system to monitor changes in cell contraction rate in response to pharmacologic agents (Kim et al. 2011).

Organ-on-a-Chip

One application of stem cell biosensors is to model tissue and organ level functions in an *in vitro* microdevice using stem cell-derived cells organized to provide the appropriate biological complexity. Real-time analysis can be facilitated by engineered reporter cell lines and integrating imaging, mechanical, electrical, and chemical probes into the system. These "organs-on-a-chip" can bridge the gap between studies on individual cells in culture and *in vivo* studies, and may supplement or replace animal models for physiological studies or drug evaluation trials.

One challenge in constructing stem cell models of tissue and organ level function is manufacturing relevant cells. An appropriate stem cell source must be chosen and efficient differentiation and purification processes must be available. Engineering function or appropriate sensing capacity into the stem cells can improve function of the construct. A physiologic microenvironment that enables survival, function, spatial arrangement, and intercellular interactions between multiple cell types must be constructed. These microenvironments often use microfluidics and biomaterials approaches to provide 3D, spatially-controlled, dynamic environments. Cell culture must be integrated with analysis, and real-time control systems often enhance construct performance.

One promising application of stem cell-based organ-on-a-chip technology is production of cardiac constructs. Efficient methods for differentiating and purifying cardiomyocytes from human pluripotent stem cells have been reported (Mummery et al. 2012). In fact, hPSC-derived cardiomyocytes are commercially available from several companies, including Cellular Dynamics International (United States) and Pluriomics (Netherlands). Contractile forces and electrophysiologic function of the stem cell-derived cardiomyocytes can be monitored in real time (Hazeltine et al. 2012; Poon et al. 2011; Fig. 4). These individual hPSC-derived cardiomyocytes lack the structural organization of heart muscle, however. A microgroove culture platform has been used to align mESC-derived cardiomyocytes into fibers that exhibit organized sarcomeres (Luna et al. 2011). More complex engineered heart tissues that comprise muscle strips or organoid chambers have been constructed from primary animal ventricular cardiomyocytes (Lee et al. 2008). Tissues of hPSC-derived cardiomyocytes, endothelial cells, and stromal cells, illustrated roles of endothelial

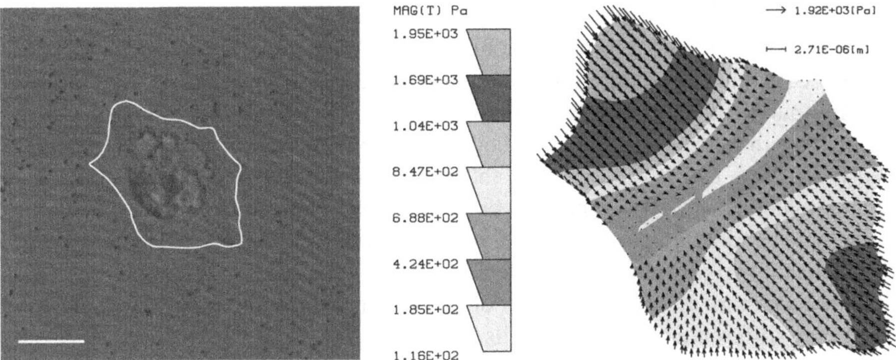

Fig. 4 Mapping of contraction stress in cardiomyocytes (Courtesy of Laurie Hazeltine). (*Left*) Phase contrast image of a contracting cardiomyocyte differentiated from a human pluripotent stem cell cultured on a flexible polyacrylamide hydrogel containing embedded fluorescent beads. The cell outline is shown in white. Scale bar = 20 μm. (*Right*) A contraction stress map of the cell shown in the phase contrast image calculated based on bead displacement during the contraction cycle

cells on cardiomyocyte proliferation, interactions between stromal and endothelial cells, vascular formation, and effects of cyclic stresses on cardiomyocyte hypertrophy and proliferation (Tulloch et al. 2011). Together, these studies demonstrate the near-term feasibility of 3D, functional stem cell-derived cardiac tissues integrated with functional readouts.

Future Directions

Stem Cell Screening

While proof-of-concept examples have demonstrated the power of high-throughput screening technologies to identify microenvironmental cues that regulate stem cell fates, the potential of these technologies is limited by structural features of HTS platforms and stem cells. The screens are typically constrained by the throughput of stem cell analysis. Thus, to screen larger libraries and increase the odds of obtaining a hit, more reliable, faster, and less expensive methods to assess cell response to the compounds in the library must be developed. Dynamic, nondestructive approaches for monitoring cell state, such as enzymatic activity or fluorescent reporter concentration, will facilitate identifying the temporal regulation of stem cells by microenvironmental cues. In addition, screening relies on observing the behavior of a single stem cell or small number of stem cells, and population heterogeneity can result in a high frequency of false positives or false negatives. An understanding of the heterogeneous responses of stem cells to microenvironmental cues and strategies to account for these differences are needed to realize the potential of HTS in stem cell applications.

An increase in screening throughput will enable higher order combinatorial screens to identify how cells process multiple cues in making fate decisions. Because of the costs and technical challenges associated with large-scale screening, it is anticipated that academic or nonprofit collaborations with industry will be productive in addressing the stem cell throughput issue. For example, researchers at I-STEM in France have collaborated with Roche to screen the effects of approximately 200,000 compounds on neural stem cell proliferation. When combined with modeling approaches and statistical analysis, these screens can provide fundamental insight into stem cell regulatory networks. Other screening outcomes may include more efficient and better defined microenvironments for expansion and controlled differentiation of stem cells.

An opportunity exists to expand HTS platforms beyond chemical libraries. Recent studies have identified important roles of microRNAs and long noncoding RNAs in developmental biology, including stem cell proliferation and differentiation as well as cell reprogramming (Pauli et al. 2011; Tiscornia and Izpisua Belmonte 2010; Yi and Fuchs 2011). Screening RNA libraries may identify new mechanisms of stem cell regulation and identify tools to control developmental programs in stem cells. Engineering platforms to screen cues such as intercellular contacts or mechanical forces in the context of a physiologically relevant chemical microenvironment would improve the potential of stem cell screening platforms. 3D screening platforms, such as microwells or biomaterials scaffolds would enable identification of factors that regulate stem cell assembly and tissue development or morphogenesis from stem cell sources.

Microfluidic Culture of Stem Cells

Microfluidic platforms offer the ability to understand the response of single stem cells or small populations of cells to defined, dynamic microenvironments. However, access to microfluidic culture systems is generally limited to labs with the fabrication capacity to produce these devices, and uniform standards for microfluidic culture of stem cells do not yet exist. To realize the potential of these devices in stem cell engineering, reliable and inexpensive sources of relatively simple microfluidic devices must be available. In addition, collaborative efforts between researchers with expertise in device manufacture and stem cell researchers posing questions that can be addressed by these devices will be important. Similarly, the use of microfluidic culture systems has been limited to the research laboratory. The opportunity exists to translate these devices to commercial or clinical applications, including production of pure populations of stem cells, stem cell-based HTS platforms, and patterned multicellular tissues for *in vitro* analysis.

The ability to construct defined microenvironments in microfluidic devices has outpaced the ability to characterize cells in these microenvironments in real time. Better integration of stem cell culture and separations platforms with cell characterization is needed. With the advent of single-cell gene expression analysis, the opportunity exists to deeply probe clonal differences in stem cell populations. Some microfluidic devices have been designed to probe gene expression and signaling activity in individual cells (Bennett and Hasty 2009; Cheong et al. 2009; Yin et al. 2010).

The opportunity exists to integrate precise microenvironmental control and cell analysis with signaling pathway modeling to obtain a systems level understanding of stem cell behavior. In fact, multiscale modeling will be needed to understand how spatial and temporal presentation of cues at the microscale can lead to longer term changes in cell and tissue level phenomena. This understanding will then in turn enable design of microenvironments that elicit the desired cell outcomes via application of chemical and physical cues to activate appropriate developmental signaling pathways.

Stem Cell Biosensors

Efforts to construct stem cell biosensors have only begun to realize the potential of stem cells in sensing applications. Next generation reporters will simultaneously monitor the activities of multiple promoters or cell processes in real time. This will require identification of sensitive and specific promoter sequences, methods to engineer stem cells, and integration of tools to monitor cell phenotypes in stem cell culture and differentiation platforms. Synthetic biology tools, which have been most widely applied to microbial cells, have the potential to redesign stem cell regulatory circuits for sensing applications.

Advances in microscale fabrication technologies will enable construction of more physiologically relevant, 3D structured cell and tissue biosensors (Gauvin and Khademhosseini 2011). For example, 3D vascular structures have been formed from microfluidic devices, direct inkjet printing, and assembly of microgels (Du et al. 2011; Wu et al. 2011; Choi et al. 2007; Fidkowski et al. 2005). These approaches will enable development of more advanced organ-on-a-chip models.

Global Assessment and Conclusions

The United States is currently a leader in development of microenvironments and platforms for using stem cells in HTS and biosensing applications. This is primarily the result of a strong and growing community of engineers trained in biomaterials and microsystem fabrication expanding their research to stem cells and collaborating with stem cell biologists. Other countries with pronounced strengths in this area include Canada, Japan, the Netherlands, and Switzerland, which have also made concerted efforts to engage engineers and stem cell biologists in research collaborations.

One area of this field that the United States does not necessarily lead is integration government and academic research with commercial efforts. Partnerships between I-STEM and pharmaceutical companies in France appear to be productive in leveraging the pharmaceutical industry's expertise in HTS with the stem cell proficiency at I-STEM. Models of commercialization in the Netherlands Institute of Regenerative Medicine and the Berlin-Brandenburg Center for Regenerative Therapies offer advantages in designing and distributing new tools that have value to the stem cell field.

It is apparent that engineered microenvironments and high-throughput methods will become more sophisticated. These approaches will provide more detailed insight into basic regulation of stem cell function, generate predictive *in vitro* tissue models, and contribute to the regenerative potential of stem cells. Efforts to engage multidisciplinary teams of stem cell engineers, biologists, and clinicians in academia, industry, and medicine will accelerate progress in this field.

References

Alves, H., K. Dechering, C. Van Blitterswijk, and J. De Boer. 2011. High-throughput assay for the identification of compounds regulating osteogenic differentiation of human mesenchymal stromal cells. *PLoS One* 6:e26678.

Andrews, P.D., M. Becroft, A. Aspegren, J. Gilmour, M J. James, S. McRae, R. Kime, R.W. Allcock, A. Abraham, Z. Jiang, R. Strehl, J.C. Mountford, G. Milligan, M.D. Houslay, D.R. Adams, and J.A. Frearson. 2010. High-content screening of feeder-free human embryonic stem cells to identify pro-survival small molecules. *Biochem. J.* 432:21–33.

Ashton, R.S., J. Peltier, C.A. Fasano, A. O'Neill, J. Leonard, S. Temple, D.V. Schaffer, and R.S. Kane. 2007. High-throughput screening of gene function in stem cells using clonal microarrays. *Stem Cells* 25:2928–2935.

Barbaric, I., P.J. Gokhale, M. Jones, A. Glen, D. Baker, and P.W. Andrews. 2010. Novel regulators of stem cell fates identified by a multivariate phenotype screen of small compounds on human embryonic stem cell colonies. *Stem Cell Res.* 5:104–19.

Bennett, M.R. and J. Hasty. 2009. Microfluidic devices for measuring gene network dynamics in single cells. *Nat. Rev. Genet.* 10:628–638.

Bichsel, C.A., S. Gobaa, S. Kobel, C. Secondini, G.N. Thalmann, M.G. Cecchini, and M.P. Lutolf. 2012. Diagnostic microchip to assay 3D colony-growth potential of captured circulating tumor cells. *Lab Chip* 12:2313–6.

Cheong, R., C.J. Wang, and A. Levchenko. 2009. High content cell screening in a microfluidic device. *Mol. Cell. Proteomics* 8:433–442.

Choi, N.W., M. Cabodi, B. Held, J.P. Gleghorn, L. J. Bonassar, and A.D. Stroock. 2007. Microfluidic scaffolds for tissue engineering. *Nat. Mater.* 6:908–915.

Davidson, K.C., A.M. Adams, J.M. Goodson, C.E. McDonald, J.C. Potter, J.D. Berndt, T.L. Biechele, R.J. Taylor, and R.T. Moon. 2012. Wnt/beta-catenin signaling promotes differentiation, not self-renewal, of human embryonic stem cells and is repressed by Oct4. *Proc. Natl. Acad. Sci. USA* 109:4485–4490.

Derda, R., S. Musah, B.P. Orner, J.R. Klim, L. Li, and L.L. Kiessling. 2010. High-throughput discovery of synthetic surfaces that support proliferation of pluripotent cells. *J. Am. Chem. Soc.* 132:1289–1295.

Desbordes, S.C., D.G. Placantonakis, A. Ciro, N.D. Socci, G. Lee, H. Djaballah, and L. Studer. 2008. High-throughput screening assay for the identification of compounds regulating self-renewal and differentiation in human embryonic stem cells. *Cell Stem Cell* 2:602–612.

Du, Y., M. Ghodousi, H. Qi, N. Haas, W. Xiao, and A. Khademhosseini. 2011. Sequential assembly of cell-laden hydrogel constructs to engineer vascular-like microchannels. *Biotechnol. Bioeng.* 108:1693–1703.

Dykes, J., A. Lenshof, I.B. Astrand-Grundstrom, T. Laurell, and S. Scheding. 2011. Efficient removal of platelets from peripheral blood progenitor cell products using a novel micro-chip based acoustophoretic platform. *PLoS One* 6:e23074.

Elliott, D.A., S.R. Braam, K. Koutsis, E.S. Ng, R. Jenny, E.L. Lagerqvist, C. Biben, T. Hatzistavrou, C.E. Hirst, Q.C. Yu, R.J. Skelton, D. Ward-van Oostwaard, S.M. Lim, O. Khammy, X. Li, S.M. Hawes, R.P. Davis, A.L. Goulburn, R. Passier, O.W. Prall, J.M. Haynes, C.W. Pouton, D.M. Kaye, C.L. Mummery, A.G. Elefanty, and E.G. Stanley. 2011. NKX2-5(eGFP/w) hESCs for isolation of human cardiac progenitors and cardiomyocytes. *Nat. Methods* 8:1037–1040.

Ellison, D., A. Munden, and A. Levchenko. 2009. Computational model and microfluidic platform for the investigation of paracrine and autocrine signaling in mouse embryonic stem cells. *Mol. Biosyst.* 5:1004–1012.

Fidkowski, C., M.R. Kaazempur-Mofrad, J. Borenstein, J.P. Vacanti, R. Langer, and Y. Wang. 2005. Endothelialized microvasculature based on a biodegradable elastomer. *Tissue Eng.* 11:302–309.

Flaim, C.J., D. Teng, S. Chien, and S.N. Bhatia. 2008. Combinatorial signaling microenvironments for studying stem cell fate. *Stem Cells Dev.* 17:29–39.

Fre, S., E. Hannezo, S. Sale, M. Huyghe, D. Lafkas, H. Kissel, A. Louvi, J. Greve, D. Louvard, and S. Artavanis-Tsakonas. 2011. Notch lineages and activity in intestinal stem cells determined by a new set of knock-in mice. *PLoS One* 6:e25785.

Gauvin, R., and A. Khademhosseini. 2011. Microscale technologies and modular approaches for tissue engineering: moving toward the fabrication of complex functional structures. *ACS Nano* 5:4258–4264.

Gerrard, L., D. Zhao, A.J. Clark, and W. Cui. 2005. Stably transfected human embryonic stem cell clones express OCT4-specific green fluorescent protein and maintain self-renewal and pluripotency. *Stem Cells* 23:124–133.

Gobaa, S., S. Hoehnel, M. Roccio, A. Negro, S. Kobel, and M.P. Lutolf. 2011. Artificial niche microarrays for probing single stem cell fate in high throughput. *Nat. Methods* 8:949–955.

Gomez-Sjoberg, R., A.A. Leyrat, D.M. Pirone, C.S. Chen, and S.R. Quake. 2007. Versatile, fully automated, microfluidic cell culture system. *Anal. Chem.* 79:8557–8563.

Hazeltine, L.B., C.S. Simmons, M.R. Salick, X. Lian, M.G. Badur, W. Han, S.M. Delgado, T. Wakatsuki, W.C. Crone, B.L. Pruitt, and S.P. Palecek. 2012. Effects of substrate mechanics on contractility of cardiomyocytes generated from human pluripotent stem cells. *Int. J. Cell Biol.* 2012:508294.

Huangfu, D., R. Maehr, W. Guo, A. Eijkelenboom, M. Snitow, A. E. Chen, and D. A. Melton. 2008. Induction of pluripotent stem cells by defined factors is greatly improved by small-molecule compounds. *Nat. Biotechnol.* 26:795–797.

Kamei, K., S. Guo, Z.T. Yu, H. Takahashi, E. Gschweng, C. Suh, X. Wang, J. Tang, J. McLaughlin, O.N. Witte, K.B. Lee, and H.R. Tseng. 2009. An integrated microfluidic culture device for quantitative analysis of human embryonic stem cells. *Lab Chip* 9:555–563.

Kamei, K., M. Ohashi, E. Gschweng, Q. Ho, J. Suh, J. Tang, Z.T. For Yu, A.T. Clark, A.D. Pyle, M.A. Teitell, K.B. Lee, O.N. Witte, and H.R. Tseng. 2010. Microfluidic image cytometry for quantitative single-cell profiling of human pluripotent stem cells in chemically defined conditions. *Lab Chip* 10:1113–1119.

Kim, S.B., H. Bae, J.M. Cha, S.J. Moon, M.R. Dokmeci, D.M. Cropek, and A. Khademhosseini. 2011. A cell-based biosensor for real-time detection of cardiotoxicity using lensfree imaging. *Lab Chip* 11:1801–1807.

Kunze, A., A. Valero, D. Zosso, and P. Renaud. 2011. Synergistic NGF/B27 gradients position synapses heterogeneously in 3D micropatterned neural cultures. *PLoS One* 6:e26187.

Kuo, Y.C., J.H. Ho, T.J. Yen, H.F. Chen, and O.K. Lee. 2011. Development of a surface plasmon resonance biosensor for real-time detection of osteogenic differentiation in live mesenchymal stem cells. *PLoS One* 6:e22382.

Lecault, V., M. Vaninsberghe, S. Sekulovic, D.J. Knapp, S. Wohrer, W. Bowden, F. Viel, T. McLaughlin, A. Jarandehei, M. Miller, D. Falconnet, A.K. White, D.G. Kent, M.R. Copley,

F. Taghipour, C.J. Eaves, R.K. Humphries, J.M. Piret, and C.L. Hansen. 2011. High-throughput analysis of single hematopoietic stem cell proliferation in microfluidic cell culture arrays. *Nat. Methods* 8:581–586.

Lee, E.J., E. Kim do, E.U. Azeloglu, and K.D. Costa. 2008. Engineered cardiac organoid chambers: toward a functional biological model ventricle. *Tissue Eng Part A* 14:215–25.

Lee, W.C., A.A. Bhagat, S. Huang, K.J. Van Vliet, J. Han, and C.T. Lim. 2011. High-throughput cell cycle synchronization using inertial forces in spiral microchannels. *Lab Chip* 11:1359–1367.

Lii, J., W.J. Hsu, H. Parsa, A. Das, R. Rouse, and S.K. Sia. 2008. Real-time microfluidic system for studying mammalian cells in 3D microenvironments. *Anal. Chem.* 80:3640–3647.

Luna, J.I., J. Ciriza, M.E. Garcia-Ojeda, M. Kong, A. Herren, D.K. Lieu, R.A. Li, C.C. Fowlkes, M. Khine, and K.E. McCloskey. 2011. Multiscale biomimetic topography for the alignment of neonatal and embryonic stem cell-derived heart cells. *Tissue Eng Part C Methods* 17:579–588.

Mei, Y., K. Saha, S.R. Bogatyrev, J. Yang, A.L. Hook, Z.I. Kalcioglu, S.W. Cho, M. Mitalipova, N. Pyzocha, F. Rojas, K.J. Van Vliet, M.C. Davies, M.R. Alexander, R. Langer, R. Jaenisch, and D.G. Anderson. 2010. Combinatorial development of biomaterials for clonal growth of human pluripotent stem cells. *Nat. Mater.* 9:768–78.

Metallo, C.M., J.C. Mohr, C.J. Detzel, J.J. de Pablo, B.J. Van Wie, and S.P. Palecek. 2007. Engineering the stem cell microenvironment. *Biotechnol. Prog.* 23:18–23.

Moledina, F., G. Clarke, A. Oskooei, K. Onishi, A. Gunther, and P.W. Zandstra. 2012. Predictive microfluidic control of regulatory ligand trajectories in individual pluripotent cells. *Proc. Natl. Acad. Sci. USA* 109:3264–3269.

Mummery, C.L., J. Zhang, E.S. Ng, D.A. Elliott, A.G. Elefanty, and T.J. Kamp. 2012. Differentiation of human embryonic stem cells and induced pluripotent stem cells to cardiomyocytes: a methods overview. *Circ. Res.* 111:344–358.

Nagrath, S., L.V. Sequist, S. Maheswaran, D.W. Bell, D. Irimia, L. Ulkus, M.R. Smith, E.L. Kwak, S. Digumarthy, A. Muzikansky, P. Ryan, U.J. Balis, R.G. Tompkins, D.A. Haber, and M. Toner. 2007. Isolation of rare circulating tumour cells in cancer patients by microchip technology. *Nature* 450:1235–1239.

Noisa, P., A. Urrutikoetxea-Uriguen, M. Li, and W. Cui. 2010. Generation of human embryonic stem cell reporter lines expressing GFP specifically in neural progenitors. *Stem Cell Rev.* 6:438–449.

Outten, J.T., X. Cheng, P. Gadue, D.L. French, and S.L. Diamond. 2011. A high-throughput multiplexed screening assay for optimizing serum-free differentiation protocols of human embryonic stem cells. *Stem Cell Res.* 6:129–42.

Pauli, A., J.L. Rinn, and A.F. Schier. 2011. Non-coding RNAs as regulators of embryogenesis. *Nat. Rev. Genet.* 12:136–149.

Plouffe, B.D., T. Kniazeva, J.E. Mayer, Jr., S.K. Murthy, and V.L. Sales. 2009. Development of microfluidics as endothelial progenitor cell capture technology for cardiovascular tissue engineering and diagnostic medicine. *FASEB J.* 23:3309–14.

Poon, E., C. W. Kong, and R. A. Li. 2011. Human pluripotent stem cell-based approaches for myocardial repair: from the electrophysiological perspective. *Mol. Pharm.* 8:1495–504.

Roccio, M., S. Gobaa, and M.P. Lutolf. 2012. High-throughput clonal analysis of neural stem cells in microarrayed artificial niches. *Integr. Biol. (Camb.)* 4:391–400.

Rosenthal, A., A. Macdonald, and J. Voldman. 2007. Cell patterning chip for controlling the stem cell microenvironment. *Biomaterials* 28:3208–3216.

Ruiz, S.A., and C.S. Chen. 2008. Emergence of patterned stem cell differentiation within multicellular structures. *Stem Cells* 26:2921–2927.

Scadden, D.T. 2006. The stem-cell niche as an entity of action. *Nature* 441:1075–9.

Soen, Y., A. Mori, T.D. Palmer, and P.O. Brown. 2006. Exploring the regulation of human neural precursor cell differentiation using arrays of signaling microenvironments. *Mol. Syst. Biol.* 2:37.

Stott, S.L., C.H. Hsu, D.I. Tsukrov, M. Yu, D.T. Miyamoto, B.A. Waltman, S.M. Rothenberg, A.M. Shah, M.E. Smas, G.K. Korir, F.P. Floyd, Jr., A.J. Gilman, J.B. Lord, D. Winokur, S. Springer,

D. Irimia, S. Nagrath, L.V. Sequist, R.J. Lee, K.J. Isselbacher, S. Maheswaran, D.A. Haber, and M. Toner. 2010. Isolation of circulating tumor cells using a microvortex-generating herringbone-chip. *Proc. Natl. Acad. Sci. USA* 107:18392–18397.

Tiscornia, G., and J.C. Izpisua Belmonte. 2010. MicroRNAs in embryonic stem cell function and fate. *Genes Dev.* 24:2732–2741.

Titmarsh, D., A. Hidalgo, J. Turner, E. Wolvetang, and J. Cooper-White. 2011. Optimization of flowrate for expansion of human embryonic stem cells in perfusion microbioreactors. *Biotechnol. Bioeng.* 108:2894–2904.

Toh, Y.C., and J. Voldman. 2011. Fluid shear stress primes mouse embryonic stem cells for differentiation in a self-renewing environment via heparan sulfate proteoglycans transduction. *FASEB J.* 25:1208–1217.

Tulloch, N.L., V. Muskheli, M.V. Razumova, F.S. Korte, M. Regnier, K.D. Hauch, L. Pabon, H. Reinecke, and C.E. Murry. 2011. Growth of engineered human myocardium with mechanical loading and vascular coculture. *Circ. Res.* 109:47–59.

Unadkat, H.V., M. Hulsman, K. Cornelissen, B.J. Papenburg, R.K. Truckenmuller, A.E. Carpenter, M. Wessling, G.F. Post, M. Uetz, M.J. Reinders, D. Stamatialis, C.A. van Blitterswijk, and J. de Boer. 2011. An algorithm-based topographical biomaterials library to instruct cell fate. *Proc. Natl. Acad. Sci. USA* 108:16565–16570.

Villa-Diaz, L.G., Y.S. Torisawa, T. Uchida, J. Ding, N.C. Nogueira-de-Souza, K.S. O'Shea, S. Takayama, and G.D. Smith. 2009. Microfluidic culture of single human embryonic stem cell colonies. *Lab Chip* 9:1749–1755.

Wu, W., A. DeConinck, and J.A. Lewis. 2011. Omnidirectional printing of 3D microvascular networks. *Adv. Mater.* 23:H178-H183.

Xu, Y., X. Zhu, H.S. Hahm, W. Wei, E. Hao, A. Hayek, and S. Ding. 2010. Revealing a core signaling regulatory mechanism for pluripotent stem cell survival and self-renewal by small molecules. *Proc. Natl. Acad. Sci. USA* 107:8129–8134.

Yi, R. and E. Fuchs. 2011. MicroRNAs and their roles in mammalian stem cells. J. *Cell Sci.* 124:1775–1783.

Yin, Z., S.C. Tao, R. Cheong, H. Zhu, and A. Levchenko. 2010. An integrated micro-electro-fluidic and protein arraying system for parallel analysis of cell responses to controlled microenvironments. *Integr. Biol. (Camb.)* 2:416–423.

Young, E.W. and D.J. Beebe. 2010. Fundamentals of microfluidic cell culture in controlled microenvironments. *Chem. Soc. Rev.* 39:1036–1048.

Zhang, Y., Z. Gazit, G. Pelled, D. Gazit, and G. Vunjak-Novakovic. 2011. Patterning osteogenesis by inducible gene expression in microfluidic culture systems. *Integr. Biol. (Camb.)* 3:39–47.

Zhao, W., S. Schafer, J. Choi, Y.J. Yamanaka, M.L. Lombardi, S. Bose, A.L. Carlson, J.A. Phillips, W. Teo, I.A. Droujinine, C.H. Cui, R.K. Jain, J. Lammerding, J.C. Love, C.P. Lin, D. Sarkar, R. Karnik, and J.M. Karp. 2011. Cell-surface sensors for real-time probing of cellular environments. *Nat. Nanotechnol.* 6:524–531.

Computational Modeling and Stem Cell Engineering

Peter W. Zandstra and Geoff Clarke

Introduction

The Need for Computational Approaches in Stem Cell Engineering

A key goal of regenerative medicine and bioengineering is the quantitative and robust control over the fate and behavior of individual cells and their populations, both *in vitro* and *in vivo*. Central to this endeavor are stem cells (SCs), which can be functionally defined as undifferentiated cells of a multicellular organism that balance the capacity for sustained self-renewal with the potential to differentiate into specialized cell types. The biology of multicellular organisms necessitates the existence and precise control of SCs to facilitate development from a single cell during embryogenesis, and tissue homeostasis in the face of continual loss of terminally differentiated cells. It is therefore not surprising that SCs have been identified and isolated from numerous adult human tissues, as well as more recently, the inner cell mass of the preimplantation human blastocyst. SCs promise a renewable source of human tissue for research, pharmaceutical testing, and cell-based therapies. Fulfilling this promise will require not only the precise control of SC self-renewal and differentiation, but also imposing this control on the formation of more functionally complex tissue-like structures.

P.W. Zandstra (✉)
Terrence Donnelly Centre for Cellular and Biomolecular Research,
University of Toronto, 160 College Street, Room 1116, Toronto, ON M5S 3E1, Canada
e-mail: peter.zandstra@utoronto.ca

G. Clarke
Peter Zandstra Lab, University of Toronto, Toronto, ON, Canada

R.M. Nerem et al. (eds.), *Stem Cell Engineering: A WTEC Global Assessment*,
Science Policy Reports, DOI 10.1007/978-3-319-05074-4_4,
© Springer International Publishing Switzerland 2014

Engineering approaches to understanding and controlling SC fate (stem cell engineering (SCE); Zandstra and Nagy 2001) will be required at multiple stages during the development and implementation of regenerative medicine (RM)-based therapy. For example, optimizing SC growth is fundamental for efforts to generate the quantities of cells (pluripotent cells and their derivatives for drug screening or somatic cells such as mesenchymal stem cells for therapy) that are projected to be required. Similarly, the rigorous control of differentiation, including the functional stabilization of stem cells and their progeny in formulations that can be used to discover RM drugs or treat disease, requires fundamental technological developments. Finally, the translation of these essential technologies into products that can be commercialized will require cost effective and robust cell generation and delivery strategies. Critically, implementing robust RM therapies, in even the simplest of tissues, will benefit from a predictive understanding of the molecular events that occur within individual SCs, and the role of the microenvironment (i.e., the SC niche; see chapter "Physical and Engineering Principles in Stem Cell Research" in this report) in perturbing these events. These molecular events, which are typically organized as cascades, include gene regulatory and intracellular signal transduction networks, cell-cell communication networks, and the mechanical, electrostatic, biochemical and cellular interactions that impinge on those networks (Guilak et al. 2009; Peerani et al. 2009; Discher et al. 2009).

Certainly understanding complex molecular processing in mammalian cells is a dominant endeavor in biomedical research. However, investigating these molecular and cellular events in SCs is made significantly more challenging by additional features of SC biology (Table 1). Ultimately new strategies and tools are needed to make progress in these areas and, as discussed in this chapter, computational methods are ideally suited to this challenge. For example, the genetic and signal transduction networks that are the focus of so much molecular analysis are responsible for stem cells fate stabilization among multiple, occasionally more preferred options (metastability—the ability of a system to remain in a state other than the most stable one). SCs are an experimentally tractable system with which to explore the biological requirements for this property. Similarly, their rarity, heterogeneity, and spatial distribution in the embryo and in adult tissues challenge us to rigorously evaluate the role of a small number of individuals on the emergence and homeostasis of much larger populations. That the SC niche, both *in vivo* and *in vitro*, is a dynamic (changing both in time and composition) and heterogeneous environment encourages us to precisely define the roles of variability on cellular behavior. Importantly, as many of these processes are nonlinear in nature and involve a large number of parameters, the degree of complexity in SCE problems demands use of computational methods to efficiently explore the putative roles of these factors in regulating SC fate.

Of course, when developing computational models of any system, one must understand its key features, make simplifying assumptions and have a rational way to evaluate options regarding the technical aspects of the modeling process. The rarity and spatially heterogeneous distribution of individual SCs requires that some degree of discrete and stochastic mathematics should be involved, as deterministic

Table 1 Features of stem cell biology relevant to modeling approaches

Stem cell property	Biological impact	Modeling impact
Rarity	Efferent signals diluted across many potential targets	Responses must be normalized for cell type ratios
	Behavior of other cells in population may overwhelm that of the stem cells	Spatial effects must be considered
		Deterministic or continuous models may not reflect underlying biology
	Stochastic responses within the small population may be important determinants of cell population behavior	Experimental validation challenging
Metastability	Dynamic responses to exogenous signals	Models must span many time and length scales
	Cells are rarely in equilibrium	Models of behavior dynamics must be utilized
Heterogeneous spatial distribution	Developmental cues originate from diverse regions	Widely varying space and time scales are important
	Afferent signals may vary widely	Population averages may not represent dominant outlier behavior
Stem cell niche is a heterogeneous and dynamic environment	Afferent signal will vary widely across the stem cell population	Population averages may not be relevant to the individual stem cell
	Different types of molecular and biophysical signals need to be integrated	Heterogeneous populations of cells should be modeled and single-cell behavior followed
		Spatial and temporal aspects should be considered

and continuous methods such as ordinary differential equations (ODEs) may not accurately describe the underlying biology. The heterogeneous nature of the niche also requires that models include descriptions of the spatial interactions and gradients among various cell types and their environment to correctly predict how any cell or cell population will behave. Finally, the metastable nature of SC fate and the hierarchical organization of differentiation suggest that guiding cell fate along specific lineages involves balancing the dynamic differentiation process and feedback from progeny and the environment. Consequently, dynamic analysis across a wide range of timescales may need to be included in accurate system-wide modeling studies in order to understand the time evolution of the cell populations.

General Computational Approaches Relevant to Stem Cell Engineering

Whereas mathematical and theoretical approaches have long been part of the arsenal of tools used in the physical sciences (Humphreys 2004), significant and widespread advances in mathematical biology have only occurred during the last century.

Additionally, while subjects such as evolutionary ecology, genetics, and epidemiology have a relatively long history of mathematical analysis, molecular and cellular biology have only recently adopted these techniques. No doubt the current appeal of modeling in cell and molecular biology has been spurred in part due to recent technological advances allowing for high-throughput screens and the generation of massive biological datasets in need of analysis. Furthermore, advances in software technologies, such as faster and more accurate numerical methods to solve ODEs and hardware advances such as parallel and graphics processing unit (GPU) computing, have made it possible to find approximate and complete solutions to the complex sets of previously intractable equations used in modeling cellular systems. This development has allowed researchers to remove simplifying assumptions from their models so they better reflect the underlying biochemical and biophysical fundamentals.

A final and perhaps more significant change that has occurred over recent decades is the dramatic shift in biology from an observational science to a *predictive* one that is able to *quantitatively* describe how a system under study will behave when exposed to conditions that have not yet been examined. Previously, biologists had been able to study biological systems only under those conditions they could observe in the environment or artificially create in a lab. With data provided by the recent high-throughput developments in biology mentioned above, researchers are now able to perform *in silico* experiments on systems under conditions that are either too expensive to monitor or create, or that have never been previously considered. This allows for the relatively rapid investigation of behaviors under conditions designed to investigate critical components of highly complex systems. In much the same manner as electrical engineers computationally design and troubleshoot new circuits, bioengineers can model known and *designed* biological systems to evaluate important hypotheses prior to performing increasingly expensive experiments. This has, in turn, led to the study of more difficult biological problems requiring expertise from many biological subdisciplines which will no doubt lead to the generation of novel tools and hypotheses to explore (National Research Council of The National Academies 2009).

There are currently three general overlapping classes of modeling approaches that are of relevance to biology. First, *mathematical biology* can be considered to include those efforts that use biological information or behavior as inspiration for the development of novel mathematical methods. These approaches usually a embrace significant degrees of abstraction in the representing biological features, and often contribute more to mathematical advances than they do to biological ones. In contrast, *theoretical biology* involves the rigorous use of mathematical concepts to identify and explore the implications of fundamental organizational principles underlying much of biology. In a sense, these theories aim to unify the complexity inherent to biology—to understand the "laws of nature," rather than explore hypotheses specific to a given set of conditions. These large-scale and unifying models benefit those methods whose goal is a more detailed and physically realistic depiction of nature, in that they can often inform and alter the conceptual thinking that is

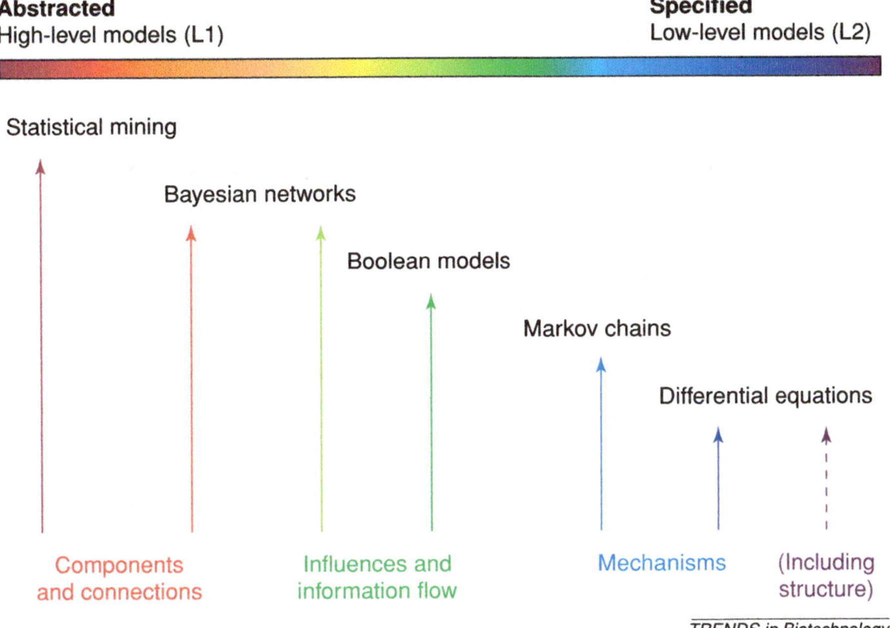

Fig. 1 Computational models in cell biology can take many forms, from abstract statistical approaches that aim to identify components and their relationships, to the more mechanistic models that explore the realistic dynamics of biophysical and biochemical systems (From Ideker and Lauffenburger 2003)

at the root of all models, whether they be detailed or abstract. Third, *computational biology*, a relative newcomer to the scene, would include aspects such as data analysis (bioinformatics, statistical methods and models), data mining, and biophysically realistic and detailed simulations of cellular systems. It would also include approaches that might be considered common in engineering; methods such as optimization or image analysis designed to enhance the ability to design and control systems to a greater degree.

While there is considerable overlap across these differing modeling philosophies, we will consider the roles of theoretical and computational biology in the remainder of this chapter. This in no way restricts us with respect to the mathematical metaphors that are applicable to stem cell engineering, as these classes provide a wealth of computational approaches (Fig. 1). For example, dynamic models of SC transcriptional and signaling networks can take many forms, including ODEs (Roeder and Glauche 2006), fuzzy logic and Boolean logic methods (Aldridge et al. 2009; Morris et al. 2010). Similarly, cell populations can be represented by ODE (Kirouac et al. 2009) or discrete-time individual-based models such as cellular automata (Fouliard et al. 2009). Moreover, hybrid combinations of approaches can be utilized to develop multiscale models of heterogeneous cell populations (Zhu et al. 2004).

Current Trends in Computational Stem Cell Engineering

Modeling approaches in SCE can not only be segregated according to the computational methodology used, but according to the cellular and molecular resolution at which the analysis is performed. In what follows, we will describe SC models at several of these levels that point to current trends and future opportunities for engineering to contribute to the understanding and control of SC biology.

Before discussing those models relevant to individual levels of SC organization, it should be noted that a somewhat new trend is emerging in models that hope to describe SC across many or all of their spatiotemporal layers. These so-called multiscale models are defined as those that span two or more levels of space or time (Deisboeck et al. 2011), and aim to shed light on emergent biological features by providing a mechanistic connection between them (for example how gene regulatory networks may control SC heterogeneity). Currently, those relating to SCE have been restricted to approximately two layers (see below for examples) due to the significant challenges of integrating disparate data types, the increasing number of parameters, and the lack of information regarding bidirectional feedback often critical to physiological function at both levels.

To address these difficulties, some have suggested utilizing a bioinformatics approach. While the majority of bioinformatics in SCE and biology utilize its powerful computational and statistical methods to study large data sets and address particular biological questions such as the structure of gene regulatory networks under various environmental conditions, bioinformatic techniques may also be used to couple large-scale data to computational methods, thereby producing tools capable of studying the more complete dynamics; such approaches have recent been used, for example in pluripotent SC (Xu et al. 2010). In these methods, data is obtained from various SC regulatory components (e.g., proteomic and gene expression networks), synthesized according to standardized formats such as the Systems Biology Markup Language (SMBL), and stored within open access databases. Many databases already in existence collate data that is applicable to SCE. For example, StemSight (www.stemsight.org) uses Bayesian network analysis to predict networks of genes that are functionally related to each other, while ESCAPE (http://www.maayanlab.net/ESCAPE/index.php) focuses on the curation and analysis of SC data related to pluripotency. Although these projects attempt to address questions at the single-cell level, they can be considered as multiscale bioinformatics projects in that the processes they explore span many different time scales, such as that required for gene expression versus that needed for a cell to change its behavior. Similar approaches that address the issues of data standardization and the linkages between spatial scales of molecules → cells → populations → tissues will require further development, but will no doubt be of extraordinary importance in the coming years.

Molecular and Single-Cell Studies

A dominant use of modeling in SC biology is that of single-cell analysis, likely due to the need to understand the endogenous mechanisms controlling SC fate, but also due to the lower degree of complexity required of biophysically realistic intracellular models when compared to those at a tissue or cell population level. In fact, one of the earliest computational studies using SC was the first to demonstrate that SCs act randomly within a population, as individuals (Till et al. 1964). In this seminal Canadian study, it was established that the distribution of colony forming cells (CFCs) within mixed populations of cells derived from the splenic colonies of cells formed after mice were injected with hematopoietic cells followed a gamma distribution. After realizing this observation was consistent with a Markov birth-death process in which individual CFCs either proliferated to form two CFCs ("birth") or underwent differentiation ("death") at fixed probabilities, a Monte-Carlo model was created which accurately fit the empirical data. This early paper demonstrates an important feature of all useful models, namely that the model results were compared directly to the empirical data in order to validate the underlying hypothesis of the model. Among more recent single-cell computational approaches to understanding SC biology, several sub-trends can be identified. Here, we will focus on a discussion of those most relevant to SC biology and SCE, specifically models of cell fate control.

Dynamic Analyses of Genetic and Signal Transduction Networks

Arguably one of the most prolific uses of modeling in cell biology is the simulation of the molecular interactions and chemical reactions responsible for gene regulation and intracellular signal transduction. The same can be said of modeling and simulation efforts in SCE, as many research groups worldwide are actively working on quantitative analyses of the biochemical networks controlling SC behavior. Likely the most active SCE field for this work addresses how cell fate—pluripotency, proliferation, quiescence, differentiation and lineage specification—is regulated. Below, we will describe a sample of current research in this area.

Many studies utilizing what could be referred to as "applied computation" have addressed the question of SC fate regulation. For example, in addition to their models on treatments for chronic myeloid leukemia (CML) (Glauche et al. 2012), members of the Institute for Medical Informatics and Biometry (IMB) at the Technical University Dresden (TUD) in Germany have been designing and analyzing mathematical models of SC biology for several years that are of direct relevance to SCE. One example of their work uses an approach commonly found in systems biology in which large and complex biochemical networks are partitioned into the modules relevant to the particular process under study. Once critical modules are identified, mathematical models and simulations are generated to study the dynamic behavior of each module independently in order to rigorously understand their behavior before progressing to the more difficult problem of modeling the complete network dynamics. In this case, the system module to be modeled was a network of three transcription factors, Oct4, Sox2, and Nanog, which are critical regulators of SC pluripotency and the initiation of differentiation.

Inspired by the observations that populations of SCs exhibit great variation in the single-cell abundance of Nanog, and that those SC expressing low levels of Nanog have a higher chance of undergoing differentiation than cells with greater amounts (Chambers et al. 2007; Kalmar et al. 2009), this study asked if the large variations in Nanog levels were the result of stochastic fluctuations in transcription or translation (i.e., "noise") or rather the result of oscillations produced by a yet unknown regulatory factor. To address these competing hypotheses, the authors developed and analyzed a model of the Oct4-Sox2-Nanog transcription factor network to determine how the two sources of variation affect the dynamic levels of Nanog (Glauche et al. 2010). The model consisted of a coupled set of ODEs describing the changes in the concentration of the Oct4-Sox2 heterodimer and Nanog. That the detailed tracking of the concentration of both Oct4 and Sox2 monomers was omitted follows from the assumption that the levels of both are maintained in dynamic equilibrium with, and are largely controlled by, the function of the Oct4-Sox2 heterodimer by positive feedback (Fig. 2). By including two sources of Nanog variation, either a Gaussian noise term in the equation describing Nanog concentration or a novel negative feedback loop that produced Nanog oscillations, the authors evaluated the contribution of each type of variability on the overall network dynamics.

Analysis of a preliminary core model lacking any variations in Nanog levels indicated that the network was bistable: Nanog levels were restricted to either high or low values, and that a cell could not switch between these alternatives unless a perturbation source was included. When network simulations included either the Gaussian noise term or the additional feedback loop, the model was able to reproduce the experimentally observed bimodal Nanog distribution. To distinguish between these two possibilities, the authors proposed a set of experiments in which cells could be sorted into Nanog-high and Nanog-low fractions, and each fraction tracked over time to determine the dynamics by which the bimodal Nanog distribution is reestablished. They predicted that stochastic fluctuations in Nanog would result in a continuous shift from the unimodal to the bimodal distribution. In contrast, if Nanog variation resulted from oscillations, then the return to a bimodal distribution should be oscillatory.

This work provides an exemplary case of a hypothesis-driven modeling effort, as the authors developed competing models representing two different mechanisms of Nanog variability and explored the consequences of each. Furthermore, it provides a prime example of the potential interactions between modeling and experiments, in that the motivation was initially provided by seemingly confusing results (known variations in Nanog levels not predicted by any current model of the regulatory network), and the results of the model analysis led directly to the suggestion of additional experiments. Finally, the work also demonstrates that certain assumptions,

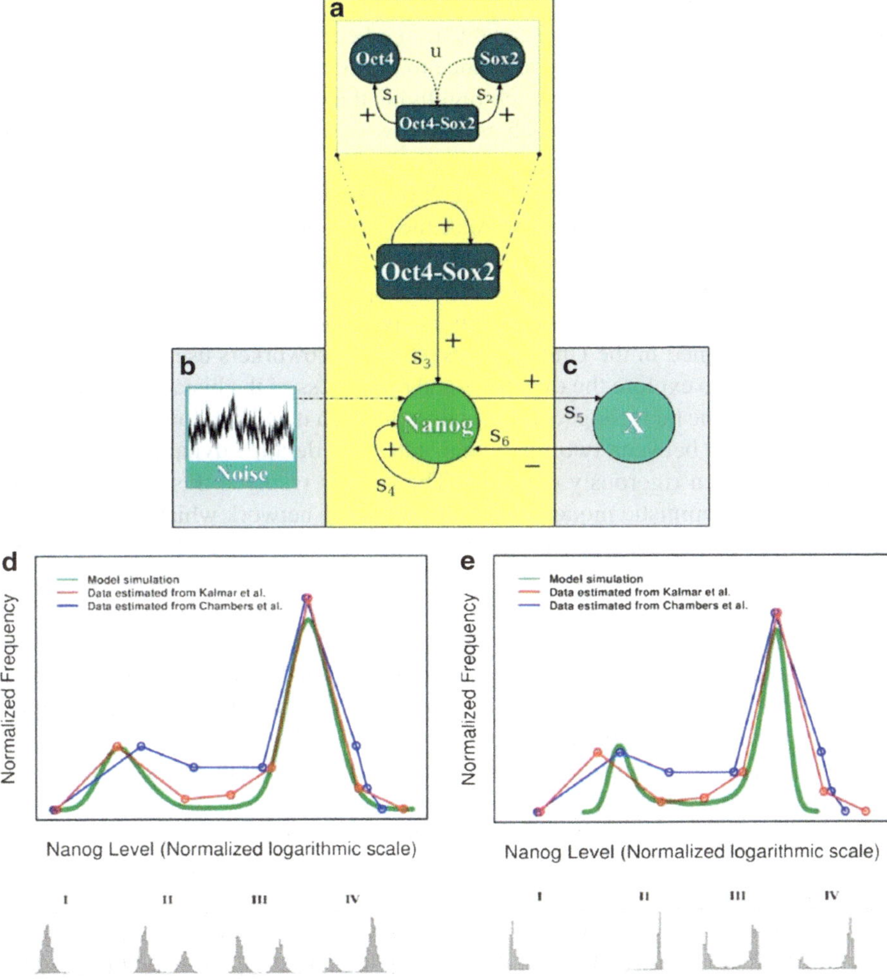

Fig. 2 Structure of the Oc4-Sox2-Nanog transcription factor network modeled by Glauche and colleagues (From Glauche et al. 2010). (**a**) The interactions between the three core components of the PSC pluripotency network were simplified by assuming that the levels of Oct4 and Sox2 are maintained in dynamic equilibrium with, and are largely controlled by, the function of the Oct4-Sox2 heterodimer by positive feedback (*inset*). Hence the individual monomers could be neglected in the model. Rate constants used in the model are indicated (S1-6). (**b**) Stochastic and oscillatory (**c**) variability in the levels of Nanog were introduced as indicated. (**d–e**) Comparisons of simulated (*green curve*) and empirical levels of Nanog, normalized to match local maxima, using the noise and oscillatory models of variability. The lower figure in each panel demonstrates the simulated dynamics by which Nanog low-expressing cells will return to the bimodal expressing stable state, suggesting that appropriate experiments should distinguish between the two variability scenarios explored

while allowing a simplification of the underlying mathematics in the model, can still be consistent with a useful model capable of predicting novel results and generating new ideas that are of immediate relevance to biological experiments.

Another example of this approach highlights the differences between deterministic and stochastic models of biochemical networks, and clearly demonstrates that these differing computational metaphors can be effectively combined to understand the role of robustness and noise (Lai et al. 2004). Deterministic models are those for which any given input will always produce the same output; no variation in response will be observed. In contrast, stochastic models allow for a variation in this input-output relationship in that a distribution of responses will be observed, with the probability of observing a specific output is always less than unity. In this interesting study (performed in the United States), Lai and coworkers used these different types of models to explore the dynamics and robustness of the bistable Gli response to exogenous Sonic hedgehog (Shh) signals. Although it had previously been known that Shh switches between two functional states, the detailed dynamics of this network had not been rigorously explored. The authors resolved this by studying an ODE-based deterministic model of the Shh signaling network which contained the major components of the Shh-Gli signaling network, including both the Gli positive feedback and the Ptc negative feedback loops (Fig. 3a).

By analyzing the stability of the network output, defined as the overall level of Gli transcription factors with respect to the magnitude of the Shh signal, the authors demonstrated that the response of this deterministic network was indeed bistable (Fig. 3b), consistent with empirical observations. Additionally, the impact of mutations in the network that are known to be associated with cancer were explored, and found to disrupt the bistability of the system so that once the Gli genes are activated, they could not be turned off.

An interesting aspect of this study was that the authors then proceeded to add a stochastic component to their model to study the impact of fluctuations in Gli that result from transcriptional and translational noise. When the Shh signal was increased to a value just below the deterministic threshold required to initiate Gli expression, the endogenous fluctuations in Gli levels produced spontaneous switches in the transition from off to on (Fig. 3c). When the average level of Gli was compared between the deterministic and stochastic simulations, it was discovered that Shh signals of intermediate magnitude, at a scale that allowed the spontaneous switching of cell states, resulted in large variations in Gli levels, although the sharpness of the switch was maintained. Finally, reasoning that positive feedback typically increases noise in a system, the authors removed the Ptc negative feedback loop, and observed a significant increase in the system noise. This result demonstrated that the negative feedback was an integral part of the robust network, required to ensure that responses to incoming signals occurred within a small window. In addition to this biological conclusion, the work of Lai and others also demonstrates that utilizing several types of computational methods when studying a biological problem can provide important insights into the underlying biological mechanism. Had the authors used only deterministic methods, the role of the second feedback loop in this network would remain unclear.

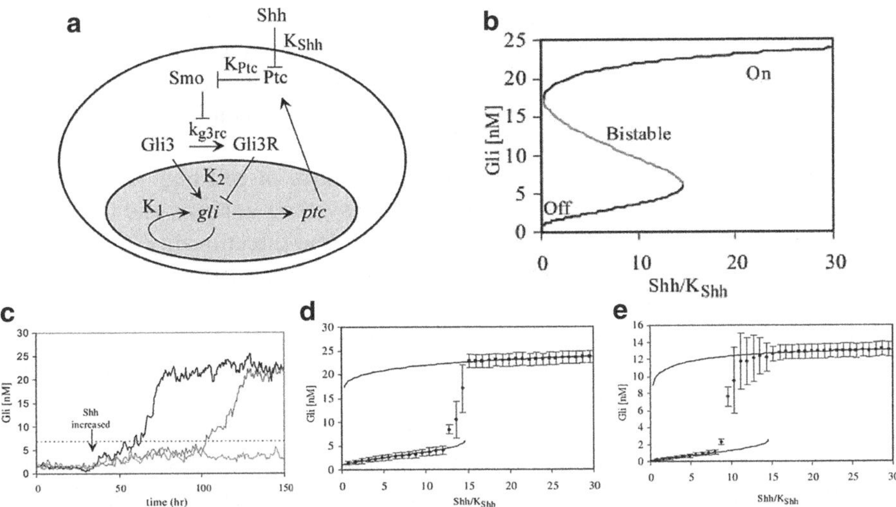

Fig. 3 Negative feedback is responsible for suppressing systems noise in the Sonic Hedgehog signaling network (From Lai et al. 2004). (**a**) Diagram representing the components of the Shh network modeled in the paper by Lai et al. (2004). *Shh* Sonic Hedgehog, *Smo* smoothened, *Ptc* patched, *Gli* glioma-associated oncogene family zinc finger. Other labels represent rate constants used in the model. (**b**) The deterministically modeled Shh network exhibits hysteresis in the abundance of the Gli transcription factors, as two stable concentrations (ON and OFF) are separated by a dynamically unstable state. Note that the transition from OFF to ON does not occur at the same parameter value as the reverse transition. (**c**) When stochastic fluctuations are included in the network, the time at which a switch from the OFF to the ON state (the "first passage time") varies considerably. Note that one Gli trajectory in this example does not switch states at all during the time simulated. (**d**) When Gli concentration is simulated stochastically as a function of Shh, variation in the Shh concentration that initiates a switch to the ON state in small, such that the stochastic switch behaves much like the deterministic system (*solid curve*). (**e**) When the Ptc negative feedback is removed from the model (by keeping Ptc constant), the Gli switch occurs at lower concentrations of Shh, and with much larger variation, indicting the negative feedback is required for robust and accurate switches in the Shh network

Theoretical Studies of Cell Fate

In contrast to these computational approaches, many theoretical studies have attempted to reveal fundamental principles of cell fate control that may be applicable across living systems. One early, yet still quite influential theoretical work was performed Stuart Kauffman, then at the University of Cincinnati. In studies of artificial and abstract "gene nets" in which the genes acted as either binary, continuous, or stochastic objects, Kauffman discovered that even randomly arranged nets consisting of only a few inputs per object could produce cycling dynamics between a small number of states, a surprisingly ordered behavior given the net's random topology (Kauffman 1969). This suggested that the organization of gene interactions into networks capable of defining discrete cell states was much more likely than previously believed. This research into random Boolean networks, which continues today (for example, see Huang et al. 2005, 2009; Markert et al. 2010; Halley et al. 2012; Morris et al. 2010), forcefully demonstrates the ability of an abstract model to highlight previously unknown fundamental biological principles, and helped initiate the use of complexity theory in biology.

More recent theoretical work addressing the control of cell fate, performed in both the United States and Canada, used dynamic systems analysis and the idea of a "potential landscape" to address the problem of "time-directionality" in SC models; that is, the concept that the apparent irreversibility of the differentiation process is not captured by standard dynamical models that allow gene expression to be reversibly controlled (Wang et al. 2010). Four primary mechanisms are thought to control this directionality: (i) avalanches of signaling within gene activation cascades, (ii) fixation of cell fate by epigenetic alterations, (iii) bifurcations in the nonlinear dynamics of the regulatory network leading to essentially irreversible switches in fate (i.e., hysteresis) or to the formation of novel attractors, both arising from alterations in system parameters, and (iv) stochastic fluctuations in the network dynamics causing the network to switch from one stable attractor to another. Of these, the first three are deterministic and require a monotonic change is some control parameter to effectively exhibit directionality, while in the case of the stochastic switching of attractors, directionality is an inherent property of the system and requires no changes to an external control parameter.

To explore the source of directionality, Wang and colleagues developed a model system composed of two mutually repressing transcription factors that can also auto-activate to induce different cell fates (Fig. 4a).

According to the standard dynamic systems approach, which has its basis in local linear stability analysis, this network behaves as a tristable system of two differentiated states separated by a single stable multipotent state (Fig. 4b, c). However, this method does not give any indication of the asymmetry in the probability of state transitions that occur in both directions; in other words, it cannot show the relative *likelihood* of state transitions between the SC and differentiated states, which would be important in addressing directionality. To address this shortcoming, the authors borrowed the concept of potential landscape analysis from the physical sciences to determine the depth and shapes of the potential surfaces surrounding the systems attractors. This approach was able to demonstrate that, as a control parameter was monotonically altered, the system's potential surface changed from one with a deep well representing the SC fate to a surface in which the SC well has been replaced with a hill-top barrier located between the now deeper differentiation-fate wells (Fig. 4d).

More detailed analyses demonstrated that the model was able to reproduce empirically-relevant observations, including a preference for one differentiated fate over another in a simulation of instructive commitment. Additionally, by comparing the relative stability of each state in the presence of noise, the authors were able to calculate the height of the potential barriers to the spontaneous reversion to the SC state (Fig. 4e) and demonstrate that as the control parameter is changed, the ratio of the SC to differentiated well depths decreases and lowers the likelihood of a differentiated cell reverting to the SC state.

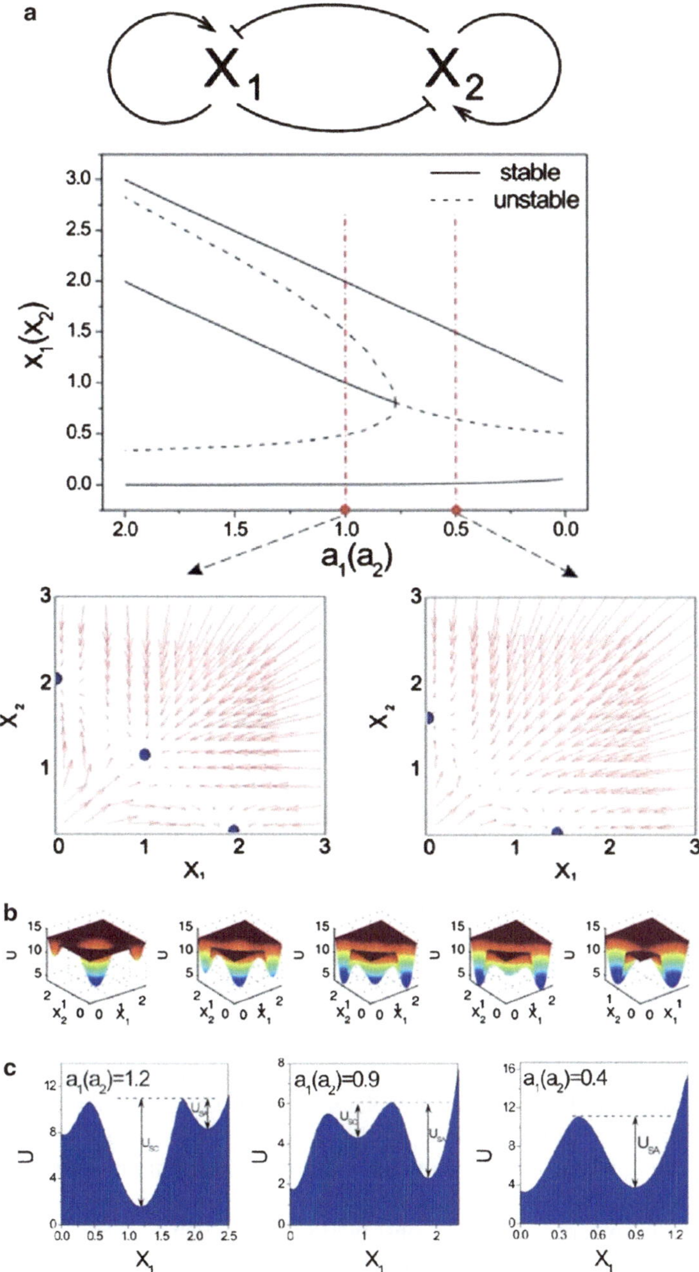

Fig. 4 Dynamics of the simplified transcription factor circuit analyzed in Wang et al. (2010). (**a**) Schematic of the circuit shows two mutually inhibitory factors that auto-activate their own transcription (*upper*). The bifurcation diagram of this network shows that as the auto-activation strength ($a1$, $a2$) decreases, one stable state disappears and is replaced by an unstable state (*middle*). Flow diagrams demonstrate the complete systems dynamics at two auto-activation values on either side of the bifurcation point (*lower*). (**b**) Potential landscapes of the systems in which the height of the landscape represents a measure of the probability that the systems will be found in a state with the given levels of each transcription factor. (**c**) Cross sections through the landscape demonstrate the changes in barrier heights as changes are made to the auto-activation parameters

This last observation therefore explains the directionality of differentiation as a change in the potential landscape representing the nonlinear dynamics of the regulatory network. In addition, this study also demonstrates several useful features of relevance to SCE opportunities. First, as the modeled network structure is a somewhat general architecture—consider several transcription factor pairs that regulate each other in a similar manner (GATA1-Pu.1, Sox2-Oct2, etc.)—it shows how a simplified representation of a known biological system can be rigorously analyzed to yield insights into fundamental mechanisms of fate control. Second, it also demonstrates the value of cross-disciplinary fertilization of techniques and concepts, as the authors were able to utilize methods from the physical sciences to great benefit in their analysis of biological networks. It also shows that detailed and precise understanding of computational methods can provide novel insights into our current understanding of SC biology, and help to further clarify our understanding of the molecular regulation of cellular systems. One further observation from the above paper was that very high levels of positive feedback significantly increased the probability of a reverse transition back to the SC fate, indicating that the modeling approach outlined above may also provide a general theoretical framework with which to understand *induced* pluripotency (Takahashi et al. 2007).

Induced Pluripotency and Priming: Bidirectional Cell Differentiation

In addition to the work by Wang and others described above, several other studies have explored induced pluripotency using models. One of the first used a model consisting of a system of six coupled ODEs to describe the dynamics of the core pluripotency genetic network (MacArthur et al. 2008). Essentially comprising a multistable network of genetic switches, this model predicted a sequence of progressive state restrictions, representing irreversible cell differentiation when exposed to increasing levels of stimulus. By converting the system to a set of coupled stochastic differential equations (SDEs), the authors were then able to analyze the effects of transcriptional noise on the reprogramming process. By essentially asking if noise might be responsible for a reversal in the unidirectional differentiation process, they observed that increasing levels of noise produced higher probabilities of a differentiated cell returning to the pluripotent state. Interestingly, the authors also noted that the levels of noise required to initiate reprogramming were much lower for Oct4 and Sox than for Nanog, suggesting that the latter gene is not required for the reprogramming process. This observation that much higher levels of Nanog (i.e., greater variation in levels) are necessary for reprogramming, provides computational insight into the previously surprising observation of the dispensability of Nanog to this process (Takahashi and Yamanaka 2006).

In a more recent study combining experiments and modeling, the fates of single pre-B cells that were reprogrammed using the "Yamanaka" factors Oct4, Sox2, Klf4, and c-Myc were followed (Hanna et al. 2009). It was observed that all cells could be reprogrammed over an extended period of time, surprising given the exceedingly low efficiencies normally described in the literature. Importantly, the time at which any cell returned to pluripotency seemed random, and the authors therefore concluded that reprogramming was both a continuous and a stochastic process. To further address this possibility, a simple one-step stochastic model was developed in which all cells had an intrinsic and constant rate of reprogramming. This basic model was able to match their experimental observations reasonably well, supporting the assertion that reprogramming was probabilistic in nature. Unfortunately, the model could not accurately predict the earliest observed cell transition times, leading the authors to suggest the reprogramming probability may be time-dependent, perhaps related to alterations in the cultures occurring during the experiments. Alternative explanations such that each cell has its own unique transition probability that results in a heterogeneous distribution across the entire population, or cell subpopulation selection occurs during adherent cell passaging, remain to be explored.

Model development has also been progressing on an issue related to cell fate (re) programming. So-called multilineage priming (Hu et al. 1997) which can be defined as a state in which gene expression patterns in a SC, present before the cell is exposed to a given fate-inducing environment, are such that the probability for the SC to differentiate along that specific trajectory is increased when the cell is exposed to that environment (Dillon 2012). Thought to allow for fast transcriptional responses to differentiation signals, priming has been studied experimentally for some time and is observed as the simultaneous low level expression of genes in a SC that are usually associated with specific differentiated states. Over the last decade, progress has also been made in the mathematical analysis of this process.

In one of the first mathematical analyses of priming (Huang et al. 2007), a model of the erythroid/myeloid switch of hematopoiesis was developed in which the interaction between the hematopoietic transcription factors PU.1—GATA1 was represented as a network of mutually inhibiting nodes which auto-activate themselves (Fig. 4a). As in the analysis of a similarly structured network discussed above (Wang et al. 2010), when no auto-activation was present in the network, it exhibited bistable behavior in which each stable state represented cells of either the myeloid or the erythroid lineage. However, when auto-activation was included, the network exhibited a novel third stable state which was interpreted as the metastable progenitor cell. Importantly, the expression levels of both PU.1 and GATA1 in this progenitor state were low and equal, consistent with the concept of lineage priming. Thus, in what is likely the simplest formal model of this system, Huang and colleagues were able to demonstrate that the presence of auto-activation was required for the tristability consistent with the PU.1—GATA1 network, and that it also could predict the low levels of each transcription factor consistent with lineage priming.

In a subsequent modeling study of the same network (Chickarmane et al. 2009), some of the assumptions of the previous models were removed, most importantly the cooperative binding between the transcription factors and the genes in the network. Without these assumptions in place, the authors demonstrated that the network, as previously modeled, was unable to exhibit bistability, and that an additional node was required in order for the network to function as a switch. In this study, the authors therefore proposed the existence of an additional factor that is upregulated by either PU.1 or GATA1, and with it, acts as a cofactor to inhibit the other factor. Thus, depending on the environmental signals, the network stabilizes in either the GATA1 or the PU.1 states. Importantly, when the novel factor is inhibited in the absence of the differentiation-inducing environmental factors, both PU.1 and GATA1 were expressed at levels low enough to retain pluripotency, consistent with the primed SC state.

These studies are interesting in their own right, and they also highlight the important fact that the experimental observations of priming and induced pluripotency indicate that models of pluripotency need to be flexible enough to exhibit bidirectional commitment, at least when experimental evidence demands it. Additionally, the two lineage priming examples considered here illustrate the interplay between modeling and experiments. Quite often models are developed that are capable of providing novel predictions of biological behavior at some "knowledge point" in time. However, frequently new experimental results invalidate the model assumptions or provide greater detail regarding the processes under study that require models be updated and assumptions reevaluated. It is only through this cyclical feedback from experiment to model to experiment that both approaches will have a maximal impact on stem cell engineering.

Data-Based Single-Cell Signaling Dynamics

The use of microfluidic technologies to analyze small biological samples has blossomed over the last decade, and its use in the study of SCs has followed that trend. An important motivating factor in this is the recognition of the importance of single SC (as opposed to population) analysis. While a more detailed discussion of these techniques and their uses in stem cell engineering is presented in chapter "High-throughput Screening, Microfluidics, Biosensors, and Real-time Phenotyping", we will describe an interesting study in which computational analyses were strongly linked and inspired by experimental observations of single-cell data obtained using microfluidic techniques. This study, performed by Dr. Savas Tay of the Basal Stem Cell Network and ETH Zurich while he was a postdoctoral fellow in Dr. Steve Quake's lab (Dept. Bioengineering, Stanford), was designed to measure the response of single mouse fibroblast cells exposed to various concentrations of TNF-α (Tay et al. 2010). In contrast to the majority of measurements of cell signaling dynamics that have been obtained from population-based bulk measurements, Tay and colleagues instead analyzed the movement of NF-κB between the cytoplasm and nucleus of individual cells in response to various levels of stimulation. A surprising finding was that the responses of individual cells were discrete in that at every concentration of TNF-α, a cell either responded or it did not. What changed as the signal levels increased was the proportion of cells responding; an observation that would be overlooked when using bulk assays. Moreover, all cell responses occurred at approximately the same magnitude regardless of the level of stimulation, indicating that the individual cell behaviors were digital. Additional experiments using multiple pulses of TNF-α further demonstrated that these responses were, at least in part, a stochastic process as individual cells varied as to whether they responded to none, one, or both stimulus pulses.

As the then-available models were unable to simultaneously reproduce both the digital cell activation and the particular dynamics of the responses, Tay colleagues revised a stochastic model of TNF-α signaling by incorporating characteristics of their experimental system such as variations in the abundance of the TNF-α receptor and modified rates of signal degradation. Importantly, they added a nonlinear activation profile for the NF-κB kinase inhibitor IKK in order to reproduce the observed digital cell responses. Surprisingly, this modified model was able to reproduce many of the experimental observations, including the probability that a cell will be activated by a given stimulus, the mean concentration of nuclear NF-κB, and the single-cell response time distributions—all using a single set of parameters.

This work highlighted the value of single-cell data when constructing models of cell signaling. Furthermore, it provides an accessible example of a multiscale model, in that even though the calculations involved events within a single cell, the model was able to reproduce observations made at a cell population level. While the difficulties involved in culturing single human pluripotent SCs currently make this type of analysis challenging, the approach nevertheless represents a significant opportunity for the SCE field in that it provides a route to understanding the complex variation observed in cultured SC populations.

Computational Design of Novel Materials and Structures

Another example of current computational work on understanding and controlling SC fate, from the van Blitterswijk and deBoer groups located at the MIRA Institute for Biomedical Technology and Technical Medicine (University of Twente, Netherlands), used mathematics in an obvious supporting role to biomaterials engineering, with a goal of creating a high-throughput method to screen for surfaces that are optimal with respect to their ability to promote specific cell behaviors (Unadkat et al. 2011). Motivated by previous work demonstrating that controlling cell spreading or shape using micro- and nanopatterns allowed for precise regulation of cell fate, the authors of this work reasoned that rational design might neglect entire classes of surfaces able to precisely regulate a cell's behavior. Consequently, they designed a "materiomics" approach in which distinct surface shapes were generated from three primitives: (1) circles, able to create large smooth areas, (2) triangles, able to produce angles, and (3) rectangles, able to form elongated structures. By computationally combining shapes of different sizes and orientations, the authors were able to generate a library consisting of over 150 million possible surfaces (Fig. 5). Assays of over 2,000 randomly selected surfaces were performed and were able to identify those surfaces most compatible with the propensity of proliferation, pluripotency, and osteogenic differentiation of human mesenchymal stromal cells. Importantly, a machine learning algorithm introduced to identify biologically important patterns was able to predict that certain spatial topographies could trigger cell proliferation (Fig. 5b). This example demonstrates that computational methods

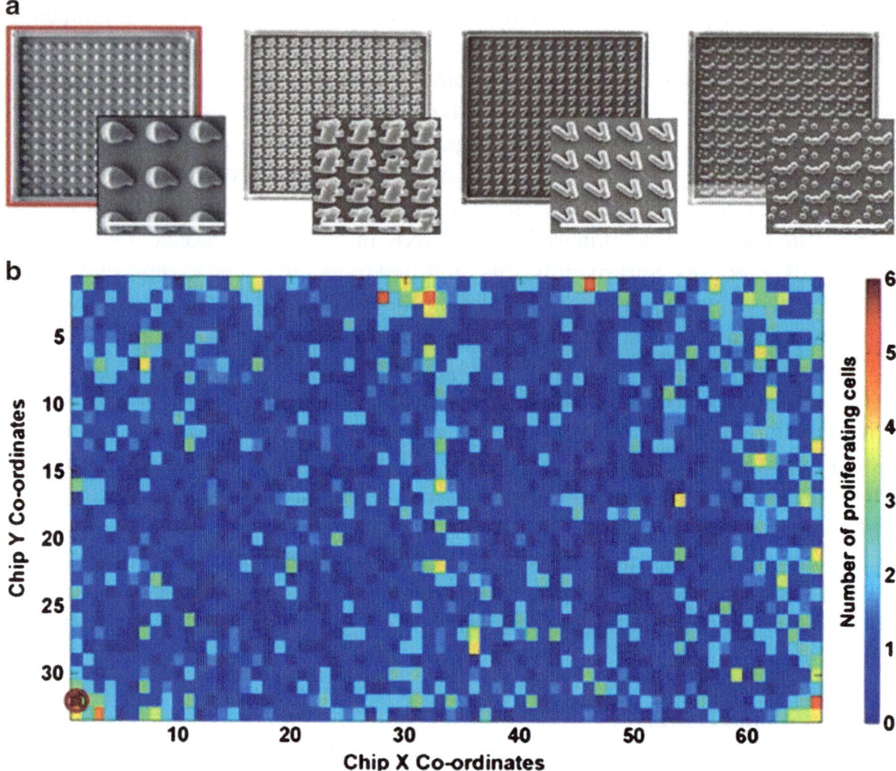

Fig. 5 Different surface structures have significant impact on the proliferation of human mesenchymal stromal cells (hMSC) (From Unadkat et al. 2011.) (**a**) Scanning electron micrographs of the four randomly designed surface structure exhibiting the highest scores in the proliferation assay. (**b**) Heat map demonstrating the complete proliferation assay. The *red circle* in the bottom left shows the structure resulting in the highest proliferation rate, corresponding to the first panel in (**a**). The entire chip is made up of a 2 cm × 2 cm array of 2,176 distinct surface patterns, in duplicate, and represents only 0.0014 % of the total number of possible surface structures that can be created from the algorithm of Unadkat et al. (2011)

can also provide important contributions to the design of reagents and experiments, thereby allowing researchers to identify important yet non-intuitive regulatory features that might otherwise be ignored.

Stem Cell Populations

The design and optimization of bioreactors and large-scale culture systems is a subject that should be very familiar to bioengineers involved with stem cell engineering, as it is of critical importance to the generation of large volumes of cells needed for both drug design and regenerative medicine. Accordingly, it is an area of active research across the world and a number of research organizations in Asia visited by the WTEC panel specialized in these approaches. For example, the Bioprocessing Technology Institute (BTI) in Singapore, until recently led by Dr. Miranda Yap, studies biomanufacturing systems focusing on the pharmaceutical and cell therapy industries. Recently, Dr. Steve Oh's group at BTI published studies in which microcarriers were used to increase yields of cardiomyocytes three fold over embryoid body controls (Lecina et al. 2010), and in which cell agitation due to shear stress in stirred cultures resulted in decreased pluripotency and an increase in differentiation-specific markers in a cell-line specific manner (Leung et al. 2011).

While the empirical studies provide a wealth of information regarding culture methodologies, a pair of papers from the Zandstra lab in Toronto, Canada provides examples of the value of a combined *in silico* and experimental approach in the analysis of cell populations and the optimization of culture technologies. Motivated by the elusive nature of techniques capable of increasing human hematopoietic stem cell (HSC) numbers *in vitro*, Kirouac et al. (2009) developed a simplified model of HSC differentiation in which cell fates were regulated by feedback from secreted molecule cell interaction networks among various cells from different points in the developmental hierarchy (Fig. 6a). Using this model, the authors were able to predict that negative feedback originating from more differentiated cells in the hierarchy is a principal regulatory mechanism controlling HSC fate.

By subsequently comparing their model predictions to experimental measurements of cell population dynamics, the authors were able to show that the model could predict the proportion of differentiated cells present under various culture manipulations (Fig. 6b). Importantly, their work suggested that modulation of these feedback loops, thereby altering the numbers of HSC and progenitors, would allow for quantitative control over cell fate.

In a recent extension of this work, Csaszar and others used these concepts to develop a strategy by which the controlled and specific inhibition of negative regulators of HSC differentiation allowed for global control of the cell population dynamics (Csaszar et al. 2012). As a preliminary step, Csaszar and coauthors identified numerous factors present in cultures with the ability to inhibit the expansion of HSC and HSC progenitor cells. In order to predict culture strategies that would maximize the abundance of HSC and progenitors, these factors were used as feedback candidates (SF1-4 in Fig. 6c) in the computational model of Kirouac et al. (2009). Of the commonly utilized fed-batch and perfusion systems, their calculations predicted that the fed-batch dilution approach would outperform other methods (Fig. 6c). When tested experimentally, the predicted protocol gave significantly higher average expansion of HSC and CD34+ progenitor cells after 12 days (Fig. 6d).

This pair of papers demonstrates the productivity achievable when even simplified models and experimental efforts are tightly linked. By integrating mathematical model outputs with validated stem cell assay outputs, the authors were able to predict the optimum culture method to maximize the yields of cell types of interest. Although this connection may not always be possible, especially for models of a

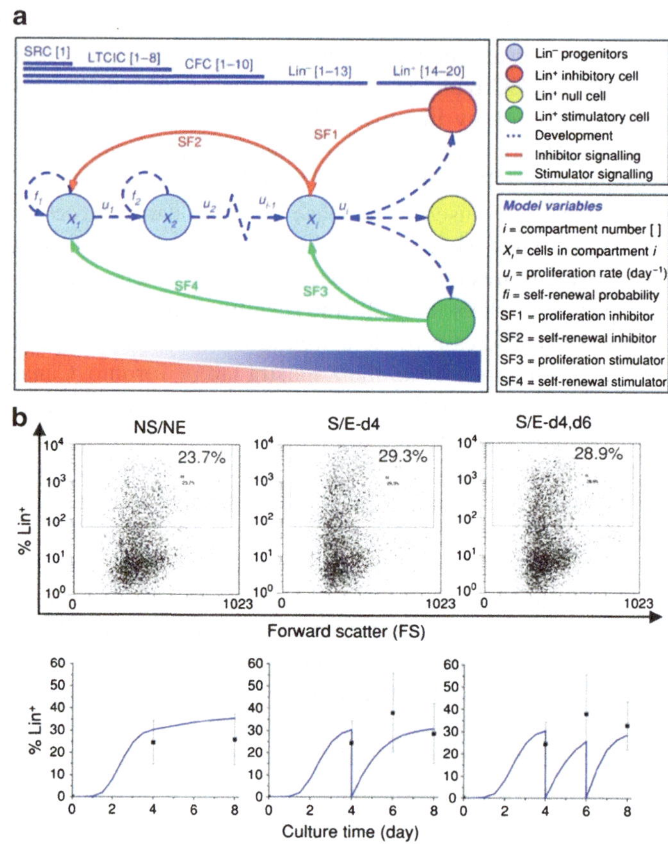

a
SRC [1]
LTCIC [1–8]
CFC [1–10]
Lin⁻ [1–13]
Lin⁺ [14–20]
SF2
SF1
f₁
u₁
f₂
u₂
u_{i-1}
u_i
X₁
X₂
X_i
SF4
SF3
Lin⁻ progenitors
Lin⁺ inhibitory cell
Lin⁺ null cell
Lin⁺ stimulatory cell
Development
Inhibitor signalling
Stimulator signalling
Model variables
i = compartment number [ ]
X_i = cells in compartment i
u_i = proliferation rate (day⁻¹)
fi = self-renewal probability
SF1 = proliferation inhibitor
SF2 = self-renewal inhibitor
SF3 = proliferation stimulator
SF4 = self-renewal stimulator
b
NS/NE
S/E-d4
S/E-d4,d6
23.7%
29.3%
28.9%
% Lin⁺
% Lin⁺
Forward scatter (FS)
Culture time (day)

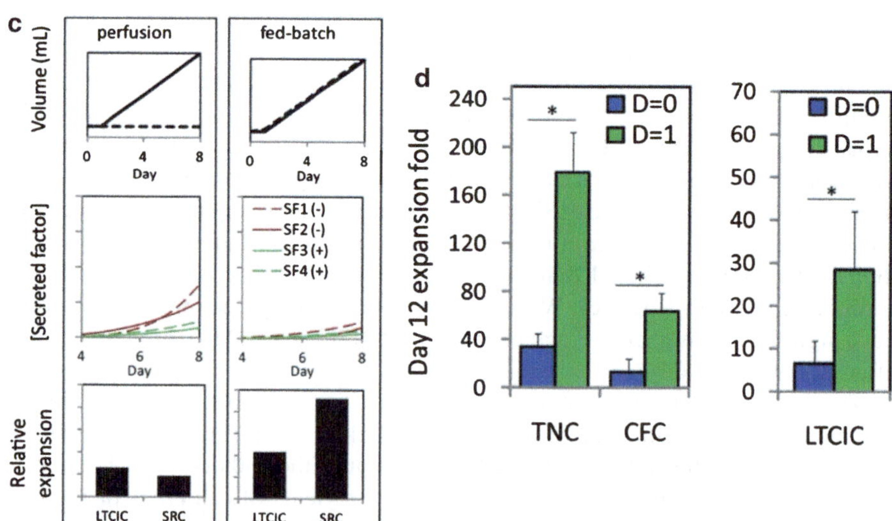

c
perfusion
fed-batch
Volume (mL)
Day
Day
[Secreted factor]
SF1 (-)
SF2 (-)
SF3 (+)
SF4 (+)
Day
Day
Relative expansion
LTCIC
SRC
LTCIC
SRC
d
Day 12 expansion fold
D=0
D=1
D=0
D=1
TNC
CFC
LTCIC

more theoretical nature, the feedback between modeling and experimental efforts should function as an iterative process to maximize the effectiveness of each, as demonstrated by these two papers.

An additional example of modeling stem cell populations studied cell differentiation dynamics using population-based methods (Task et al. 2012). Adapting a model based on previous studies on HSC development (Roeder and Loeffler 2002), the authors used cell-based stochastic simulations in which a heterogeneous population evolves according to user-defined rules to explore the mechanism of lineage commitment during pluripotent cell endoderm differentiation. As in the examples above, this study linked the available assays of cell type and number to the model outputs so that the mechanisms represented in their model system could be experimentally validated.

Briefly, this model assumed two cellular compartments representing a differentiating state (DS) in which cells actively age, proliferate, and differentiate, and another quiescent SC state. Transfer of cells between these states was probabilistic, with the likelihood of transfer from the SC to the DS remaining constant, while the reverse probability declines over time. Once the transition probability of any cell falls below a set threshold, that cell irreversibly enters the DS, where it has a unique and finite lifetime. Once this commitment to the DS has occurred, the likelihood of differentiating increases over time, and when these probabilities exceed a set threshold, the cell is irreversibly committed to a particular lineage.

To study the mechanisms of endoderm differentiation, two culture conditions were modeled and validated: differentiation using Activin A alone, or Activin A in combination with BMP4 and FGF2. Various mechanisms containing combinations of three features were then modeled to determine which was most consistent with data: (1) the presence of a mesendoderm (ME) intermediate germ layer (i.e., a 2-stage differentiation mechanism), (2) the presence of the marker CXCR4 in mesoderm (motivated by the known non-linearity of CXCR4 expression and conflicting reports of its expression), and (3) whether one cell type demonstrates preferential

---

Fig. 6 A cell interaction model of the HSC differentiation hierarchy is able to predict improved culture conditions that maximize cell expansion. (**a**) A schematic of the interactions included in the simplified model of the HSC hierarchy in which compartments representing SCs (X1) and progenitors (X2 … Xi) progressively lead to the formation of differentiated progeny(*red, yellow,* and *green circles*). *Red* and *green arrows* represent negative and positive feedback, respectively. The resulting mathematical model contains 24 state variables and 16 internal parameters in a system of ODEs. (**b**) *Upper panel* exhibits fluorescence activated cell sorting analysis of cell cultures shows the increased percentage of Lin+ cells from 8-day cultures that have undergone cell selection and media exchange, as predicted by the model. *NS/NE* no selection or media exchange, *S/E-d4* selection and media exchange at d4, *S/E-d4,d6* selection and exchange at both d4 and d6. *Lower panel* shows a comparison of predicted and measured values over the course of the entire culture period (From Kirouac et al. 2009; used under Creative Commons license from *Molecular Systems Biology*). (**c**) Simulated predictions of secreted factor concentration and compartmental expansion using perfusion and fed-batch culture methods. (**d**) Experimental measures of the 12-day expansion of various cell compartments in the perfusion (D=0) and fed-batch (D=1) culture systems (From Csaszar et al. 2012)

proliferation over the others. Depending on the specifics of each of the 12 mechanisms analyzed, the dynamics with which populations of SC, visceral endoderm (VE), ME, and endoderm cells arise will vary over time. Interestingly, a comparison of model predictions to data suggested a single mechanism was consistent with data obtained under both culture conditions examined, and suggested a mechanism of lineage commitment involving: (1) a ME intermediate, (2) inclusion of a CXCR4-mesoderm population, and (3) a higher proliferation of ESC, ME and endoderm over other cell populations. The cell dynamics were also able to demonstrate that the addition of FGF2 and BMP4 was consistent with both a lower threshold for commitment beyond the ME stage and a higher transition rate from ME to the proliferative meso- and endoderm states, accounting for higher rates of differentiation observed under these conditions. One surprising observation, given the inherent non-linearity of the differentiation process, was that the fits to this single mechanism was strong enough that not even a detailed parameter search for alternatives could provide an improvement of the predicted mechanism.

This work highlights the ability of models to rapidly and efficiently explore many competing hypotheses, and provide useful predictions regarding experimental results. Furthermore, it demonstrates a burgeoning trend in SCE models: using population-based methods to understand the dynamics of heterogeneous populations of cells evolving under different conditions. While not considering the cellular feedback included in the papers discussed previously (Kirouac et al. 2009; Csaszar et al. 2012), this work was able to provide mechanistic insights into what previously had not been understood. Additionally, it showcases another novel use of a computational method in SCE that had been pioneered in alternate fields of computational biology (Sneddon et al. 2011; Yang et al. 2010; Maus et al. 2011), and thus points to future opportunities for modeling in SCE.

Tissues and Development

Far fewer computational studies have been performed specifically exploring the role of SCs in whole tissues and development than have been performed for individual or SC populations. One area in which models have been utilized is in the study of the vertebrate gut, specifically the development of the intestinal crypt. Two recent papers from the University of Leipzig illustrate the use of models in this area, and highlight the importance of biomechanical forces in regulating cell fate and tissue morphogenesis.

Motivated by the idea that the traditional view of hierarchical tissue organization in which SC differentiate into progenitors, which subsequently terminally differentiate may not be required in order to explain aspects of tissue development, the authors of these papers first developed a model to determine if cell-cell and cell-environment interactions alone could result in the self-organization of the intestinal crypt (Buske et al. 2011). This multiscale model treated cells as elastic objects capable of growing, dividing, moving, and making contacts with other cells and the surrounding extracellular matrix. Cell fate was modeled to be dependent on the internal activity of the cell itself, alone with that of its neighboring cells. Accordingly, individual cell differentiation, and therefore tissue development, was assumed to be dependent on the curvature of the crypt basal membrane (BM) and the types and number of contacts it makes to its surroundings. The critical prediction made using this model was that the robustness of this tissue to cell loss was made possible by the flexibility in the cell fate decision process—any population could be removed from the simulated crypt without impacting the long-term tissue organization because the cell fate transitions experienced by the progenitors in the model were reversible, and the development occurred within an externally-imposed Wnt gradient.

In a recent extension of this model motivated by the recent success in establishing long-term intestinal organoids in culture, the self-organization concept was extended to the intestinal basal epithelium itself (Buske et al. 2012). In contrast to the rigid triangular network used to represent the BM in the previous model, this extension allowed for a flexible membrane whose shape was determined by interactions with a proliferating and heterogeneous cell population. With this modification, the authors observed that organoid shape changes result from differences in cell proliferation in that "buckled organoids" were formed as long as cell division continued. Once proliferation was eliminated, the organoids relaxed to spherical shapes. Furthermore, the model predicted that structures with stable geometries resembling that of the observed *in vitro* organoids would only form when three conditions were met: (1) differentiated Paneth cells are mixed with SCs, (2) Paneth cells become localized to regions of high curvature, and (3) Paneth cells induce further curvature in the BM of their local environment. Finally, the model also predicted that the expansion of the intestinal SC population depended on the rigidity of the BM network. This work therefore provides a nice example of how the incorporation of cell and tissue mechanics, along with realistic 3D tissue geometries can provide valuable insight into the connections between cells and their surroundings, whether during development or within a tissue exhibiting continual cell turnover, such as the intestine.

Future Opportunities

The preceding discussion covers only a tiny fraction of the research of relevance to SCE that utilizes computational approaches, as our goal was not to be encyclopedic but to highlight some interesting approaches and trends in the field. While these methods demonstrate that much progress has been made on integrating mathematics in the last several years, there are many opportunities for advancing computational SCE. Due to their quantitative and experimental training, bioengineers are remarkably well situated to seize these opportunities and drive SCE (and regenerative medicine) forward.

One type of opportunity relevant to all computational biologists includes advances made to the fundamental tools used in generating and analyzing models. For example, most biochemical models used in SCE are deterministic, and therefore assume that the system has an infinite volume (so concentrations can be depicted by continuous functions) and are well mixed (so the spatial heterogeneity can be ignored). These assumptions are unrealistic for all biological systems, but are routinely accepted because the mathematical tools required to analyze deterministic models under these conditions are well known. In order to increase the accuracy and predictability of SCE models, stochastic and spatial models—along with the appropriate tools required to analyze them—are needed. Although software packages are available which provide some of the necessary methods (Cowan et al. 2012; Stewart-Ornstein and El-Samad 2012; Hepburn et al. 2012; Drawert et al. 2012), they are generally not well used in the SCE field. Once further developed and adapted by the SCE community, these types of software packages could become a standard tool for both experimental and computational SC engineers.

Improved Links Between Experiment and Computational Efforts

A theme highlighted by several of the papers discussed above is that of the iterative and integrated nature of the successful computational-experimental approach to SCE. Many examples can be found in the literature that demonstrate the use of empirical data for the development and validation of models of SC biology, yet the majority of cases are examples of data sharing rather than the ideal of a long-term interaction between research groups with differing expertise. Examples of this collaborative effort can be found within individual research groups (for example, see Kirouac et al. 2009; Csaszar et al. 2012; Scherf et al. 2012; Tay et al. 2010), and between labs with very different expertise. The latter case is represented by a recent single-cell lineage analysis in which cell proliferation and differentiation were modeled as a stochastic multi-type branching process (Nordon et al. 2012). First described several years ago (Eilken et al. 2009), single-cell video microscopy using cell type specific markers in live cells has allowed experimentalists to follow the proliferation, division, and differentiation of individual hematopoietic SCs in culture to explicitly link related cell types in the HSC developmental hierarchy. When these original single-cell methods were combined with a powerful computation method, Nordon and colleagues were able to predict the ratios of cell types present in large-scale cultures. By likewise involving both computational and empirically-based researchers in the design of experiments to ensure that the data produced is easily integrated into modeling efforts (as in Kirouac et al. 2009 and Csaszar et al. 2012), the iterative process required to advance SCE will be more likely to occur. Ideally, this approach of combining modeling and experiment should be considered a standard one in the SCE toolbox, although it could be considered a long-term goal as it would necessitate a considerable increase in the degree of mathematical and modeling training provided to all those within the SC research community.

Of particular interest and relevance to the bioengineering community are computational methods used to optimize large-scale cell production in bioreactors. The environment within stirred-suspension bioreactors is remarkably complex environment dependent on many physical parameters (Kinney et al. 2011), yet many optimization attempts use a purely empirical approach to maximize yields. Consequently, one important opportunity for computational SCE is to take advantage of the many mathematical optimization procedures available (Banga 2008). Some examples exist in the literature in which computational fluid dynamics (Hidalgo-Bastida et al. 2011), cell population dynamics (Kresnowati et al. 2011), or various Bayesian selection and regression techniques (Winkler and Burden 2012) have been used to evaluate and predict the results of modifying bioreactor parameters. In general, however, these questions remain under explored by mathematically-inclined research labs. By improving the links between engineers familiar with the mathematical aspects of large-scale process design and those focusing on maximizing SC yields, significant progress can be made.

Improved Links Between Computational Methods

Due to recent technological advances, we are now gaining the ability to analyze events within individual cells and measure the distributions of behavior within cell populations rather than restricting our analyses to time and population-averaged bulk measurements. While the culture of single human pluripotent SC remains problematic, the adaptation of these innovative experimental approaches to other aspects of SCE will not only provide the much needed deconvolution of bulk and single-cell measurements, but will also allow for the use and validation of relatively new computational approaches addressing multiple scales of cellular organization (i.e., intracellular, whole cell, cell population and tissues). As described above for mouse fibroblasts (Tay et al. 2010) and for HSC (Luni et al. 2011) elsewhere, the prediction of response distributions across a population of cells can provide great insights into the mechanisms by which population-based bulk measurements arise.

Large-scale modeling efforts will also provide further opportunities for SCE. However, in comparison to other branches of computational biology, SCE lags far behind in such efforts. For example, a recent paper modeling the life cycle of the pathogen *Mycoplasma genitalium* represents the state of the art in these types of endeavors as it represents a whole cell model of an organism (Karr et al. 2012). Although the genome of *M. genitalium* only consists of 525 genes, this work represents a truly remarkable effort by a relatively small group of researchers, who utilized a modular design including all known gene and protein interactions, described using a variety of mathematical approaches, to simulate the complete cell cycle of the organism. Such a large-scale, SC-specific modeling project will be even more difficult due to the increased complexity of the eukaryotic cell, and will doubtless require considerable coordination across many labs and funding institutions. Indeed recent progress in synthetic biology, and the use of synthetic biology tools to model and control stem cells fate, is an exciting future opportunity. Furthermore, the continuing development of large-scale databases and standardization of data description, visualization and mathematical language approaches—not dissimilar to efforts in the bioinformatics community (Demir et al. 2010)—will also be a requirement for achieving goals of this magnitude. However, the gains to our understanding of biological systems in general, and to clinically relevant SC application in particular will no doubt be enormous.

Incorporation of Non-standard Computational Approaches from Additional Disciplines to Gain Novel Insight

Although there are many differences across biological scales that require computational exploration, there are also many similarities across both biological and physical systems. Arguably the greatest opportunity for computational SCE therefore lies in the adaptation of methods pioneered in other scientific fields. While relatively obvious connections exist between SCE and bioinformatics, molecular biology, and genetics, other less obvious relationships can be identified.

For example, SC biology has much in common with evolutionary ecology, from terminology to the types of problems that are posed (Mangel and Bonsall 2007; Powell 2005). Stem cells and organisms both occupy a spatially discrete niche (individual-environment interactions) and change their behavior in response to changes in the surroundings (adaptation), proliferate, die, and interact with individuals of both similar and different types (population dynamics). Whereas ecology makes productive use of models addressing these issues (for example, see Evans et al. 2012), their use in SCE has largely been ignored. Accordingly, many of the mathematical techniques that have been used in evolutionary ecology to explore these processes will undoubtedly find fruitful application in fundamental and translational SC biology and SCE.

It is interesting to note the fact that many ecological modeling approaches have been utilized in other areas of cell biology. A particularly active area is the study of heterogeneous population dynamics in cancer progression (e.g., Cleveland et al. 2012; Rodriguez-Brenes et al. 2011; Dingli et al. 2009; Chen and Pienta 2011; Bowler and Kelly 2012; Tieu et al. 2012). The goal in these efforts is to explain and predict the evolution of cancer cell populations within tumors by treating them as communities of interacting agents. Another area of activity is in the study of population dynamics of microorganisms (Gore et al. 2009; Cremer et al. 2011; Huang and Wu 2012). In contrast, little integration of ecological principles has filtered into SCE modeling efforts. As a paper published several years ago focused on developing a quantitative foundation for the ecological view of SCs (Mangel and Bonsall 2008), this may change in the near future.

In addition to the insight that can be provided by looking to other biological subdisciplines, much can also be learned from exploring the physical sciences. Stochastic simulations (Stewart-Ornstein and El-Samad 2012) of biochemical systems provide an illustrative example of this, as the approaches commonly used in biology today were originally developed to describe the coalescence of clouds (Gillespie 1975), demonstrating that biologically productive ideas can arise in unlikely places. Similar to evolutionary methods, these approaches are already being productively adapted for use in cancer research (Michor et al. 2011), but applications in SCE have been lacking. Numerous current opportunities for incorporation into SCE models could be listed, including network theory (Barabási 2011; Newman 2011; Motter and Albert 2012), single molecule dynamics (Barkai et al. 2012), bioelectricity, and biomechanics (Buske et al. 2012).

Finally, computational SCE can also look to other engineering subdisciplines for inspiration. A prime example of this is an analysis of intestinal crypt dynamics using optimal control theory (Itzkovitz et al. 2012). This approach uses a control function to continually modify system variables in order to reach a desired goal. By analyzing hypothetical intestinal cell lineages, the authors were able to identify "rules" by which the crypt produces the appropriate numbers of SC and differentiated cells in the shortest period of time. Interestingly, they were able to demonstrate computationally, and subsequently validate empirically, that cell dynamics in the crypt exhibit an efficient two-stage process consisting of a primary symmetric division of SC followed by a secondary burst of asymmetric division that produced the final differentiated cells.

Global Assessment and Conclusions

As evidenced by the publications described above, activity in computational SCE is a global effort, with important publications coming from labs around the world. While small pockets of particularly productive research can be found in the United States, Canada, and Germany, there does not appear to be any national region that is significantly more productive than the others. One aspect that is clear from the analysis of the above work is that mathematical approaches are increasingly being used to gain fundamental insight into the mechanistic underpinnings of complex biological systems, that many of these studies benefit from interdisciplinary approaches and data sharing, and that engineers are particularly well suited to take a leadership soles roles in the area. A further prediction is that the increasing use of team-based science to fund interdisciplinary efforts around target problems is likely important to supporting success in the area of stem cell mathematical modeling. Strategies in the United States to continue the growth of stem cell modeling based approaches, for example, through specific research initiatives, dedicated conferences and organizations, should be developed and implemented.

As outlined above, computational modeling can provide a solid foundation on which to study and understand the complex system made up of the SC and its environment. Although great progress has been achieved in the 50 years since the first mathematical treatment of SC biology was published (Till et al. 1964), significant prospects remain for advancing the field of computational SCE. By consolidating efforts among modelers and experimentalists of various scientific backgrounds, we can expect to make more rapid progress in the years to come. This is especially true given the incredible rate at which new and larger data sets are generated, and novel technologies produced that allow for increasingly sophisticated questions to be addressed. With backgrounds spanning experimental biology, computer science, physics, chemistry, and design, bioengineers are in an enviable position to rapidly and efficiently advance our knowledge of SCE, leading to innovations in fundamental biology and its clinical applications.

References

Aldridge, B.B., J. Saez-Rodriguez, J.L. Muhlich, P.K. Sorger, and D.A. Lauffenburger 2009. Fuzzy logic analysis of kinase pathway crosstalk in TNF/EGF/insulin-induced signaling. *PLoS Computational Biology* 5(4):e1000340. Available at: http://www.pubmedcentral.nih.gov/articlerender.fcgi?artid=2663056&tool=pmcentrez&rendertype=abstract.

Banga, J.R. 2008. Optimization in computational systems biology. *BMC Systems Biology* 2:47. Available at: http://www.pubmedcentral.nih.gov/articlerender.fcgi?artid=2435524&tool=pmc entrez&rendertype=abstract.

Barabási, A.-L. 2011. The network takeover. *Nature Physics* 8(1):14–16. Available at: http://www.nature.com/doifinder/10.1038/nphys2188.

Barkai, E., Y. Garini, and R. Metzler. 2012. Strange kinetics of single molecules in living cells. *Physics Today* 65(8):29. Available at: http://link.aip.org/link/PHTOAD/v65/i8/p29/s1&Agg=doi.

Bowler, M.G., and C.K. Kelly. 2012. On the statistical mechanics of species abundance distributions. *Theoretical Population Biology* 82(2):85–91. Available at: http://www.ncbi.nlm.nih.gov/pubmed/22683489.

Buske, P., J. Galle, N. Barker, G. Aust, H. Clevers, and M. Loeffler. 2011. A comprehensive model of the spatio-temporal stem cell and tissue organisation in the intestinal crypt. *PLoS Computational Biology* 7(1):e1001045. Available at: http://www.pubmedcentral.nih.gov/articlerender.fcgi?artid=3017108&tool=pmcentrez&rendertype=abstract.

Buske, P., J. Przybilla, M. Loeffler, N. Sachs, T. Sato, H. Clevers, and J. Galle. 2012. On the biomechanics of stem cell niche formation in the gut—modelling growing organoids. *FEBS Journal* 279(18):3475–3487, doi:10.1111/j.1742-4658.2012.08646.x. Available at: http://www.ncbi.nlm.nih.gov/pubmed/22632461.

Chambers, I., J. Silva, D. Colby, J. Nichols, B. Nijmeijer, M. Robertson, J. Vrana, K. Jones, L. Grotewold, and A. Smith. 2007. Nanog safeguards pluripotency and mediates germline development. *Nature* 450(7173):1230–1234. Available at: http://www.ncbi.nlm.nih.gov/pubmed/18097409.

Chen, K.-W., and K.J. Pienta. 2011. Modeling invasion of metastasizing cancer cells to bone marrow utilizing ecological principles. *Theoretical Biology & Medical Modelling* 8(1):36. Available at: http://www.ncbi.nlm.nih.gov/pubmed/21967667.

Chickarmane, V., T. Enver, and C. Peterson. 2009. Computational modeling of the hematopoietic erythroid-myeloid switch reveals insights into cooperativity, priming, and irreversibility. *PLoS Computational Biology* 5(1):e1000268. Available at: http://www.pubmedcentral.nih.gov/articlerender.fcgi?artid=2613533&tool=pmcentrez&rendertype=abstract.

Cleveland, C., D. Liao, and R. Austin. 2012. Physics of cancer propagation: a game theory perspective. *AIP Advances* 2(1):011202. Available at: http://link.aip.org/link/AAIDBI/v2/i1/p011202/s1&Agg=doi.

Cowan, A.E., I.I. Moraru, J.C. Schaff, B.M. Slepchenko, and L.M. Loew. 2012. Spatial modeling of cell signaling networks. *Methods Cell Biology* 110:195–221. Available at: http://www.ncbi.nlm.nih.gov/pubmed/22482950.

Cremer, J., A. Melbinger, and E. Frey. 2011. Evolutionary and population dynamics: A coupled approach. *Physical Review E* 84(5):58–60. Available at: http://link.aps.org/doi/10.1103/PhysRevE.84.051921.

Csaszar, E., D.C. Kirouac, M. Yu, W. Wang, W. Qiao, M.P. Cooke, A.E. Boitano, C. Ito, and P.W. Zandstra. 2012. Rapid expansion of human hematopoietic stem cells by automated control of inhibitory feedback signaling. *Cell Stem Cell* 10(2):218–229. Available at: http://www.ncbi.nlm.nih.gov/pubmed/22305571.

Deisboeck, T.S., Z. Wang, P. Macklin, and V. Cristini. 2011. Multiscale cancer modeling. *Annual Review of Biomedical Engineering* 13:127–155. Available at: http://www.ncbi.nlm.nih.gov/pubmed/21529163.

Demir, E., M.P. Cary, S. Paley, K. Fukuda, C. Lemer, I. Vastrik, G. Wu, et al. 2010. The BioPAX community standard for pathway data sharing. *Nature Biotechnology* 28(9):935–42. Available at: http://www.pubmedcentral.nih.gov/articlerender.fcgi?artid=3001121&tool=pmcentrez&rendertype=abstract.

Dillon, N. 2012. Factor mediated gene priming in pluripotent stem cells sets the stage for lineage specification. BioEssays : news and reviews in molecular, cellular and developmental biology, 34(3):194–204. Available at: http://www.ncbi.nlm.nih.gov/pubmed/22247014.

Dingli, D., F.A.C.C. Chalub, F.C. Santos, S. Van Segbroeck, and J.M. Pacheco. 2009. Cancer phenotype as the outcome of an evolutionary game between normal and malignant cells. *British Journal of Cancer* 101(7):1130–1136. Available at: http://www.pubmedcentral.nih.gov/articlerender.fcgi?artid=2768082&tool=pmcentrez&rendertype=abstract.

Discher, D.E., D.J. Mooney, and P.W. Zandstra. 2009. Growth factors, matrices, and forces combine and control stem cells. *Science* 324(5935):1673–1677. Available at: http://www.pubmedcentral.nih.gov/articlerender.fcgi?artid=2847855&tool=pmcentrez&rendertype=abstract.

Drawert, B., S. Engblom, and A. Hellander. 2012. URDME: a modular framework for stochastic simulation of reaction-transport processes in complex geometries. *BMC Systems Biology* 6(1):76. Available at: http://www.ncbi.nlm.nih.gov/pubmed/22727185.

Eilken, H.M., S.-I. Nishikawa, and T. Schroeder. 2009. Continuous single-cell imaging of blood generation from haemogenic endothelium. *Nature* 457(7231):896–900. Available at: http://www.ncbi.nlm.nih.gov/pubmed/19212410.

Evans, M.R., K.J. Norris, and T.G. Benton. 2012. Predictive ecology: systems approaches. *Philosophical Transactions of the Royal Society of London, Series B, Biological Sciences* 367(1586):163–169. Available at: http://rstb.royalsocietypublishing.org/cgi/doi/10.1098/rstb.2011.0191.

Fouliard, S., S. Benhamidaa, N. Lenuzzab, F. Xaviera. 2009. Modeling and simulation of cell populations interaction. *Mathematical and Computer Modelling* 49(11–12):2104–2108. Available at: http://linkinghub.elsevier.com/retrieve/pii/S0895717708002501.

Gillespie, D.T. 1975. An exact method for numerically simulating the stochastic coalescence process in a cloud. *Journal of the Atmospheric Sciences* 32:1977–1989.

Glauche, I., K. Horn, M. Horn, L. Thielecke, M.A. Essers, A. Trumpp, and I. Roeder. 2012. Therapy of chronic myeloid leukaemia can benefit from the activation of stem cells: simulation studies of different treatment combinations. *British Journal of Cancer* 106(11):1742–1752. Available at: http://www.ncbi.nlm.nih.gov/pubmed/22538973.

Glauche, I., M. Herberg, and I. Roeder. 2010. Nanog variability and pluripotency regulation of embryonic stem cells—insights from a mathematical model analysis. *PloS One* 5(6):e11238. Available at: http://www.pubmedcentral.nih.gov/articlerender.fcgi?artid=2888652&tool=pmcentrez&rendertype=abstract.

Gore, J., H. Youk, and A. van Oudenaarden. 2009. Snowdrift game dynamics and facultative cheating in yeast. *Nature* 459(7244):253–256. Available at: http://www.pubmedcentral.nih.gov/articlerender.fcgi?artid=2888597&tool=pmcentrez&rendertype=abstract.

Guilak, F., D.M. Cohen, B.T. Estes, J.M. Gimble, W. Liedtke, and C.S. Chen. 2009. Control of stem cell fate by physical interactions with the extracellular matrix. *Cell Stem Cell* 5(1):17–26, doi:10.1016/j.stem.2009.06.016. Available at: http://www.ncbi.nlm.nih.gov/pmc/articles/PMC2768283/.

Halley, J.D., K. Smith-Miles, D.A. Winkler, T. Kalkan, S. Huang, and A. Smith. 2012. Self-organizing circuitry and emergent computation in mouse embryonic stem cells. *Stem Cell Research* 8(2):324–333. Available at: http://www.ncbi.nlm.nih.gov/pubmed/22169460.

Hanna, J., K. Saha, B. Pando, J. van Zon, C.J. Lengner, M.P. Creyghton, A. van Oudenaarden, and R. Jaenisch. 2009. Direct cell reprogramming is a stochastic process amenable to acceleration. *Nature* 462(7273):595–601. Available at: http://www.pubmedcentral.nih.gov/articlerender.fcg i?artid=2789972&tool=pmcentrez&rendertype=abstract.

Hepburn, I., W. Chen, S. Wils, and E. De Schutter. 2012. STEPS: efficient simulation of stochastic reaction-diffusion models in realistic morphologies. *BMC Systems Biology* 6(1):36. Available at: http://www.ncbi.nlm.nih.gov/pubmed/22574658.

Hidalgo-Bastida, L.A., S. Thirunavukkarasu, S. Griffiths, S.H. Cartmell, and S. Naire. 2011. Modeling and design of optimal flow perfusion bioreactors for tissue engineering applications. *Biotechnology and Bioengineering* 109(4):1095–1099, doi:10.1002/bit.24368. Available at: http://www.ncbi.nlm.nih.gov/pubmed/22068720.

Hu, M., D. Krause, M. Greaves, S. Sharkis, M. Dexter, C. Heyworth, and T. Enver. 1997. Multilineage gene expression precedes commitment in the hemopoietic system. *Genes & Development* 11(6):774–785. Available at: http://www.genesdev.org/cgi/doi/10.1101/gad.11.6.774.

Huang, A.C., L. Hu, S.A. Kauffman, W. Zhang, and I. Shmulevich. 2009. Using cell fate attractors to uncover transcriptional regulation of HL60 neutrophil differentiation. *BMC Systems Biology* 3:20. Available at: http://www.pubmedcentral.nih.gov/articlerender.fcgi?artid=2652435&tool =pmcentrez&rendertype=abstract.

Huang, S. Y.P. Guo, G. May, and T. Enver. 2007. Bifurcation dynamics in lineage-commitment in bipotent progenitor cells. *Developmental Biology* 305(2):695–713. Available at: http://www.ncbi.nlm.nih.gov/pubmed/17412320.

Huang, S., G. Eichler, Y. Bar-Yam, and D.E. Ingber. 2005. Cell fates as high-dimensional attractor states of a complex gene regulatory network. *Physical Review Letters* 94(12):1–4. Available at: http://link.aps.org/doi/10.1103/PhysRevLett.94.128701.

Huang, Y., and Z. Wu. 2012. Game dynamic model for yeast development. *Bulletin of Mathematical Biology* Available at: http://www.ncbi.nlm.nih.gov/pubmed/22434448.

Humphreys, P. 2004. *Extending ourselves: computational science, empiricism, and scientific method.* Toronto: Oxford University Press.

Ideker, T., and D. Lauffenburger. 2003. Building with a scaffold: emerging strategies for high- to low-level cellular modeling. *Trends in Biotechnology* 21(6):255–262. Available at: http://idekerlab.ucsd.edu/Documents/idekerTiB2003.pdf.

Itzkovitz, S., I.C. Blat, T. Jacks, H. Clevers, and A. van Oudenaarden. 2012. Optimality in the development of intestinal crypts. *Cell* 148(3):608–619. Available at: http://linkinghub.elsevier.com/retrieve/pii/S0092867412000128.

Kalmar, T., C. Lim, P. Hayward, S. Muñoz-Descalzo, J. Nichols, J. Garcia-Ojalvo, and A. Martinez Arias. 2009. Regulated fluctuations in nanog expression mediate cell fate decisions in embryonic stem cells. *PLoS Biology* 7(7):e1000149. Available at: http://www.pubmedcentral.nih.gov/articlerender.fcgi?artid=2700273&tool=pmcentrez&rendertype=abstract.

Karr, J.R., J.C. Sanghvi, D.N. Macklin, M.V. Gutschow, J.M. Jacobs, B. Bolival, N. Assad-Garcia, J.I. Glass, and M.W. Covert. 2012. A whole-cell computational model predicts phenotype from genotype. *Cell* 150(2):389–401. Available at: http://www.sciencedirect.com/science/article/pii/S0092867412007763.

Kauffman, S.A. 1969. Homeostatic and differentiation in random genetic control networks. *Nature* 224:177–178. Available at: http://www.nature.com/nature/journal/v224/n5215/pdf/224177a0.pdf.

Kinney, M.A., C.Y. Sargent, and T.C. McDevitt. 2011. The multiparametric effects of hydrodynamic environments on stem cell culture. *Tissue Engineering. Part B, Reviews* 17(4):249–262. Available at: http://www.ncbi.nlm.nih.gov/pubmed/21491967.

Kirouac, D.C., G.J. Madlambayan, M. Yu, E.A. Sykes, C. Ito, and P.W. Zandstra. 2009. Cell-cell interaction networks regulate blood stem and progenitor cell fate. *Molecular Systems Biology* 5(293):293, doi:10.1038/msb.2009.49. Available at: http://www.pubmedcentral.nih.gov/articlerender.fcgi?artid=2724979&tool=pmcentrez&rendertype=abstract.

Kresnowati, M.T., G.M. Forde, and X.D. Chen. 2011. Model-based analysis and optimization of bioreactor for hematopoietic stem cell cultivation. *Bioprocess and Biosystems Engineering* 34(1):81–93. Available at: http://www.ncbi.nlm.nih.gov/pubmed/20652600.

Lai, K., M.J. Robertson, and D.V. Schaffer. 2004. The sonic hedgehog signaling system as a bistable genetic switch. *Biophysical Journal* 86(5):2748–2757. Available at: http://www.pubmedcentral.nih.gov/articlerender.fcgi?artid=1304145&tool=pmcentrez&rendertype=abstract.

Lecina, M., S. Ting, A. Choo, S. Reuveny, and S. Oh. 2010. Scalable platform for human embryonic stem cell differentiation to cardiomyocytes in suspended microcarrier cultures. *Tissue Engineering, Part C, Methods* 16(6):1609–1619. Available at: http://www.ncbi.nlm.nih.gov/pubmed/20590381.

Leung, H.W., A. Chen, A.B. Choo, S. Reuveny, and S.K. Oh. 2011. Agitation can induce differentiation of human pluripotent stem cells in microcarrier cultures. *Tissue Engineering, Part C, Methods* 17(2):165–172. Available at: http://www.ncbi.nlm.nih.gov/pubmed/20698747.

Luni, C., F.J. Doyle, and N. Elvassore. 2011. Cell population modelling describes intrinsic heterogeneity: a case study for hematopoietic stem cells. *IET Systems Biology* 5(3):164–173. Available at: http://www.ncbi.nlm.nih.gov/pubmed/21639590.

MacArthur, B.D., C.P. Please, and R.O.C. Oreffo. 2008. Stochasticity and the molecular mechanisms of induced pluripotency. *PloS One* 3(8):e3086. Available at: http://www.pubmedcentral.nih.gov/articlerender.fcgi?artid=2517845&tool=pmcentrez&rendertype=abstract.

Mangel, M., and M.B. Bonsall. 2007. The evolutionary ecology of stem cells and their niches - the time isnow.*Oikos*116(11):1779–1781.Availableat:http://doi.wiley.com/10.1111/j.2007.0030-1299.16248.x.

Mangel, M., and M.B. Bonsall. 2008. Phenotypic evolutionary models in stem cell biology: replacement, quiescence, and variability. *PloS One* 3(2):e1591. Available at: http://www.pubmedcentral.nih.gov/articlerender.fcgi?artid=2217616&tool=pmcentrez&rendertype=abstract

Markert, E.K., N. Baas, A.J. Levine, and A. Vazquez. 2010. Higher order Boolean networks as models of cell state dynamics. *Journal of Theoretical Biology* 264(3):945–951. Available at: http://www.ncbi.nlm.nih.gov/pubmed/20303985.

Maus, C., S. Rybacki, and A.M. Uhrmacher. 2011. Rule-based multi-level modeling of cell biological systems. *BMC Systems Biology* 5(1):166. Available at: http://www.ncbi.nlm.nih.gov/pubmed/22005019.

Michor, F., J. Liphardt, M. Ferrari, and J. Widom. 2011. What does physics have to do with cancer? *Nature Reviews, Cancer* 11(9):657–70, doi:10.1038/nrc3092. Available at: http://www.ncbi.nlm.nih.gov/pubmed/21850037.

Morris, M.K., J. Saez-Rodriguez, P.K. Sorger, and D.A. Lauffenburger. 2010. Logic-based models for the analysis of cell signaling networks. *Biochemistry* 49(15):3216–3224. Available at: http://www.pubmedcentral.nih.gov/articlerender.fcgi?artid=2853906&tool=pmcentrez&rendertype=abstract.

Motter, A.E., and R. Albert. 2012. Networks in motion. *Physics Today* 65(4), p.43. Available at: http://link.aip.org/link/PHTOAD/v65/i4/p43/s1&Agg=doi.

National Research Council of The National Academies. 2009. *A New Biology for the 21st Century: Ensuring the United States Leads the Coming Biology Revolution*, Washington D.C.: The National Academies Press. Available at: http://www.nap.edu/catalog/12764.html.

Newman, M.E.J. 2011. Communities, modules and large-scale structure in networks. *Nature Physics* 8(1):25–31. Available at: http://www.nature.com/doifinder/10.1038/nphys2162.

Nordon, R.E., K.H. Ko, R. Odell, and T. Schroeder. 2012. Multi-type branching models to describe cell differentiation programs. *Journal of Theoretical Biology* 277(1):7–18. Available at: http://www.ncbi.nlm.nih.gov/pubmed/21333658.

Peerani, R., K. Onishi, A. Mahdavi, E. Kumacheva, and P.W. Zandstra. 2009. Manipulation of signaling thresholds in "engineered stem cell niches" identifies design criteria for pluripotent stem cell screens. *PloS One* 4(7):e6438. Available at: http://www.pubmedcentral.nih.gov/articlerender.fcgi?artid=2713412&tool=pmcentrez&rendertype=abstract.

Powell, K. 2005. It's the ecology, stupid! *Nature* 435(May):268–270.

Rodriguez-Brenes, I.A., N.L. Komarova, and S. Wodarz. 2011. Evolutionary dynamics of feedback escape and the development of stem-cell-driven cancers. *Proceedings of the National Academy of Sciences of the USA* 108(47):18983–18988. Available at: http://www.ncbi.nlm.nih.gov/pubmed/22084071.

Roeder, I. and I. Glauche. 2006. Towards an understanding of lineage specification in hematopoietic stem cells: a mathematical model for the interaction of transcription factors GATA-1 and PU.1. *Journal of Theoretical Biology* 241(4):852–865. Available at: http://www.ncbi.nlm.nih.gov/pubmed/16510158.

Roeder, I., and M. Loeffler. 2002. A novel dynamic model of hematopoietic stem cell organization based on the concept of within-tissue plasticity. *Experimental Hematology* 30(8):853–861. Available at: http://www.ncbi.nlm.nih.gov/pubmed/12160836.

Scherf, N., K. Franke, I. Glauche, I. Kurth, M. Bornhäuser, C. Werner, T. Pompe, and I. Roeder. 2012. On the symmetry of siblings: automated single-cell tracking to quantify the behavior of hematopoietic stem cells in a biomimetic setup. *Experimental Hematology* 40(2):119–130.e9. Available at: http://www.ncbi.nlm.nih.gov/pubmed/22085452.

Sneddon, M.W., J.R. Faeder, and T. Emonet. 2011. Efficient modeling, simulation and coarse-graining of biological complexity with NFsim. *Nature Methods* 8(2):177–183. Available at: http://www.ncbi.nlm.nih.gov/pubmed/21186362.

Stewart-Ornstein, J., and H. El-Samad. 2012. Stochastic modeling of cellular networks. *Methods in Cell Biology* 110:111–37. Available at: http://www.ncbi.nlm.nih.gov/pubmed/22482947.

Takahashi, K., K. Tanabe, M. Ohnuki, M. Narita, T. Ichisaka, K. Tomoda, and S. Yamanaka. 2007. Induction of pluripotent stem cells from adult human fibroblasts by defined factors. *Cell* 131(5):861–872. Available at: http://www.ncbi.nlm.nih.gov/pubmed/18035408.

Takahashi, K., and S. Yamanaka. 2006. Induction of pluripotent stem cells from mouse embryonic and adult fibroblast cultures by defined factors. *Cell* 126(4):663–676. Available at: http://www.ncbi.nlm.nih.gov/pubmed/16904174.

Task, K., M. Jaramillo, and I. Banerjee. 2012. Population based model of human embryonic stem cell (hESC) Differentiation during endoderm induction. *PloS One* 7(3):e32975. Available at: http://www.pubmedcentral.nih.gov/articlerender.fcgi?artid=3299713&tool=pmcentrez&rendertype=abstract.

Tay, S., J.J. Hughey, T.K. Lee, T. Lipniacki, S.R. Quake, and M.W. Covert. 2010. Single-cell NF-kappaB dynamics reveal digital activation and analogue information processing. *Nature* 466(7303):267–271. Available at: http://www.pubmedcentral.nih.gov/articlerender.fcgi?artid=3105528&tool=pmcentrez&rendertype=abstract.

Tieu, K.S., R.S. Tieu, J.A. Martinez-Agosto, and M.E. Sehl. 2012. Stem cell niche dynamics: from homeostasis to carcinogenesis. *Stem Cells International* 2012:367567. Available at: http://www.pubmedcentral.nih.gov/articlerender.fcgi?artid=3289927&tool=pmcentrez&rendertype=abstract.

Till, J.E., E.A. McCulloch, and L. Siminovich. 1964. A stochastic model of stem cell proliferation, based on the growth of spleen colony-forming cells. *Proceedings of the National Academy of Sciences of the USA* 51:29–36.

Unadkat, H.V., M. Hulsman, K. Cornelissen, B.J. Papenburg, R.K. Truckenmüller, A.E. Carpenter, M. Wessling, G.F. Post, M. Uetz, M.J.T. Reinders, D. Stamatialis, C.A. van Blitterswijk, and J. de Boer 2011. An algorithm-based topographical biomaterials library to instruct cell fate. *Proceedings of the National Academy of Sciences of the USA* 108(40):16565–16570. Available at: http://www.pubmedcentral.nih.gov/articlerender.fcgi?artid=3189082&tool=pmcentrez&rendertype=abstract.

Wang, J., L. Xu, E. Wang, and S. Huang. 2010. The potential landscape of genetic circuits imposes the arrow of time in stem cell differentiation. *Biophysical Journal* 99(1):29–39. Available at: http://www.pubmedcentral.nih.gov/articlerender.fcgi?artid=2895388&tool=pmcentrez&rendertype=abstract.

Winkler, D.A., and F.R. Burden. 2012. Robust, quantitative tools for modelling ex-vivo expansion of haematopoietic stem cells and progenitors. *Molecular BioSystems* 8(3):913–920. Available at: http://www.ncbi.nlm.nih.gov/pubmed/22282302.

Xu, H., C. Schaniel, I.R. Lemischka, and A. Ma'ayan. 2010. Toward a complete *in silico*, multi-layered embryonic stem cell regulatory network. *Wiley Interdisciplinary Reviews. Systems Biology and Medicine* 2(6):708–733. Available at: http://www.pubmedcentral.nih.gov/articlerender.fcgi?artid=2951283&tool=pmcentrez&rendertype=abstract.

Yang, J., X. Meng, and W.S. Hlavacek. 2010. Rule-based modelling and simulation of biochemical systems with molecular finite automata. *IET Systems Biology* 4(6):453–466. Available at: http://www.pubmedcentral.nih.gov/articlerender.fcgi?artid=3070173&tool=pmcentrez&rendertype=abstract.

Zandstra, P.W., and A. Nagy. 2001. Stem cell bioengineering. *Annual Review of Biomedical Engineering* 3:275–305. Available at: http://www.ncbi.nlm.nih.gov/pubmed/11447065.

Zhu, H., P.Y.H. Pang, Y. Sun, and P. Dhar. 2004. Asynchronous adaptive time step in quantitative cellular automata modeling. *BMC Bioinformatics* 5:85. Available at: http://www.pubmedcentral.nih.gov/articlerender.fcgi?artid=459211&tool=pmcentrez&rendertype=abstract.

Stem Cell Bioprocessing and Biomanufacturing

Todd C. McDevitt

Introduction

The ability to manufacture products from stem cells is required to deliver on the many envisioned potential applications of these cells and will ultimately dictate the translational success of stem cells. The technologies and processes for bioprocessing are dependent on insights gained from the physical science principles, screening technologies, and computational analysis approaches described in each of the previous chapters. Ultimately insights from each of these different approaches will contribute significantly to the different elements of biomanufacturing schemes including culture platforms, monitoring techniques, and quality control/assurance. The development and implementation of bioprocessing technologies will require collaborations between academic institutions and a rapidly growing global industry seeking to commercialize stem cell products, as well as interactions with regulatory agencies providing oversight of these processes. This chapter will highlight the most common current approaches and remaining challenges facing the development of bioprocessing technologies necessary for the scalable and robust manufacturing of stem cells and stem cell-derived products.

The terms "bioprocessing" and "biomanufacturing" are often used interchangeably and are largely synonymous, but a subtle distinction between the two does exist. "Bioprocessing" refers to the engineering of individual technology components and overall systems capable of producing a biological product—in this case, stem cells and stem cell-derived products. "Biomanufacturing" refers to the use of bioprocessing technologies for manufacturing purposes in industry, in this

T.C. McDevitt (✉)
Biomedical Engineering, Stem Cell Engineering Center Georgia Institute of Technology, 315 Ferst Dr. NW, 30332 Atlanta, GA, USA
e-mail: todd.mcdevitt@bme.gatech.edu

R.M. Nerem et al. (eds.), *Stem Cell Engineering: A WTEC Global Assessment*, Science Policy Reports, DOI 10.1007/978-3-319-05074-4_5,
© Springer International Publishing Switzerland 2014

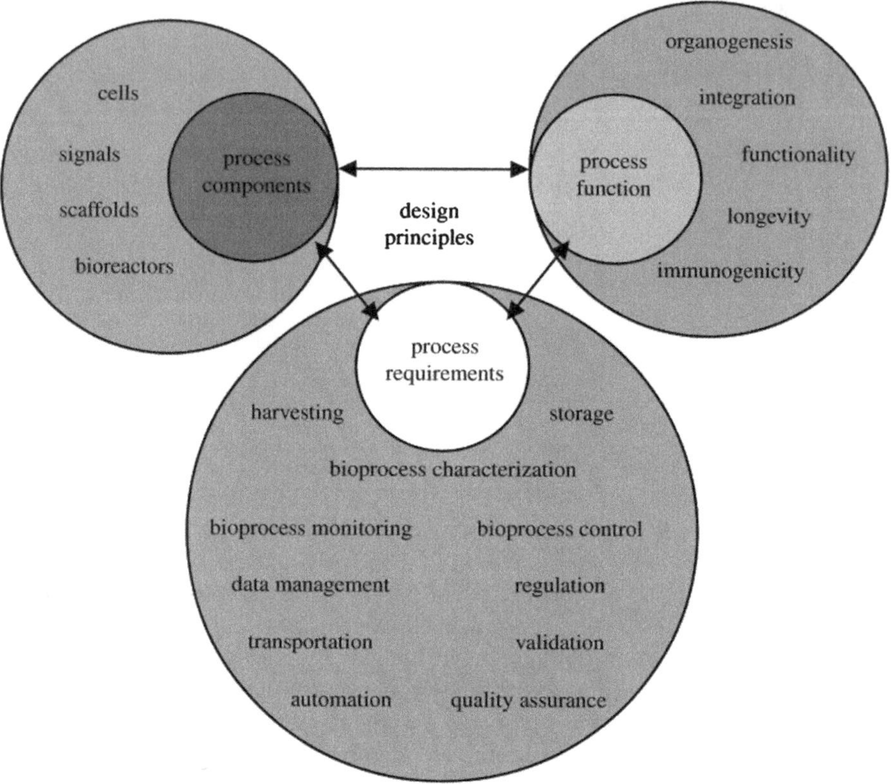

Fig. 1 Design principles for stem cell bioprocesses (From Placzek et al. 2009)

case the biotechnology sector. In this context, "bioprocessing" encompasses the technologies and systems processes whereas "biomanufacturing" reflects the implementation of bioprocessing practices to produce stem cells and derivatives of stem cells for specific applications.

Every bioprocessing scheme consists of three primary design considerations that need to be taken into account initially: the materials to be used (process components), the various manipulations and assessments of the materials (process requirements), and the performance of the output product(s) (process function) (Fig. 1). The process components include the choice of cells, media, chemicals and biomolecules, substrates, vessels, and sensors that are used in the overall process for the manufacturing of a biological product. The process requirements entail everything needed for the integration of the components to successfully produce and deliver a well-defined and robust desired product. The process function is then the assessment of how well the product of the bioprocessing scheme attains an application-specific purpose. Altogether, consideration of these critical principles at the outset of designing a bioprocess dictates the efficacy of the overall system and its ultimate utility.

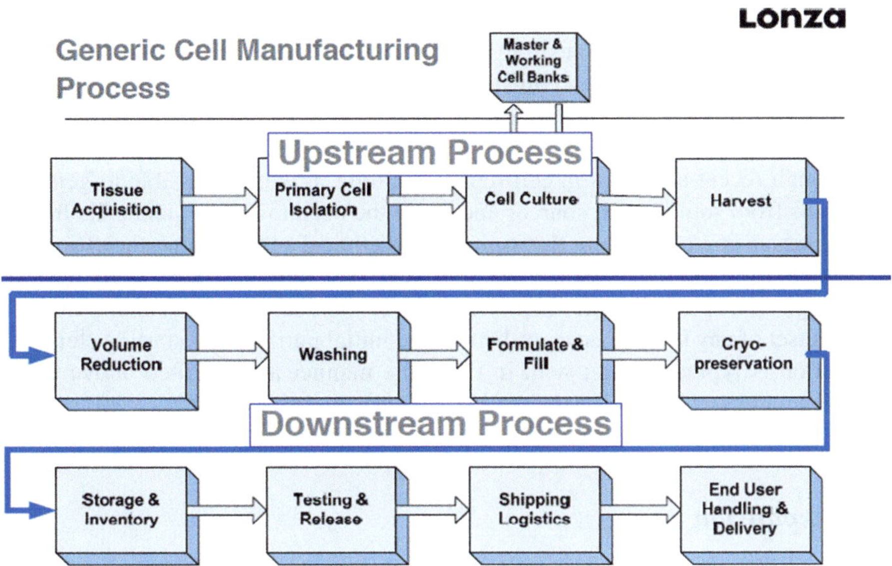

Fig. 2 Cell manufacturing process. The manufacturing of cell products can be divided into upstream and downstream processing steps (Courtesy of Lonza Group, Ltd.)

Biomanufacturing strategies can generally be divided into "upstream" and "downstream" bioprocesses (Fig. 2). The upstream portions encompass everything from the initial acquisition of the starting material through to the production of the final crude product, whereas downstream processing then entails all of the remaining necessary steps to deliver a well-defined, purified, and packaged final product. Within this scope, any bioprocess for the manufacturing of cell products consists generally of five stages in sequential order: isolation/derivation, expansion, differentiation, separation/purification and preservation; the former three are considered part of the upstream processes, whereas the latter two are considered part of the downstream stages. Specific considerations for each of these stages are described below.

Cell Isolation/Derivation

The initial requirement for the bioprocessing of any stem cell population is a starting cell population. In most cases this requires the isolation of stem cells from a tissue source in order to separate the stem cells of interest from the remaining non-stem cell population. Typically this requires a combination of enzymatic and physical dissociation methods to break down a tissue into its smallest functional units, individual cells, before positive and/or negative selection methods can be employed to obtain an enriched population of stem cells relative to the initial starting heterogeneous population. The starting purity of any stem cell population is a quantifiable metric that defines the input stream to any bioprocessing approach. Since contaminating cells (i.e., non-stem cells) can adversely affect the efficiency of bioprocesses, it is desirable to remove them as effectively as possible prior to and during subsequent steps.

Through recent advances in cell reprogramming, it is now possible to generate stem cells from somatic cell sources and avoid the need to isolate stem cells from a starting tissue source. Despite this significant technical achievement, the efficiency of most reprogramming methods remains low, and thus the same challenges of enriching and/or purifying the stem cells from contaminating cells remains an issue at the outset of any bioprocessing scheme. The initial purity of the starting stem cell population is typically dealt with in the same manner as described above using positive and/or negative selection methods.

Cell Expansion

A critical component of any biomanufacturing scheme is amplification of the initial starting cell population to increase the quantity of cells from the starting pool of stem cells. It is imperative for successful expansion strategies to maintain a relatively homogeneous population of stem cells in their undifferentiated state. This is generally attempted by maintaining optimal conditions that favor stem cell self-renewal and limit differentiation. The increasing density of cells over time and increasing consumption of reagents necessary for maintenance of cells in an appropriate undifferentiated state requires serially splitting cells and/or increasing the volume of media and surface area (for adherent cell types) to generate sufficiently large quantities of stem cells.

The most critical factors to successful stem cell expansion are often the addition of biochemical reagents that stimulate self-renewal and inhibit differentiation of the cells. As is true for any mammalian cell in culture, it is also necessary to provide sufficient oxygen and basic nutrients to the cells, as well as maintain the correct salt concentration and pH in the media to maintain optimal cell viability. In addition to providing factors to the cells, it is also necessary to remove factors that can adversely affect stem cell phenotype and growth, such as the buildup of excess waste by-products, by exchanging some or all of the media on a periodic basis.

Evaluation of cell expansion should consider the quantity as well as the quality of cells. The quantity of cells can be readily obtained by counting the numbers of cells accumulating over time; this can be performed over short periods of time, such as for a single passage, as well as over long periods of time, such as over multiple passages, to determine the expansion potential. The quality of expanded stem cell populations can be examined by the use of different phenotypic markers; most commonly these would be cell surface markers, but could also include intracellular markers, and can be readily assessed by flow cytometry analysis.

Cell Differentiation

In many cases, the target cell of interest is not necessarily the undifferentiated stem cell, but some derivative(s) of a stem cell that can be used therapeutically or for diagnostic *in vitro* applications. Similar to cell expansion, the primary considerations for effectively directing differentiation of stem cells, typically to a single lineage of interest, are the biochemical and biophysical elements of the culture system that promote loss of a "stem" phenotype and commitment to a more differentiated and lineage-restricted cell fate. In most cases, the analysis of differentiation is discretized into a series of cell states that are described in a binary manner by the loss of certain cell markers and acquisition of others. In reality, differentiation of most stem cells is more accurately appreciated as a continuum of cell states that result from the relative magnitudes of multiple phenotypic markers changing over time.

Most differentiation protocols involve the sequential administration and removal of different soluble factors to the culture media as a function of time. Media composition changes are often performed in concert with manipulating the cell culture format once or more throughout the process to expose the cells to different combinations of instructive cell-cell and cell-extracellular matrix adhesive cues. The series of environmental changes is intended to favor the stem cells commitment to specific lineages at critical junctures of a hierarchical decision-making tree to eventually arrive at a target cell type of interest. For this reason, stem cells with greater potency and differentiation of more mature cell types generally require lengthier and more complicated differentiation protocols.

Often the most challenging aspect and critical determinant of the success of various differentiation protocols is their ability to efficiently and reproducibly yield a population of cells of a well-defined and uniform phenotype. This can be difficult to attain due to a lack of synchrony among the population of cells and heterogeneities that exist within most culture systems with regard to spatial and temporal integration of differentiation inducing signals. The efficacy of most differentiation protocols is assessed primarily on the final purity of the cell population of interest and yield of differentiated cells produced relative to the starting number of stem cells. Although gene expression analyses can provide average phenotypic information on cell populations for a number of markers at once, flow cytometry with a small panel of markers (usually less than 4–6) can provide an accurate assessment of differentiated cell purity and yield, at least for a singular, defined target cell type of interest.

Cell Separation/Purification

At the conclusion of expansion and/or differentiation stages of stem cell culture, the resulting population of cells usually contains some degree of heterogeneous cell phenotypes. Thus, prior to packaging and storage in a final product form, it is often necessary to separate the desired cell type(s) away from contaminating cells present in the final gross product. Separation at this latter stage is typically handled in much the same manner as during the initial isolation steps at the outset of the overall process to enhance stem cell purity. Positive and/or negative selection methods are used to physically remove either the stem cells or stem cell-derived products from other unintended cell phenotypes.

As stem cell bioprocesses increase in scale, one of the biggest problems becomes the handling of the relatively large volumes of media and liquid waste needing to be disposed of in order to concentrate the final cell product. This challenge stands to be the biggest bottleneck in the overall biomanufacturing process if new technologies capable of addressing this critical issue are not developed.

Cell Preservation

Preservation of cells at the output of bioprocessing systems is critical to the success of biomanufacturing as a means of stably storing and distributing the cell products for their ultimate application(s). Cryopreservation techniques are the preferred method of choice to suspend cells in a viable state for long-term storage. Cryo-protective agents are typically added to a high-density volume of the cells in culture media supplemented with an excess of growth factors and freezing is controlled at a rate intended to limit ice crystal formation which can damage cell membranes and induce cell death. Once frozen, cell stocks can be useful over long periods of time if maintained appropriately at low-temperature (i.e., liquid nitrogen vapor) conditions. In a frozen format, the cells can also be shipped on ice without thawing in order to distribute them over long distances.

Stem Cell Bioprocessing Culture Technologies

Stem cell culture technologies include two primary design considerations: the use of sterile vessels to contain the cells during processing steps and the format in which the cells are cultured within the vessels. These parameters directly affect the scalability of biomanufacturing processes, namely whether the processes can be "scaled-out" or "scaled-up" (Fig. 3). "Scale-out" reflects massive parallelization of the process to increase the final yield due to a linear correlation between the output cell number and vessel parameters, such as surface area for adherent stem cell populations. In contrast, "scale-up" refers to the ability to exponentially enhance cell output without changing physical volume or size of the system thereby more efficiently increasing the yield of a system and is more commonly associated with suspension culture systems. "Scaling-up" usually begins by optimizing a cell process on a small scale and then progressively increasing the volume of the system without adversely affecting the efficiency of the process. These strategies directly impact the ultimate size of cell lots that can be attained by either method, as well as the related cost-of-goods necessary for production.

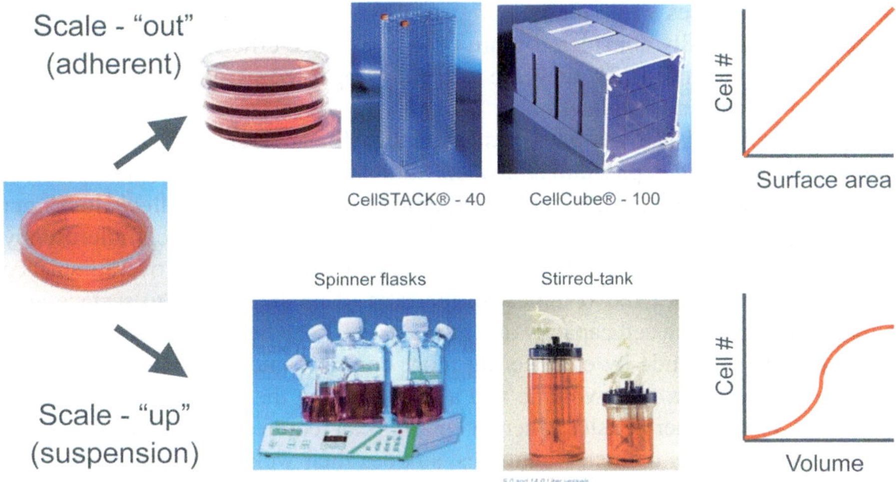

Fig. 3 Different strategies for scalable biomanufacturing. Scaling "out" typically refers to expanding the quantity of cells in a linear manner and is most relevant to increasing surface area for adherent cell types. Scaling "up," on the other hand, reflects an ability to increase cell yields in a non-linear fashion and is more related to suspension culture systems (Courtesy of the author)

In addition to the reagents needed to expand, differentiate, and preserve stem cells and the vessels to house them in during these processes, routine monitoring of the cells is required to know when to proceed from one stage of the process to the next. In conventional adherent flask cultures, this is most often performed by simply viewing the cells directly by microscopy and making decisions about how and when to proceed based upon the appearance and density of the cells; typically only a portion of any cell culture is monitored in this manner and random sampling is performed to qualitatively assess the uniformity of the cultures and inform subsequent decision-making. Slightly more advanced means of monitoring cell cultures rely upon the integration of physical sensor elements to quantify cell metabolites and/or media conditions (i.e., pH, pO_2) embedded directly within the housing vessel or integrated in-line so as to monitor and potentially introduce feedback control into the processes.

Culture Vessels

Culture vessels used for stem cell biomanufacturing can be generally divided into two categories: planar surfaces, such as flasks, for adherent cell culture and suspension bioreactors. Flasks are closed systems that are typically cultured in a static manner and are limited to use in batch processes. Suspension bioreactors, on the other hand, almost always involve dynamic fluid environments internally to maintain well-mixed conditions and can be configured to operate in batch, semi-batch or even continuous processes while maintaining sterility. Additionally, suspension culture systems are generally considered more cost- and space-efficient to accommodate larger media volumes, thus they are generally regarded as being preferable for scale-up and industrial applications.

Flasks provide large amounts of surface area for adherent cells to attach and multiply while being exposed to equivalent soluble media components independent of their spatial location. The planar configuration of flask cultures allows for direct visual inspection of cell morphology and density using simple optical microscopy techniques. Multi-stack vessels have been created to increase the overall surface area density within a fixed volume, but the ratio of media volume to surface area remains constant. The cell capacity and ultimate yield can be significantly increased by orders of magnitude by engineering more complex internal geometries to add surface area within a fixed volume and also increasing the overall volume/size of culture vessels. Higher-density adherent cultures can require less processing of individual vessels, but as their size increases, so does their weight, making them more cumbersome to manipulate and prone to larger spill volumes, hence the need for greater secondary containment. As the internal complexity of culture flask systems increases and they continue to be designed for single-use, the manufacturing costs skyrocket, making them less attractive from a cost-efficiency standpoint.

The simplest of suspension culture systems currently used for stem cell biomanufacturing are clear, disposable bags into which cells and media can be introduced via sterile ports, but thereafter cultured as a "closed" system. The bags are then placed onto a rocking instrument to create a gentle and continuous wave of media moving back-and-forth, hence the name "wave" bioreactors. Originally described in 1999 (Singh 1999), the WAVE Bioreactor™ systems consisting of bags (Cellbag™ bioreactors) and rocking devices (WAVE Mixers™) are now commercially available from GE Healthcare Life Sciences. The gas permeable bags facilitate better oxygen transport than solid wall culture systems and can be used over a wide range of volumes (up to 100 L) without the need to modify hardware needs (i.e., the rocking device). The culture bags have been developed to be compatible with single-use and reusable sensor technologies for process controls and can be run in perfusion culture mode with special filters attached. This system is mostly commonly used to culture non-adherent cells in suspension, but can support adherent cells on microcarriers.

Bioreactors are the preferred vessel for industrial biotechnology manufacturing platforms due to their cost-efficiency and ability to provide well-mixed, relatively homogeneous culture environments. Furthermore, most bioreactors are amenable to the inclusion of various types of sensor technologies useful for continuous monitoring of the bioprocesses. Fixed-wall bioreactors, such as spinner flasks and stirred tanks, that rely upon the creation of mixing with impeller systems contained within the bioreactor to ensure continuous mixing of the media and dispersion of the cells throughout the vessel have been the two most commonly used such systems for stem cell culture to date because of their historical use in bioprocessing. Rotating vessel systems, originally designed for simulating microgravity conditions, can be used to culture encapsulated stem cell populations to produce different differentiated lineages (Hwang et al. 2009; Siti-Ismail et al. 2012). Comparisons between the different types of stirred or rotating systems have indicated that the choice of vessel can influence the profile of differentiated cells (Fridley et al. 2010).

The addition of perfusion to bioreactors is commonly done to enable better long-term maintenance of culture environments with the intent of providing more uniform conditions for the cells over time. Frequently perfusion is used to maintain not just the appropriate supply of nutrients for cell viability, but also provide stable oxygen tension levels when this is used as a process requirement consideration. The yield of differentiated cells from either encapsulated stem cells or stem cells on microcarriers has been demonstrated to be increased by controlling oxygen tension in perfusion bioreactors (Bauwens et al. 2005; Serra et al. 2010). Perfusion also permits the continuous removal of cell by-product wastes that increasingly accumulate with increasing cell density in order to obtain higher overall cell yields.

As a result of their larger size (usually starting at ~100 mL and progressively increasingly to tens of L volumes) compared to flasks and culture dishes, most suspension bioreactor systems are not inherently capable of readily permitting visual monitoring of individual cells and morphology, but sampling of the media and even a portion of cells regularly can be performed to quantitatively assess changes in cell density and media composition as a function of time. Routine monitoring of the cells or culture media permits the opportunity to introduce feedback control into the processes, a significant advantage over closed, flask systems, which can lead more readily to optimized system designs for manufacturing. Bioreactors can be designed to be reusable depending on the materials they are made of, but many experimental and low-volume systems (<1 L) are commonly produced as disposable vessels manufactured from cheap, sterile plastic materials.

Culture Formats

The format in which stem cells are cultured affects the choice(s) of culture vessels that can be used for bioprocessing. Several different formats for stem cell expansion and differentiation have been examined by a variety of investigators over the past decade or so. Many of the current formats are derived from bioprocesses originally used for non-stem cells, but easily adapted to accommodate stem cells. Generally speaking, the primary formats include culturing stem cells on materials, as aggregates (with or without materials), or encapsulated within materials (Fig. 4).

Planar substrates for stem cell culture remain the most commonly used platform by experimental researchers and the current biomanufacturing industry for adherent cell types. Surface treatment and adhesive coating technologies originally designed to promote non-stem cell attachment and growth work well for several types of stem cells, but not all. Innovations in planar culture technologies for stem cells have largely focused on the development of adhesive chemistries capable of specifically supporting stem cell growth. In many cases, pre-existing products have been examined and repurposed for stem cell culture. For example, Matrigel, a complex basement

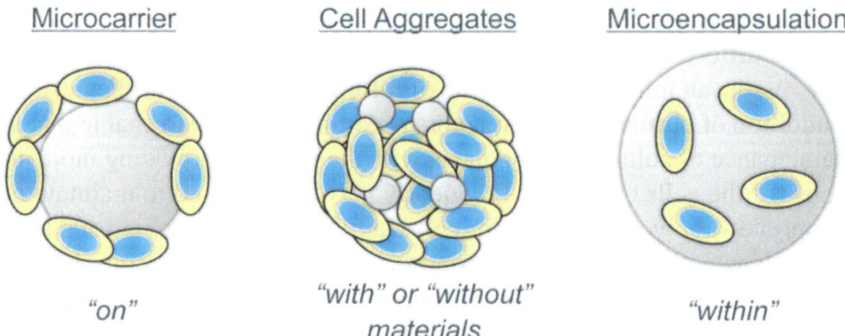

Fig. 4 Suspension culture formats. Stem cells in suspension systems can be cultured in several different forms, including on the surface of microcarriers, as aggregates either with or without incorporated materials, or encapsulated within hydrogel materials (Courtesy of the author)

membrane derived from mouse EHS sarcoma cells, was the first commercially available product to culture human pluripotent stem cells in a "feeder-free" manner and it is now available in a qualified lot format that is screened by Becton Dickinson *a priori* to support hPSC growth. More recently, an E-cadherin domain containing protein, sold under the trade name of "StemAdhere" by Primorigen has been used as a more reproducible and well-defined surface adhesive coating for hPSC growth. Corning has also developed a synthetic hydrogel-coated cultureware product (Synthemax) that can be used for stable growth of undifferentiated human pluripotent stem cells and other stem cell/progenitor types.

Microcarriers are small particles (~100's of μm in diameter) that support adherent cell growth and can be introduced into suspension culture systems. Due to their radius of curvature, microcarrier technologies can significantly increase the ratio of surface area for growth to volume of culture media compared to planar culture. Many different types of microcarriers have been used successfully for the expansion of various types of stem cells in spinner flasks (Abranches et al. 2007; Alfred et al. 2011; Fok and Zandstra 2005; Serra et al. 2009) and scaled-up to larger volume stirred tank reactors (Fernandes-Platzgummer et al. 2011). Differentiation of stem cells on microcarriers has also been demonstrated following initial expansion and without having to dissociate and transfer the cells to different carrier materials (Bardy et al. 2012; Lock and Tzanakakis 2009; Rodrigues et al. 2011). Stem cells on microcarriers can also be recovered from cryopreservation with better survival (Nie et al. 2009), further demonstrating how microcarriers can be used throughout various stages of stem cell bioprocessing in suspension culture systems.

Many types of stem cells can self-assemble in non-adherent, suspension culture conditions to form 3D multicellular aggregates via intercellular adhesion molecules. Pluripotent (ES and iPS), multipotent (MSCs, neural stem cells) and even unipotent/progenitor (skeletal myoblasts) can be coaxed to form such aggregates, and in this form can be subsequently cultured in suspension culture for prolonged periods of time. Multicellular aggregation formation is typically controlled by the inoculation density of single-cell suspensions and the hydrodynamic conditions created by stirred suspension culture systems. In many cases 3D aggregate format can stimulate differentiation of the stem cells and therefore has often been used primarily for such purposes, however aggregates of undifferentiated cells can be maintained under appropriate conditions. Long-term culture of stable undifferentiated stem cells as aggregates without adversely affecting their phenotype can be achieved (Amit et al. 2011; Cormier et al. 2006; Zweigerdt et al. 2011), even for long periods of time under appropriate conditions (zur Nieden et al. 2007). For undifferentiated expansion, the aggregates typically need to be dissociated and re-associated at regular intervals in order to limit spontaneous differentiation that increases with larger aggregate size. Furthermore, it has even become possible to derive cells, expand them, and differentiate all within the same suspension culture system by controlling the aggregation of the cells via hydrodynamic conditions in concert with culture media and environmental parameters (Fluri et al. 2012; Steiner et al. 2010).

Somewhat analogous to microcarrier culture, microparticles ($\sim$1–10 µm in diameter) can be integrated within aggregates of stem cells to deliver differentiation factors and present immobilized signals locally to the cells. This concept was first demonstrated by Mahoney and Saltzman with PLGA microparticles delivering nerve growth factor to fetal brain cell aggregates to promote maturation and survival *in vitro* and *in vivo* (Mahoney and Saltzman 2001). Similarly, delivery of differentiation factors from PLGA or gelatin microparticles incorporated within 3D stem cell aggregates can promote differentiation equally or even more effectively than soluble treatment methods (Carpenedo et al. 2009; Purpura et al. 2012; Solorio et al. 2010). The primary advantages of delivering molecules in this manner to stem cells are the controlled release that can be achieved over different durations as a function of material properties and the significant reduction in total amount of the molecules (usually an order of magnitude at least) that can used to direct cell fate. Interestingly, microparticles from different types of "unloaded" materials can also provoke differences in the differentiation of stem cells (Bratt-Leal et al. 2011), suggesting a simple means by which various types of material properties alone might be used to engineer properties of stem cell microenvironments.

Stem cells can also be encapsulated, either as single cells or as 3D cell aggregates, within microbeads in order to control their local microenvironment and culture the cells in higher density formats. Hydrogel materials, such as alginate or agarose, are commonly used to physically entrap stem cells before they are introduced into suspension culture (Dang et al. 2004; Siti-Ismail et al. 2008). Microencapsulation shields the cells directly from fluid shear forces created by mixed/stirred suspension cultures that can negatively impact viability or even perhaps directly affect stem cell phenotype. Transport of molecules into and out from the microbeads can be controlled by the density of the hydrogels and coating of the exterior surface. Microcarrier and encapsulation techniques can be combined for stem cell expansion and improve the recovery of cell yield following cryopreservation (Serra et al. 2011).

Automation Platforms for Stem Cell Culture and Preservation

Automation of established procedures is a common element of nearly all mature manufacturing practices whereby machines are introduced to substitute for manual labor. Automated processes handled by robotic instrumentation can increase throughput, reproducibility and reliability of manufacturing at a reduced long-term cost. Robotic instrumentation can achieve repeated precision more confidently than even the most highly skilled person and can completely remove the potential for human error.

Stem cell culture, whether for research purposes or in biomanufacturing, is currently performed almost exclusively by humans at all stages of the process and it is clear that subtle environmental changes or even minor variations in processing steps can affect cell phenotype. Thus, in order to improve the quality and quantity of stem cell products in the future it will become necessary to increasingly introduce automated processes into the stem cell biomanufacturing pipeline.

Flask Passaging

The first introduction of automation into stem cell culture practices has been for the purpose of re-feeding and passaging adherent cells during expansion and differentiation stages. Current first-generation systems consist of a single-arm robotic unit housed within a sterile environment that is capable of grasping different size culture flasks and performing liquid-handling steps to rinse, trypsinize, re-plate, and feed the cells. The robotic compartment is physically distinct but adjacent to one or more cell incubator environments housing the cells. Peripheral equipment, such as cell counters and/or microscopes, can be added to assess and record cell metrics throughout the process for quality control and system validation purposes.

The most commonly used platform thus far for automation of adherent stem cell culture is the Compact Select instrument developed by TAP Biosystems. This multi-unit system is quite large and handles a single T-flask at a time, so all of the culture processes are conducted in a serial manner. Several Compact Select units are in existence in academic research environments conducting cell manufacturing and tissue engineering research, such as at the University of Loughborough and the University of Nottingham, for education and training purposes. The Compact Select technology is also being used within the context of more industrial research environments, such as at I-STEM in France. The units are embedded within existing cell culture suites to enable trainees and investigators to begin to directly compare automation culture to normal manual practices and also allow for optimization studies to be performed in a highly controlled and reproducible manner.

In addition to the use of commercially developed automation platforms, researchers are now beginning to develop new types of automation platforms with the potential of offering greater flexibility and overall throughput. At Tokyo Women's Medical

University in Japan, researchers have devised a modular approach whereby a central, single-arm robotic system can be interchangeably connected to various types of adjacent units for housing, processing or analyzing cells cultured in flasks. As compared to a rigidly pre-designed system, such a modular approach can be compatible with numerous types of currently existing or future technologies and therefore provide an extremely flexible system in addition to being more space-efficient than a fixed processing unit. Another approach being developed by researchers at the Fraunhofer in Leipzig, Germany is an "assembly-line" approach to cell manufacturing processes. Similar to many of the types of operating systems at modern manufacturing facilities, automation technologies can remain fixed in place while simple conveyor systems can allow samples to be processed in a continuous sequential manner. Assembly line systems require multiple, independent automated tasks that can be individually simpler than robotic handling arms used for the aforementioned approaches, so throughput can be increased and feedback control can be more easily integrated.

Suspension Culture Systems

Continuous stirred tank reactors (CSTRs) are perhaps the simplest and most prevalent form of automated suspension culture system that has been used for a long time as a precursor to larger volume scale-up for the manufacturing of biologics in the biotechnology industry. The well-stirred environment within the tank is created by an impellor that is driven by an external controller unit. Perfusion of stirred tank reactors can be achieved with external pumping systems to recycle or continuously introduce fresh media at different rates by running the system in different modes. The reactors are also designed to accommodate different types of sensor probes that can be inserted into the vessel to directly and continuously monitor culture environment conditions. An additional advantage of completely closed systems, like the stirred tank bioreactors, is that they can be run outside of an incubated environment due to the fact that the temperature and gas are internally controlled within the tank and the reservoir of excess media connected to the tank.

The typical large volumes (>100 mL) of even the simplest suspension culture systems and need for individual controller units for each culture has precluded systematic screening of all of the various culture parameters that can affect stem cell expansion and phenotype. As a result, comparative studies of two conditions have been the most commonly reported findings of experimental studies. The broad parameter space for suspension cultures requires higher throughput systematic screening platforms that accurately mimic large volume systems in order to perform optimization studies prior to scale-up. TAP Biosystems recently introduced a bench-top "ambr" instrument capable of operating up to 48 parallel micro-scale bioreactors, each on the order or 10 mL culture volumes. Although the number of samples per experiment is only half of conventional adherent screening platforms (i.e., 96-well plates), this system provides continuous monitoring of pH and pO_2 with sensors integrated into the culture vessels and can operate at a range of different agitation speeds, culture media, cell densities, and gas environments, all within a single experiment and with sufficient replicates for meaningful statistical analyses. Such a system represents a major step forward in "scaling down" culture systems for high-throughput screening prior to scaling up for manufacturing purposes.

Cryopreservation

Most of emphasis on automation of stem cell bioprocessing to date has focused on the upstream manufacturing of the cells and some downstream processing related to harvesting and purifying/enriching the final cell product. The Fraunhofer IBMT, in collaboration with the University Hospital Zurich, has actually developed an automated platform for cryopreserving cells. Controlled rate freezing is a common practice intended to promote cell recovery post-thaw by precisely regulating the freeze rate of the cells prior to cryopreservation. To ensure greater consistency, the process of freezing and cell retrieval can be completely controlled in an automated fashion and the history of the temperature of individual cells can be recorded with sensors embedded into the cryovials and the bottoms of the cryostorage boxes. The ability to record temperature helps not only with storage, but also provides a unique ability to detect any potential problems with temperature stability that could occur during shipping of cryopreserved cells.

Future Directions

Despite the increasing amount of activity in the nascent field of stem cell bioprocessing over the past decade, numerous opportunities remain to further improve existing processes and develop innovative and disruptive technologies to significantly enhance the scalability and purity of manufacturing stem cell products. Many of the current challenges and limitations will likely be addressed by increasing amounts of interdisciplinary activities that leverage the expertise of tangential fields that have solved similar types of problems in the past. However, some of the most challenging problems will also need to be solved by engineers and scientists with a significant understanding of the fundamentals of stem cell biology and the myriad of factors that can affect cell behavior, namely fate decisions.

Culture Technologies

Culture media has been the most active area of commercial development of stem cell culture reagents for experimental and manufacturing purposes. The emphasis has primarily been on providing serum-free defined media and supplements capable of adequately supporting cell expansion and differentiation while reducing the risk of exposure to potential xenogenic pathogens in anticipation of safety concerns of national and international agencies likely to regulate stem cell products. Many varieties of such culture media now exist for nearly all types of stem cells, but further reductions in the cost of these specialized media are necessary as volumes increase by orders of magnitude to keep the cost of goods low enough to produce affordable stem cell therapies and diagnostics. The identification and commercialization of small molecules that can specifically affect stem cell growth and differentiation (Chambers et al. 2012; Yao et al. 2006) are an attractive substitute for more expensive growth factors commonly used in current stem cell culture. The spatial and temporally controlled presentation of key factors for stem cell expansion and differentiation, using perfusion systems or controlled release materials, could also use molecules more efficiently without sacrificing the quality of the cultures.

Engineering of materials and surface chemistries that contact stem cells at different stages of bioprocesses offers tremendous opportunities for enhancement. Recently, surface coatings derived from native ECM molecules (Rodin et al. 2010) and synthetic chemistries (Melkoumian et al. 2010; Villa-Diaz et al. 2010) have been shown to be capable of supporting stem cell growth and directed differentiation. Furthermore, engineered surface properties can be designed and optimized to work in concert with specific culture media in a synergistic manner (Irwin et al. 2011). Tethering to surfaces of molecules that can promote stem cell growth (Alberti et al. 2008) and/or influence the differentiation of stem cells can regulate cell behaviors in a localized manner. Engineering of molecular groups, similar to small molecule development for culture media, can also influence the fate decisions of encapsulated stem cells (Benoit et al. 2008). Physical properties of materials may also be used to tune stem cell responses in addition to chemical/biochemical traits.

Moving beyond the use of simply the current bioreactors and culture vessels also represents an area of rich technological development. The design of smaller scale vessels that can accurately mimic the hydrodynamics and culture environments of larger bioreactors would allow for screening of conditions and true optimization studies to identify key parameters and process requirements that could be directly implemented in larger scale systems without sacrificing cell yield or purity of the bioprocess. Vessel geometries and the resulting hydrodynamic conditions that can affect potential phenotypic responses of the cells (compared to static culture conditions) should be taken into consideration and modifications of existing suspension culture systems or design of new types of reactors with new capabilities should be pursued.

Sensor technologies, routine in most manufacturing systems, have not been developed expressly for stem cell characterization purposes to be capable of monitoring environmental parameters and also cell phenotype simultaneously. Nearly all forms of current phenotype analysis require removing cells from the culture environment, thus disrupting the process, and/or sacrificing cells to assess the average state of the cell population. Although meaningful correlations between media readings and cell phenotype could be used to monitor processes, these links have yet to be established. Examining all of the cells at once in increasingly large culture volumes is infeasible, so sampling a portion of cells needs to represent an entire population. Label-free or transient labeling methods that don't affect cell phenotype significantly could permit repeated measures of cells without destruction. In-line or off-line sensors integrated into culture vessels should become capable of monitoring cells directly without significantly reducing cell yield of the process by non-destructively analyzing their state.

Automation Platforms

Automation will continue to be integrated into stem cell biomanufacturing platforms to serve a variety of different purposes. The major efforts will be focused on development of robotics capable of replacing some or all of the manual labor required for routine culture practices, such as feeding, passaging, and seeding of cells. Compared to existing technologies, there will be an increasing need for greater flexibility of robotic capabilities and introduction of new types of robotic automation into stem cell bioprocessing schemes. In addition to building completely new systems, an as-of-yet unexplored possibility is for the integration of robotics into existing cell culture environments consisting of standard biosafety cabinets and incubators. Robotic operations should accommodate different types of vessels of different geometries and sizes and not be designed solely for singular culture formats. In addition to hardware changes, modification of processing algorithms via software engineering will also be required. The overall throughput and speed of bioprocessing may also be enhanced by parallel processing methods conducted by robotics and based on non-intuitive process design algorithms that can't be attained by manual labor. Decision-making algorithms based on feedback control of cultures and processes could be a unique opportunity to introduce artificial intelligence into stem cell biomanufacturing.

Global Assessment and Conclusions

The United States has traditionally been a global leader in the development of bioprocessing technologies and the biomanufacturing industry. Nations in Europe and Asia with similar highly-trained work forces have also developed strengths in these areas. For most of the past half century, the field of bioprocessing has largely been occupied by chemical and process engineers developing the systems in collaboration with microbiologists engineering the cells used to manufacture biological products. Based on this rich history, it is not surprising therefore that most engineers who have ventured into the field of stem cell bioprocessing come from the field of chemical engineering and that globally, those nations with previous bioprocessing expertise and industries are the current leaders in stem cell biomanufacturing.

Nations such as Germany and Japan, with a long-standing history of robotics and automation, are currently leading efforts to implement these practices into stem cell biomanufacturing platforms. The U.K., Portugal, and Singapore have invested heavily in research laboratories conducting stem cell bioprocessing work, and as a result are looked to globally as intellectual leaders in this field. For the most part though, the bioprocessing technologies for culture and monitoring of cells have not changed significantly to accommodate the specific needs and concerns of stem cells, so there are tremendous opportunities for technology development in these areas. Major U.S. initiatives launched in recent years in advanced manufacturing and robotics are directly relevant to many of needs of the burgeoning stem cell biomanufacturing industry and should therefore address this within their scope.

Although current platforms and practices can clearly produce sufficient quantities of some stem cells to manufacture first generation technologies for therapeutic, diagnostic, and research use, the anticipated future demand for stem cell products will necessitate significantly greater innovation in this field. The biggest challenges are anticipated to be developing flexible systems capable of handling variable inputs, efficiently producing well-defined cell populations, monitoring the processes in real-time in a non-destructive manner, and delivering reproducible and reliable products at the point of use. It is expected that to meet all of these goals will require interdisciplinary teams of engineers and scientists working collaboratively with industry partners and regulatory agencies to produce commercially viable and readily adoptable bioprocesses for the successful biomanufacturing of stem cells. Ultimately, the responsibility to deliver on the widespread hope for translation of stem cell potency into real and impactful societal benefits to human health will fall upon the biomanufacturing industry to produce sufficient quality and quantities of such products.

References

Abranches, E., E. Bekman, D. Henrique, and J.M. Cabral. 2007. Expansion of mouse embryonic stem cells on microcarriers. *Biotechnology and Bioengineering* 96:1211–1221.

Alberti, K., R.E. Davey, K. Onishi, S. George, K. Salchert, F.P. Seib, M. Bornhauser, T. Pompe, A. Nagy, C. Werner, and P. W. Zandstra. 2008. Functional immobilization of signaling proteins enables control of stem cell fate. *Nature Methods* 5:645–650.

Alfred, R., J. Radford, J. Fan, K. Boon, R. Krawetz, D. Rancourt, and M.S. Kallos. 2011. Efficient suspension bioreactor expansion of murine embryonic stem cells on microcarriers in serum-free medium. *Biotechnology Progress* 27:811–823.

Amit, M., I. Laevsky, Y. Miropolsky, K. Shariki, M. Peri, and J. Itskovitz-Eldor. 2011. Dynamic suspension culture for scalable expansion of undifferentiated human pluripotent stem cells. *Nature Protocols* 6:572–579.

Bardy, J., A.K. Chen, Y.M. Lim, S. Wu, S. Wei, H. Weiping, K. Chan, S. Reuveny, and S.K. Oh. 2012. Microcarrier suspension cultures for high-density expansion and differentiation of human pluripotent stem cells to neural progenitor cells. *Tissue Engineering. Part C, Methods* EPub ahead of print, September 4, 2012.

Bauwens, C., T. Yin, S. Dang, R. Peerani, and P.W. Zandstra. 2005. Development of a perfusion fed bioreactor for embryonic stem cell-derived cardiomyocyte generation: oxygen-mediated enhancement of cardiomyocyte output. *Biotechnology and Bioengineering* 90:452–461.

Benoit, D. S., M.P. Schwartz, A.R. Durney, and K.S. Anseth. 2008. Small functional groups for controlled differentiation of hydrogel-encapsulated human mesenchymal stem cells. *Nature Materials* 7:816–823.

Bratt-Leal, A. M., R.L. Carpenedo, M.D. Ungrin, P.W. Zandstra, and T.C. McDevitt. 2011. Incorporation of biomaterials in multicellular aggregates modulates pluripotent stem cell differentiation. *Biomaterials* 32:48–56.

Carpenedo, R. L., A.M. Bratt-Leal, R.A. Marklein, S.A. Seaman, N.J. Bowen, J.F. McDonald, and T.C. McDevitt. 2009. Homogeneous and organized differentiation within embryoid bodies induced by microsphere-mediated delivery of small molecules. *Biomaterials* 30:2507–2515.

Chambers, S. M., Y. Qi, Y. Mica, G. Lee, X.J. Zhang, L. Niu, J. Bilsland, L. Cao, E. Stevens, P. Whiting, S.H. Shi, and L. Studer. 2012. Combined small-molecule inhibition accelerates developmental timing and converts human pluripotent stem cells into nociceptors. *Nature Biotechnology* 0:715–720.

Cormier, J. T., N.I. zur Nieden, D.E. Rancourt, and M.S. Kallos. 2006. Expansion of undifferentiated murine embryonic stem cells as aggregates in suspension culture bioreactors. *Tissue Engineering* 12:3233–3245.

Dang, S. M., S. Gerecht-Nir, J. Chen, J. Itskovitz-Eldor, and P.W. Zandstra. 2004. Controlled, scalable embryonic stem cell differentiation culture. *Stem Cells* 22:275–82.

Fernandes-Platzgummer, A., M.M. Diogo, R.P. Baptista, C.L. da Silva, and J.M. Cabral. 2011. Scale-up of mouse embryonic stem cell expansion in stirred bioreactors. *Biotechnology Progress* 27:1421–1432.

Fluri, D. A., P.D. Tonge, H. Song, R.P. Baptista, N. Shakiba, S. Shukla, G. Clarke, A. Nagy, and P.W. Zandstra. 2012. Derivation, expansion and differentiation of induced pluripotent stem cells in continuous suspension cultures. *Nature methods* 9:509–516.

Fok, E. Y. and P.W. Zandstra. 2005. Shear-controlled single-step mouse embryonic stem cell expansion and embryoid body-based differentiation. *Stem Cells* 23:1333–1342.

Fridley, K.M., I. Fernandez, M.T. Li, R.B. Kettlewell, and K. Roy. 2010. Unique differentiation profile of mouse embryonic stem cells in rotary and stirred tank bioreactors. *Tissue Engineering. Part A* 16:3285–98.

Hwang, Y. S., J. Cho, F. Tay, J.Y. Heng, R. Ho, S.G. Kazarian, D.R. Williams, A.R. Boccaccini, J.M. Polak, and A. Mantalaris. 2009. The use of murine embryonic stem cells, alginate encapsulation, and rotary microgravity bioreactor in bone tissue engineering. *Biomaterials* 30:499–507.

Irwin, E.F., R. Gupta, D.C. Dashti, and K.E. Healy. 2011. Engineered polymer-media interfaces for the long-term self-renewal of human embryonic stem cells. *Biomaterials* 32:6912–9.

Lock, L.T. and E.S. Tzanakakis. 2009. Expansion and differentiation of human embryonic stem cells to endoderm progeny in a microcarrier stirred-suspension culture. *Tissue Engineering. Part A* 15:2051–2063.

Mahoney, M.J. and W.M. Saltzman. 2001. Transplantation of brain cells assembled around a programmable synthetic microenvironment. *Nature biotechnology* 19:934–939.

Melkoumian, Z., J.L. Weber, D.M. Weber, A.G. Fadeev, Y. Zhou, P. Dolley-Sonneville, J. Yang, L. Qiu, C.A. Priest, C. Shogbon, A.W. Martin, J. Nelson, P. West, J.P. Beltzer, S. Pal, and R. Brandenberger. 2010. Synthetic peptide-acrylate surfaces for long-term self-renewal and cardiomyocyte differentiation of human embryonic stem cells. *Nature Biotechnology* 28:606–610.

Nie, Y., V. Bergendahl, D.J. Hei, J.M. Jones, and S.P. Palecek. 2009. Scalable culture and cryopreservation of human embryonic stem cells on microcarriers. *Biotechnology Progress* 25:20–31.

Placzek, M.R., I.M. Chung, H.M. Macedo, S. Ismail, T. Mortera Blanco, M. Lim, J.M. Cha, I. Fauzi, Y. Kang, D.C. Yeo, C.Y. Ma, J.M. Polak, N. Panoskaltsis, and A. Mantalaris. 2009. Stem cell bioprocessing: fundamentals and principles. *Journal of the Royal Society Interface* 6:209–232.

Purpura, K.A., A.M. Bratt-Leal, K.A. Hammersmith, T.C. McDevitt, and P.W. Zandstra. 2012. Systematic engineering of 3D pluripotent stem cell niches to guide blood development. *Biomaterials* 33:1271–1280.

Rodin, S., A. Domogatskaya, S. Strom, E.M. Hansson, K.R. Chien, J. Inzunza, O. Hovatta, and K. Tryggvason. 2010. Long-term self-renewal of human pluripotent stem cells on human recombinant laminin-511. *Nature Biotechnology* 28:611–615.

Rodrigues, C.A., M.M. Diogo, C.L. da Silva, and J.M. Cabral. 2011. Microcarrier expansion of mouse embryonic stem cell-derived neural stem cells in stirred bioreactors. *Biotechnology and Applied Biochemistry* 58:231–242.

Serra, M., C. Brito, S.B. Leite, E. Gorjup, H. von Briesen, M.J. Carrondo, and P.M. Alves. 2009. Stirred bioreactors for the expansion of adult pancreatic stem cells. *Annals of Anatomy = Anatomischer Anzeiger: official organ of the Anatomische Gesellschaft* 191:104–115.

Serra, M., C. Brito, M.F. Sousa, J. Jensen, R. Tostoes, J. Clemente, R. Strehl, J. Hyllner, M.J. Carrondo, and P.M. Alves. 2010. Improving expansion of pluripotent human embryonic stem cells in perfused bioreactors through oxygen control. *Journal of Biotechnology* 148:208–215.

Serra, M., C. Correia, R. Malpique, C. Brito, J. Jensen, P. Bjorquist, M.J. Carrondo, and P.M. Alves. 2011. Microencapsulation technology: a powerful tool for integrating expansion and cryopreservation of human embryonic stem cells. *PloS One* 6:e23212.

Singh, V. 1999. Disposable bioreactor for cell culture using wave-induced agitation. *Cytotechnology* 30:149–158.

Siti-Ismail, N., A.E. Bishop, J.M. Polak, and A. Mantalaris. 2008. The benefit of human embryonic stem cell encapsulation for prolonged feeder-free maintenance. *Biomaterials* 29:3946–3952.

Siti-Ismail, N., A. Samadikuchaksaraei, A.E. Bishop, J.M. Polak, and A. Mantalaris. 2012. Development of a novel three-dimensional, automatable and integrated bioprocess for the differentiation of embryonic stem cells into pulmonary alveolar cells in a rotating vessel bioreactor system. *Tissue Engineering. Part C, Methods* 18:263–272.

Solorio, L. D., A.S. Fu, R. Hernandez-Irizarry, and E. Alsberg. 2010. Chondrogenic differentiation of human mesenchymal stem cell aggregates via controlled release of TGF-beta1 from incorporated polymer microspheres. *Journal of Biomedical Materials Research. Part A* 92:1139–1144.

Steiner, D., H. Khaner, M. Cohen, S. Even-Ram, Y. Gil, P. Itsykson, T. Turetsky, M. Idelson, E. Aizenman, R. Ram, Y. Berman-Zaken, and B. Reubinoff. 2010. Derivation, propagation and controlled differentiation of human embryonic stem cells in suspension. *Nature Biotechnology* 28:361–364.

Villa-Diaz, L.G., H. Nandivada, J. Ding, N.C. Nogueira-de-Souza, P.H. Krebsbach, K.S. O'Shea, J. Lahann, and G.D. Smith. 2010. Synthetic polymer coatings for long-term growth of human embryonic stem cells. *Nature biotechnology* 28:581–583.

Yao, S., S. Chen, J. Clark, E. Hao, G.M. Beattie, A. Hayek, and S. Ding. 2006. Long-term self-renewal and directed differentiation of human embryonic stem cells in chemically defined conditions. *Proceedings of the National Academy of Sciences of the USA* 103:6907–6912.

zur Nieden, N.I., J.T. Cormier, D.E. Rancourt, and M.S. Kallos. 2007. Embryonic stem cells remain highly pluripotent following long term expansion as aggregates in suspension bioreactors. *Journal of Biotechnology* 129:421–432.

Zweigerdt, R., R. Olmer, H. Singh, A. Haverich, and U. Martin. 2011. Scalable expansion of human pluripotent stem cells in suspension culture. *Nature Protocols* 6:689–700.

Appendices

Appendix A: Delegation Biographies

Robert M. Nerem (Panel Chair), Georgia Institute of Technology

Robert M. Nerem joined Georgia Tech in 1987 as the Parker H. Petit Distinguished Chair for Engineering in Medicine. He is an Institute Professor and Parker H. Petit Distinguished Chair Emeritus. He currently serves as the Director of the Georgia Tech/Emory Center (GTEC) for Regenerative Medicine, a center established with an NSF—Engineering Research award. He also is a part-time Distinguished Visiting Professor at Chonbuk National University in Korea. He received his Ph.D. in 1964 from Ohio State University and is the author of more than 200 publications. He is a Fellow and was the founding President of the American Institute of Medical and Biological Engineering (1992–1994), and he is past President of the Tissue Engineering Society International, the forerunner of the Tissue Engineering and Regenerative Medicine International Society (TERMIS). In addition, he was the part-time Senior Advisor for Bioengineering in the new National Institute for Biomedical Imaging and Bioengineering at the National Institutes of Health (2003–2006). In 1988 Professor Nerem was elected to the National Academy of Engineering (NAE), and he served on the NAE Council (1998–2004). In 1992 he

R.M. Nerem et al. (eds.), *Stem Cell Engineering: A WTEC Global Assessment,*
Science Policy Reports, DOI 10.1007/978-3-319-05074-4,
© Springer International Publishing Switzerland 2014

was elected to the Institute of Medicine of the National Academy of Sciences and in 1998 a Fellow of the American Academy of Arts and Sciences. 1994 he was elected a Foreign Member of the Polish Academy of Sciences, and in 1998 he was made an Honorary Fellow of the Institution of Mechanical Engineers in the United Kingdom. In 2004 he was elected an honorary foreign member of the Japan Society for Medical and Biological Engineering and in 2006 a Foreign Member of the Swedish Royal Academy of Engineering Sciences. Professor Nerem holds honorary doctorates from the University of Paris, Imperial College London, and Illinois Institute of Technology. In 2008 he was selected by NAE for the Founders Award.

Peter W. Zandstra, University of Toronto

Peter W. Zandstra, graduated with a Bachelor of Engineering degree from McGill University in the Department of Chemical Engineering, obtained his Ph.D. degree from the University of British Columbia in the Department of Chemical Engineering and Biotechnology (working with Jamie Piret and Connie Eaves). Finally, he did a postdoctoral fellowship in the laboratory of Douglas Lauffenburger at the MIT before moving to the University of Toronto in 1999. Research in the Zandstra Laboratory is focused on the generation of functional tissue from adult and pluripotent stem cells. His group's quantitative, bioengineering-based approach strives to gain new insight into the fundamental mechanisms that control stem cell fate and to develop robust technologies for the use of stem cells and their derivatives to treat disease. Specific areas of research focus include blood stem cell expansion and the generation of cardiac tissue and endoderm progenitors from pluripotent stem cells. Dr. Zandstra is a Professor in the Institute of Biomaterials and Biomedical Engineering, the Department of Chemical Engineering and Applied Chemistry, and the Donnelly Centre at the University of Toronto. He is also a member of the McEwen Centre for Regenerative Medicine and the Heart and Stroke/Richard Lewar Centre of Excellence. He currently acts as Chief Scientific Officer for the Centre for the Commercialization of Regenerative Medicine (http://www.ccrm.ca/). Dr. Zandstra's accomplishments have been recognized by a number of awards and accolades including a Guggenheim Fellowship and the McLean Award. Dr. Zandstra's strong commitment to training the next generation of researchers is evidenced by his role as the Director of the undergraduate Bioengineering Program.

David V. Schaffer, University of California, Berkeley

David V. Schaffer is a Professor of Chemical Engineering, Bioengineering, and Neuroscience at the University of California, Berkeley, where he also serves as the codirector of the Berkeley Stem Cell Center. He graduated from Stanford University with a B.S. degree in Chemical Engineering in 1993. Afterward, he attended Massachusetts Institute of Technology and earned his Ph.D. also in Chemical Engineering in 1998 with Professor Doug Lauffenburger. Finally, he did a postdoctoral fellowship in the laboratory of Fred Gage at the Salk Institute for Biological Studies in La Jolla, CA before moving to UC Berkeley in 1999. At Berkeley, Dr. Schaffer applies engineering principles to enhance stem cell and gene therapy approaches for neuroregeneration. This work includes mechanistic investigation of stem cell control, as well as molecular evolution and engineering of viral gene delivery vehicles. David Schaffer has received an NSF CAREER Award, Office of Naval Research Young Investigator Award, Whitaker Foundation Young Investigator Award, and was named a Technology Review Top 100 Innovator. He was also awarded the Biomedical Engineering Society Rita Shaffer Young Investigator Award in 2000, the American Chemical Society BIOT Division Young Investigator Award in 2006, and was inducted into the College of Fellows of the American Institute of Medical and Biological Engineering in 2010.

Todd C. McDevitt, Georgia Institute of Technology/Emory University

Todd C. McDevitt is an Associate Professor in the Wallace H. Coulter Department of Biomedical Engineering at the Georgia Institute of Technology and Emory University, and a Petit Faculty Fellow of the Parker H. Petit Institute for Bioengineering and Bioscience at Georgia Tech. In 2009, Dr. McDevitt was appointed the founding Director of the Stem Cell Engineering Center at Georgia Tech (http://scec.gatech.edu/), an interdisciplinary initiative to advance stem cell translation and enhance stem cell biology research through multi-investigator collaborative efforts. The McDevitt Laboratory for the Engineering of Stem Cell Technologies (http://mcdevitt.bme.gatech.edu/) is focused on developing enabling technologies for the directed differentiation and morphogenesis of stem cells for regenerative medicine therapies and *in vitro* diagnostic applications. Dr. McDevitt's research program has been supported by funding from the National Institutes of Health, National Science Foundation, American Heart Association and Georgia Research Alliance, among other agencies. Dr. McDevitt graduated cum laude with a Bachelor of Science in Engineering (B.S.E.) from Duke University in 1997 double majoring in Biomedical and Electrical Engineering and he received the Howard Clark Award for undergraduate research. He received his Ph.D. in Bioengineering from the University of Washington in 2001, and conducted postdoctoral research in the Department of Pathology at the University of Washington 2002–2004 before starting as an Assistant Professor at Georgia Tech in August 2004. Dr. McDevitt has received several honors, including the Society for Biomaterials Young Investigator Award (2010), the Georgia Tech Junior Faculty Outstanding Undergraduate Research Mentor Award (2010), the Petit Institute Interdisciplinary Research and Education Award (2009), and an American Heart Association New Investigator Award (2004).

Sean P. Palecek, University of Wisconsin—Madison

Sean P. Palecek is a Professor of Chemical and Biological Engineering at the University of Wisconsin—Madison. He is also affiliated with the Department of Biomedical Engineering, the Stem Cell and Regenerative Medicine Center, and WiCell Research Institute. Prof. Palecek received his B.Ch.E. in Chemical Engineering from the University of Delaware, M.S. in Chemical Engineering from the University of Illinois at Urbana-Champaign, and Ph.D. in Chemical Engineering from MIT. He is a recipient of a National Science Foundation CAREER award. Prof. Palecek's research identifies chemical and mechanical cues that regulate human pluripotent stem cell self-renewal and differentiation, then uses those principles to design culture systems that apply those cues in the appropriate spatial and temporal manner. He has made contributions to human pluripotent stem cell expansion and differentiation to cardiac myocyte, vascular endothelial, and epidermal cell lineages.

Jeanne Loring, The Scripps Research Institute

Jeanne Loring is a professor and the Director of the Center for Regenerative Medicine at The Scripps Research Institute. Dr. Loring has a B.S. in Molecular

Biology and a Ph.D. in Developmental Neurobiology. She was on the faculty of the University of California at Davis, and has held research and management positions at biotechnology companies including Hana Biologics, GenPharm International, Incyte Genomics, and Arcos BioScience. She joined the faculty of Sanford-Burnham Medical Research Institute as a principal investigator in January 2004 and served as codirector of the institute's NIH Exploratory Center for Human Embryonic Stem Cell Research and Director of the NIH Human Embryonic Stem Cell Training Course. In 2007, Dr. Loring joined The Scripps Research Institute as founding director of the stem cell regenerative medicine program. Dr. Loring's current research focuses on the genomics and epigenomics of human pluripotent stem cells (embryonic and induced pluripotent stem cells), with the major goal of ensuring the safety of stem cell therapies and accuracy of models of human disease. Dr. Loring is also developing practical applications for these cells for drug discovery, drug delivery, and cell therapy. She works with collaborators to develop stem cell applications for Alzheimer's disease, multiple sclerosis, and arthritis, and is using stem cells to investigate autism. To improve the drug development process, her laboratory is building an ethnically diverse cell bank of iPSCs for drug toxicity screening.

Appendix B: Site Visit Reports

Site visit reports are arranged in alphabetical order by organization name.

Academy of Military Medical Sciences, Tissue Engineering Research Center

Site Address:	27 Tai Ping Road, Haidian District, Beijing 100850
	http://www.itelab.com
Date Visited:	November 14, 2011
WTEC Attendees:	S. Palecek (report author), N. Moore, P. Zandstra, F. Huband
Host(s):	Prof. Chang Yong Wang
	Tel.: 86-10-68166874
	Fax: 86-10-68166874
	wcy2000@yahoo.com
	Haibin Wang
	whb_zzgc@yahoo.com.cn

Overview

Prof. Chang Yong Wang is the director of the Tissue Engineering Research Center at the Academy of Military Medical Sciences (AMMS). He has an M.D. and a background in clinical medicine. His research effort focuses on tissue engineering and regenerative medicine. Dr. Haibin Wang is a postdoctoral researcher in the Tissue Engineering Research Center.

R.M. Nerem et al. (eds.), *Stem Cell Engineering: A WTEC Global Assessment,*
Science Policy Reports, DOI 10.1007/978-3-319-05074-4,
© Springer International Publishing Switzerland 2014

Research and Development Activities

The Tissue Engineering Research Center at AMMS consists of 30 researchers working on a diverse array of tissues, including heart, brain, liver, kidney, lung, and uterus. The center has significant activity in using stem cells, including embryonic stem cells (ESCs), somatic cell nuclear transfer ESCs (scNT-ESCs), induced pluripotent stem cells (iPSCs), and mesenchymal stem cells (MSCs) in engineered tissues.

The center performs basic studies on stem cell expansion, stem cell differentiation, and cell-material interactions that will facilitate its translational efforts. Cells and engineered tissues for clinical application are under development, including cardiomyocytes for myocardial infarction, cartilage and bone, and biomaterials for cell delivery. Nonclinical products are also being developed, including models for development and drug screening or evaluation and cellular biochips models.

Researchers in the center have differentiated ESCs and scNT-ESCs to cardiomyocytes. Mechanical and electrical stimulation have been used to culture and mature the cells. The cells were implanted in a rat infarction model, and large grafts of these cells have been formed. Injectable hydrogels for delivery of stem cells into the infarct wall have also been developed.

Efforts in cartilage engineering have used MSCs and porous bioceramics. Chitosan hydrogels have been engineered to delivery growth factors that aid cell survival and engraftment *in vivo*.

Challenges faced in developing engineered tissues include efficient cell seeding, appropriate scaffold composition and structure, cell delivery, and maintaining tissue survival and function after implantation.

Translation

The Tissue Engineering Research Center at AMMS is focused on tissue development for use in humans, and translation is a large part of their effort. Development is designed with translation in mind, but with a vision of substantially improving on current technology. For example, acellular materials are easier to translate but cellularized constructs will likely provide superior functionality. In addition, development of whole organs is a goal of the center.

Sources of Support

The government funds early stages of product translation. Private enterprise partnerships with the government are also possible for commercializing products. Tissue engineering product approval is regulated by the Chinese FDA. The mechanism for approving these products is currently being developed.

Collaboration Possibilities

The projects at the Tissue Engineering Research Center at AMMS are highly collaborative. The researchers implement a "virtual lab" model in which they provide their expertise and platforms to researchers at other institutions. This effort is funded by the government, and outcomes are shared among researchers. The center has collaborations with experts in materials manufacturing, materials characterization, and biochip assembly. Collaboration partners include Rice University, Drexel University, Tsinghua University, Peking University, and CAS.

The AMMS has a similar research environment to CAS, although limitations on visitations in both directions exist. The hosts expressed particular interest in collaborating with stem cell scientists and engineers working on mechanistic problems, in addition to application-oriented researchers.

Summary and Conclusions

The Tissue Engineering Research Center at the AMMS has a very active research program using several different stem cell sources in a wide variety of engineered tissues and organs. The developmental and translational efforts are facilitated by the ease of animal trials and clinical studies in China. This multidisciplinary center is an example of how advances in stem cell engineering can advance the field of tissue engineering. Additional collaborations with stem cell biologists and engineers would benefit the tissue engineering efforts of the center.

Selected References

Gao, J., R. Liu, J. Wu, Z. Liu, J. Li, J. Zhou, T. Hao, Y. Wang, Z. Du, C. Duan, and C. Wang. 2012. The use of chitosan based hydrogel for enhancing the therapeutic benefits of adipose-derived MSCs for acute kidney injury. *Biomaterials* 33: 3673–3681.

Liu, Z., H. Wang, Y. Wang, Q. Lin, A. Yao, F. Cao, D. Li, J. Zhou, C. Duan, Z. Du, Y. Wang, and C. Wang. 2012. The influence of chitosan hydrogel on stem cell engraftment, survival and homing in the ischemic myocardial microenvironment. *Biomaterials* 33: 3093–3106.

Lu, S., H. Wang, W. Lu, S. Liu, Q. Lin, D. Li, C. Duan, T. Hao, J. Zhou, Y. Wang, S. Gao, and C. Wang. 2010. Both the transplantation of somatic cell nuclear transfer and fertilization-derived mouse embryonic stem cells with temperature-responsive chitosan hydrogels improve myocardial performance in infarcted rat hearts. *Tissue Engineering Part A* 16:1303–1315.

Lu, S., Y. Li, S. Gao, S. Liu, H. Wang, W. He, J. Zhou, Z. Liu, Y. Zhang, Q. Lin, C. Duan, X. Yang, and C. Wang. 2010. Engineered heart tissue graft derived from somatic cell nuclear transfer embryonic stem cells improve myocardial performance in infarcted rat heart. *Journal of Cellular and Molecular Medicine* 14:2771–2779.

Wang, H., J. Zhou, Z. Liu, and C. Wang. 2010. Injectable cardiac tissue engineering for the treatment of myocardial infarction. *Journal of Cellular and Molecular Medicine* 14: 1044–1055.

Basel Stem Cell Network (BSCN), University Hospital Basel and University of Basel

Site Address:	[Meeting held with Basel Stem Cell Network (BSCN) investigators at University of Zurich, Center for Regenerative Medicine] Moussonstrasse 13 CH-8091 Zürich http://www.baselstemcells.ch/network.html
Date Visited:	February 29, 2012
WTEC Attendees:	D. Schaffer (report author), T. McDevitt, P. Zandstra, N. Moore, H. Sarin
Host(s):	Prof. Verdon Taylor Laboratory of Embryology and Stem Cell Biology Department of Biomedicine, University of Basel Mattenstrasse 28 CH-4058 Basel, Switzerland Tel.: +41 61 695 30 91 Fax :+41 61 695 30 90 verdon.taylor@unibas.ch
	Prof. Dr. Savas Tay Department of Biosystems Science and Engineering ETH Zurich Swiss Federal Institute of Technology Mattenstrasse 26 4058 Basel, Switzerland Tel.: +41 61 387 31 57 savas.tay@bsse.ethz.ch

Overview

The Basel Stem Cell Network (BSCN) is a consortium of investigators working at different sites across Basel, including at the Department of Biomedicine at the University of Basel, ETH Department of Biosystems Science and Engineering in Basel, Friedrich Miescher Institute for Biomedical Research (FMI) in Basel, various clinical departments at the University Hospital of Basel, as well as investigators working at Novartis and Roche. The mission of the network is to foster collaborative opportunities arising from basic science and clinical research to answer a wide variety of questions related to stem cells. Stem cell research has a long-standing tradition in Basel, with origins in the study of the hematopoietic system for clinical application. The ongoing focus of network investigators is finding clinical applicability of stem cell-based therapies in regenerative medicine beyond the hematopoietic system, in addition to answering more fundamental questions in developmental biology. The research activities of the network are funded by the University of Basel, Swiss National Science Foundation (SNSF), European Union Framework Programme 7 (FP7), University Hospital Basel, and the pharmaceutical industry.

Research and Development Activities

Prof. Taylor is British, was educated at King's College, London, the University of Basel and at the ETH, has worked as an independent group leader at Max Planck and senior lecturer (associate professor) in Sheffield, and recently accepted a Professorship position at the University of Basel. The goals of his research program are to identify neural stem cells in the mammalian brain, to identify and understand niche derived signals that control stem cell maintenance and fate *in vivo* (using a mouse model and relying on conditional lineage tracing), to uncover transcriptional and post-transcriptional networks that are controlled by the niche and determine cell fate, to examine changes in transcriptional and post-transcriptional networks under pathophysiological conditions at the single-cell level, and to uncover mechanisms of stem cell aging leading to dormancy with an aim toward rejuvenation. In recent work, he has been investigating the role of Notch signaling, and the cells in which such signaling is active, in adult subventricular zone and hippocampal neural stem cell function and neurogenesis. As a reporter of Notch activation, they generated Hes5::GFP mice, which enabled the identification of two populations of stem cells (radial and horizontal) that respond differently to exercise, seizure, and aging (Lugert et al. 2010, Fig. B.1). They are also using Hes5:CreER^{T2} mice for lineage tracing and recently found that while the immature NSC and a later stage neuroblast undergo extensive proliferation, the intermediate progenitor cells are not highly mitotic. Furthermore, it can take considerable lengths of time (up to 100 days) for the immature cells to fully convert to differentiated neurons (Lugert et al. 2012). In addition to this fundamental work, he has recently been collaborating with Prof. Matthias Lutolf at EPFL to conduct high-throughput screening of biofunctionalized hydrogel microenvironments that influence NSC fate and function.

Prof. Tay has arrived at ETH in Basel relatively recently, after having conducted a postdoctoral fellowship with Prof. Steve Quake. He is now applying his strong experience in microfluidics and quantitative analysis to problems in stem cells.

Fig. B.1 The adult hippocampus contains three distinct populations of neural stem cells: quiescent radial, quiescent horizontal and active horizontal. These three stem cell populations respond selectively to pathophysiological stimuli (From Lugert et al. 2010)

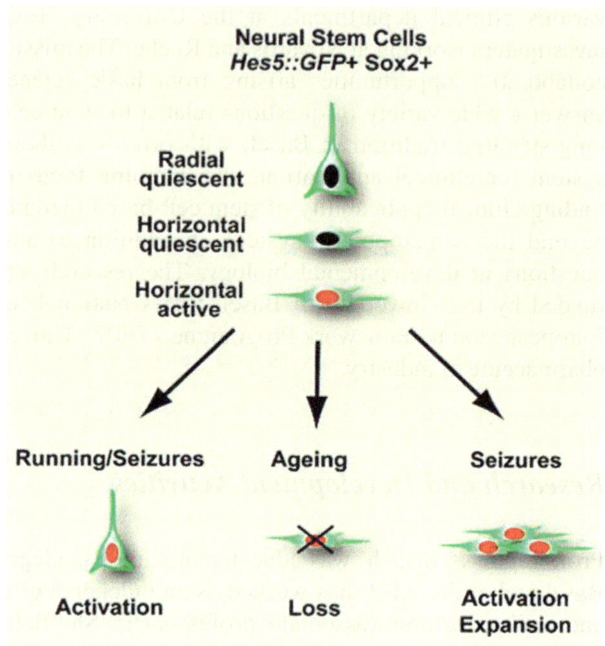

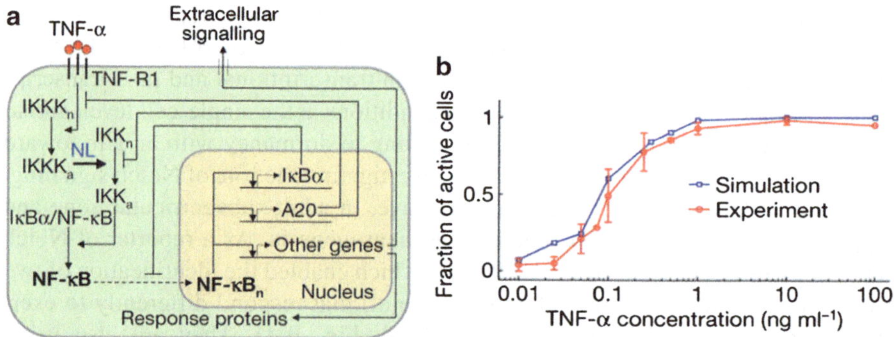

Fig. B.2 Mathematical model development and simulations (From Tay et al. 2010, Figure 3a, b) (*Left*) Model architecture is based on stochastic description of receptor and gene activity, quadratic representation of IKK activation, and negative feedback via IkBa and A20. (*Right*) Simulated (*blue*) and measured (*red*) fraction of activated cells (error bars indicate standard error of the mean)

In prior work, he used a microfluidic cell culture system to investigate NF-κB signaling in thousands of individual cells (fibroblasts) in response to TNF-α (Tay et al. 2010, Fig. B.2). In contrast to what bulk culture measurements would indicate, they find that the response is digital, with different fractions of cells responding or not as a function of ligand concentration. However, the nature of the response could vary from cell to cell in "analog properties" including response time, amplitude, and number of oscillations. Finally, they developed a stochastic model that could predict a number of cellular outcomes.

Translation

As described below, the BSCN is a broad network of investigators that blends basic with translational and clinical work. Given their growing resources and investigator strengths, the network has very strong translational potential.

Sources of Support

The BSCN is funded by the Swiss National Science Foundation (SNSF), European Union Framework Programme 7 (FP7), University Hospital Basel, and the pharmaceutical industry. In addition, they have recently submitted a large application to a SNSF proposal call for centers of excellence, which would provide 20–30 million CHF over 4 years. The proposal, entitled Re2stem (for regulation of and regeneration by stem cells), would involve basic investigation, the use of bioengineering approaches to create artificial niches, and clinical translation. Also, there would be efforts in early embryo stem cells (germ, pluripotent), hematopoietic stem cells (with a GMP facility), and neural stem cells.

Collaboration Possibilities

The BSCN is a dynamic entity with the potential to grow rapidly. They have a considerable amount of expertise within the network, and they have growing efforts in both materials research and mathematical modeling. Additional opportunities for collaborations in materials science and quantitative analysis likely exist.

Summary and Conclusions

These two investigators, and the network in general, have a broad range of expertise including basic investigation of stem cell function using *in vivo* models, biomaterials research, microfluidics, and mathematical modeling. A major strength of Switzerland is that it has a broad range of academic, medical, and pharmaceutical expertise, all within a relatively small country that encourages tight networking and collaboration.

Selected References

Lugert, S., O. Basak, P. Knuckles, U. Haussler, K. Fabel, M. Götz, C.A. Haas, G. Kempermann, V. Taylor, and C. Giachino. 2010. Quiescent and active hippocampal neural stem cells with distinct morphologies respond selectively to physiological and pathological stimuli and aging. *Cell Stem Cell* 6:445–456.

Lugert, S., M. Vogt, J.S. Tchorz, M. Müller, C. Giachino, and V. Taylor. 2012. Homeostatic neurogenesis in the adult hippocampus does not involve amplification of Ascl1^{high} intermediate progenitors. *Nat. Commun.* 3:670, doi:10.1038/ncomms1670.

Tay, S., J.J. Hughey, T.K. Lee, T. Lipniacki, S.R. Quake, and M.W. Covert. 2010. Single-cell NF-kappaB dynamics reveal digital activation and analogue information processing. *Nature* 466:267–271.

Berlin-Brandenburg Center for Regenerative Therapies

Site Address:	Charite-Campus Virchow Clinic Augustenburger Platz 1 D-13353 Berlin, Germany http://bcrt.charite.de/
Date Visited:	February 29, 2012
WTEC Attendees:	R.M. Nerem (report author), J. Loring, S. Palecek, H. Ali
Host(s):	Professor Dr.-Ing. Georg Duda, Director of the Julius Wolff Institute, Vice-Director BCRT, Board Member of the CSSB Tel.: +49 30 450 55 90 79 Fax: +49 30 450 55 99 69 georg.duda@charite.de Dr. Frank-Roman Lauter, Head of Business Development Tel.: +49 30 450 539 413 Fax: +49 30 450 539 904 frank-roman.lauter@charite.de Professor Dr. med. Hans-Dieter Volk, Director, Institute for Medical Immunology Vice-Director BCRT, Board Member of the CSSB Tel.: +49 30 450 524 062 Fax: +49 30 450 524 932 hans-dieter.volk@charite.de Dr. Manfred Gossen, Research Group Leader, Genetic Engineering Tel.: +49 30 450 539 491 Fax: +49 30 450 539 991 manfred.gossen@charite.de

(continued)

Dr. Nan Ma, Center for Biomaterial Development, Helmholtz-Zentrum
Geesthacht [Teltow]
Tel.: +49 3328 352 0
nan.ma@hzg.de

Dr.-Ing. Jochen Ringe, Junior Research Group Leader
Laboratory for Tissue Engineering
Tel.: +49-(0)30-450 513 293
jochen.ringe@charite.de

Dr. Tobias Winkler, Center for Musculoskeletal Surgery
and Julius Wolff Institute
Tel.: +49 30 450 6 109
tobias.winkler@charite.de

Professor Dr. Petra Reinke, Professor of Nephrology
Dept. Nephrology and Internal Intensive Care, CVK
Medical Director Kidney Transplant Program
Platform leader "Immunology" at the BCRT
Tel.: +49 (0)30 450653490
petra.reinke@charite.de

Prof. Dr. Christof Stamm, German Heart Institute, Berlin
Tel.: +49 (0)30 4593 2109
stamm@dhzb.de

Dipl.-Ing. Sophie Van Linthout, Department of Cardiology
Tel.: +49 (0)30 8445 2715
Sophie.Van-linthout@charite.de

Overview

The Berlin-Brandenburg Center for Regenerative Therapies (BCRT) was established as a translational center in 2006 with a 4-year grant from the German Federal Ministry for Education and Research. This funding has been renewed, and the BCRT is now in its second 4-year period of support. The BCRT is located in a newly reconstructed building at the Charité Campus Virchow Clinic and hosts to 26 newly implemented research groups. This center has been established on the belief that regenerative medicine requires new translational and educational structures. It is an interdisciplinary center, not simply a multidisciplinary center, and thus the classical disciplinary groups have been replaced by project teams.

There are three platforms, and these are the immune system, the cardiovascular system, and the musculoskeletal system. These build on cell development and characterization and on polymer-based biomaterials. There also is a clinical open

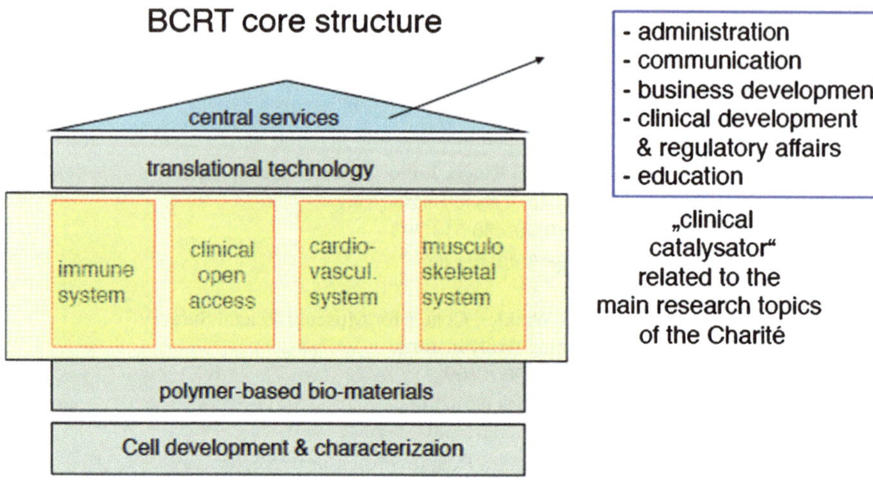

Fig. B.3 Organization of the Berlin-Brandenburg Center for Regenerative Therapies (BCRT; Courtesy of BRCT)

access platform, which is available to assist others and a business development unit headed by Dr. Frank-Roman Lauter. The BCRT is unique in that it brings together biology, engineering, and clinical activities. This is not only apparent in its research and development activities, but also in its educational program. This core structure is illustrated in Fig. B.3.

The BCRT has as its focus endogenous regeneration. This strategy is based on the fact that classical tissue engineering has had limited clinical success, but there are lessons from which one can learn. Some key lessons include that inflammation is essential to the regenerative cascades needed, that there is a lack of the potent cells needed to overcome the hurdles in regeneration, and that material science can contribute to the endogenous formation of tissues during regeneration.

Research and Development Activities

During the site visit the WTEC attendees heard a series of presentations covering the different platforms. Dr. Manfred Gossen discussed the cell differentiation and molecular characterization activities at the BCRT. This included his presenting on "high end" cell engineering and transposon based chromosomal integration. Goals for the immediate future are focused on the establishment of gene transfer protocols for iPSCs. This includes the derivation of patient specific iPS cells.

This was followed by a presentation by Dr. Nan Ma, head of the department "Biocompatibility" at the Centre of Biomaterial Development of Helmholtz-Zentrum Geesthacht in Teltow, entitled "Stem cell-Biomaterial Interactions." Of interest is the project using magnetically-controlled gene delivery in the cardiovascular system. This involves the injection of magnetic bead/gene complexes that are "captured" by an external magnet resulting in gene expression in a desired site that is defined by the external magnet. This is one example of the many projects that are ongoing in the biomaterials group.

Then there were two presentations from the musculoskeletal platform. One was from Dr. Tobias Winkler, an orthopedic surgeon, and the other from Dr. Jochen Ringe, an engineer. Dr. Winkler noted that there is no method for regenerating skeletal muscle today; however, initial results indicate that an MSC-based cell therapy may be able to improve muscle function. Dr. Ringe presented data on cartilage repair using both autologous cell implantation and also matrix-assisted autologous cell implantation. He also indicated that the next generation therapy would be a cell-free, regenerative approach using a matrix-based tissue inductive material, one to which bioactive factors could be incorporated. So far, there are four spin-off companies, and the products on the market include OralBone, BioSeed, and ChondroTissue. By 2015 there could be a new product, Chondrokine, a cell-free approach that actively recruits MSCs using chemokines. It should be noted that Dr. Ringe is the research director of the Tissue Engineering Group. This group was established in 1994 by Professor Michael Sittinger, who not only still heads this group but also heads translational technology research.

The technical presentations in the afternoon included Dr. Petra Reinke who discussed immune cell therapy, i.e., an approach for reshaping the immune response, and two presentations in the cardiovascular area. The first of these was by Dr. Christof Stamm discussing the use of cell-based therapy for ischemic heart disease. This activity is joint between the BCRT and the German Heart Center. Dr. Stamm reviewed the history of cell therapy clinical studies in this area, and indicated that today, if a patient is of an age less than 73, a cell-based therapy may be of some help. The question of the use of a cardiac patch came up, and Dr. Stamm indicated that it was not clear that the use of such a patch as a cell delivery vehicle would help. The final presentation was from Dr. Sophie van Linthout who discussed the use of mesenchymal stromal cells for the treatment of inflammatory cardiomyopathy. The idea that inflammation could be a therapeutic target for heart disease is an intriguing one. The question is can MSCs modulate the inflammatory response and in doing so reduce cardiac damage. In conclusion, these seven presentations on research and development activities at BCRT provided evidence of the exciting projects being conducted in the center.

Translation

There also was a presentation by Frank-Roman Lauter who heads business development at BCRT. The mission of this unit is to identify high potential products and to increase translational efficiency. There are five members of his team, and out of the center's 130 total projects, 12 have been identified as key projects. Of these 12, three are entering the clinic. It should be noted that, whereas it is normal at a university to do what might be called an opportunity analysis once there is a patent and thus defined intellectual property, at BCRT such an opportunity analysis is carried out much earlier, in some cases at the start of the research and development program for a specific project. There also is regulatory and health-economic analysis expertise available. It should also be noted that there are a variety of partnerships with industry. These include more strategic ones, co-developments, the spinning off of companies, and joint ventures, BCRT also does conduct contract research. Partnerships have been established with several companies such as Miltenyi Biotec, B. Braun, Pharmicell Europe, and Pluristem.

Sources of Support

The primary sources of support are the German Federal Ministry of Education and Research and the Helmholtz Association. Additional support comes from the states of Berlin and Brandenburg as well as Charité-Universitätsmedizin Berlin. There also are the more conventional single investigator grants and industry support.

Collaboration Possibilities

BCRT has a variety of ongoing collaborations. The BCRT core groups closely interact with the research groups of the institutions of the principal investigators. It is a founding member of the Regenerative Medicine Coalition that has the goal of accelerating the delivery of regenerative therapies to patients. It also is a member of TERM (Tissue Engineering Regenerative Medicine), which is a European collaboration within regenerative medicine, whose objective is to strengthen the cooperation between regional research clusters in Europe in the field of tissue engineering and regenerative medicine. Finally, a unique feature is the appointment of visiting fellows. Such individuals have a laboratory at BCRT as well as research support, with one such visiting fellow being Professor David Mooney from Harvard in the United States.

Summary and Conclusions

The BCRT is a unique organization in at least two ways. The first of these is its translational nature with a business development unit including activities in the regulatory affairs area and health economic analysis. Second is it being organized to bring together biology, engineering, and clinical activities. In this regard Berlin has first-rate biologists and clinicians, from the Freie and Humboldt Universities; the Technical University of Berlin has some excellent individuals complementing other disciplines. In the context of the engineering expertise within BCRT, there is a strong biomaterials group and there are also biomechanicians. Also, the research director of the Tissue Engineering Group, Dr. Jochen Ringe, is an engineer and this group is largely made up of engineers. One of our hosts, Professor Georg Duda, has a mechanical engineering education and is a vice-director of BCRT. Engineering is integrated within this clinically dominated research center and is central to its technology-driven approach toward clinical translation.

Selected References

Alexander, T., L. Templin, S. Kohler, C. Groß, A. Sattler, A. Meisel, G.-R. Burmester, A. Radbruch, A. Thiel, and F. Hiepe. 2012. Helios+ Foxp3+ naturally occurring regulatory t cells are peripherally expanded in active systemic lupus erythematosus. *Annals of the Rheumatic Diseases* 71:A41–A42, doi:10.1136/annrheumdis-2011-201234.

Cipitria, A., C. Lange, H. Schell, W. Wagermaier, J.C. Reichert, D.W. Hutmacher, and P. Fratzl, and G.N. Duda. 2012. Porous scaffold architecture guides tissue formation. *Journal of Bone and Mineral Research* 27:1275–1288, doi:10.1002/jbmr.1589.

Heinrich, V., J. Stange, T. Dickhaus, P. Imkeller, U. Krüger, S. Bauer, S. Mundlos, P.N. Robinson, J. Hecht, and P.M. Krawitz. 2012. The allele distribution in next-generation sequencing data sets is accurately described as the result of a stochastic branching process. *Nucleic Acids Research* 40:2426–2431, doi:10.1093/nar/gkr1073.

Hutmacher, D.W., G. Duda, and R. E. Guldberg. 2012. Endogenous musculo-skeletal tissue regeneration. *Cell and Tissue Research* 347:485–488, doi:10.1007/s00441-012-1357-0.

Klopocki, E., S. Lohan, S.C. Doelken, S. Stricker, C.W. Ockeloen, R. S.T. de Aguiar, K. Lezirovitz, R.C. Mingroni-Netto, A. Jamsheer, H. Shah, I. Kurth, R. Habenicht, M. Warman, K. Devriendt, U. Kordaß, M. Hempel, A. Rajab, O. Mäkitie, M. Naveed, U. Radhakrishna, S.E. Antonarakis, D. Horn, S. Mundlos. 2012. Duplications of BHLHA9 are associated with ectrodactyly and tibia hemimelia inherited in non-Mendelian fashion. *Journal of Medical Genetics* 49:119–125, doi:10.1136/jmedgenet-2011-100409.

Kurtz, A., and S. J. Oh. 2012. Age related changes of the extracellular matrix and stem cell maintenance. *Preventive Medicine* 54(Suppl.):S50-S56.

Leutz, A., and J.J. Smink. 2012. A TORway to osteolytic disease. *Cell Cycle* 11:637–638.

Liman, P., N. Babel, T. Schachtner, N. Unterwalder, J. König, J. Hofmann, P. Reinke, and P. Nickel. 2012. Mannose-binding lectin deficiency is not associated with increased risk for polyomavirus nephropathy. *Transplant Immunology* 26(2–3):123–127.

Poller, W., M. Rother, C. Skurk, and C. Scheibenbogen. 2012. Endogenous migration modulators as parent compounds for the development of novel cardiovascular and anti-inflammatory drugs. *British Journal of Pharmacology*165:2044–2058, doi:10.1111/j.1476-5381.2011.01762.x.

Schachtner, T. 2012. Measurement of interferon-gamma induced protein 10 in serum: a risk assessment approach for bkv-associated nephropathy. *American Journal of Transplantation* 12:112 (May 2012).

Schmueck, M. 2012. Preferential expansion of virus-specific multifunctional central-memory T cells. *American Journal of Transplantation* 12: 461. (May 2012)

Schmueck, M., A.M. Fischer, B. Hammoud, G. Brestrich, H. Fuehrer, S.-H. Luu, K. Mueller, N. Babel, H.-D. Volk, and P. Reinke. 2012. Preferential expansion of human virus-specific multifunctional central memory T Cells by partial targeting of the IL-2 receptor signaling pathway: the key role of CD4+ T cells. *Journal of Immunology* 188:5189–5198.

Schwele, S., A.M. Fischer, G. Brestrich, M.W. Wlodarski, L. Wagner, M. Schmueck, A. Roemhild, S. Thomas, M.H. Hammer, N. Babel, A. Kurtz, J.P. Maciejewski, P. Reinke, and H.-D. Volk. 2012. Cytomegalovirus-specific regulatory and effector T cells share TCR clonality—possible relation to repetitive CMV infections. *American Journal of Transplantation* 12:669–681, doi:10.1111/j.1600-6143.2011.03842.x.

Stricker, S., S. Mathia, J. Haupt, P. Seemann, J. Meier, and S. Mundlos.2012. Odd-skipped related genes regulate differentiation of embryonic limb mesenchyme and bone marrow mesenchymal stromal cells. *Stem Cells and Development* 21(4):623–633.

Van Linthout, S. 2012. Human cardiac biopsy-derived cells improve angiotensin II-induced heart failure. *Cardiovascular Research* 93:S14 (Mar 15, 2012).

von Haehling, S., J.C. Schefold, E.A. Jankowska, J. Springer, A. Vazir, P.R. Kalra, A. Sandek, G. Fauler, T. Stojakovic, N, Trauner, P. Ponikowski, H.-D. Volk, W. Doehner, A.J. Coats, P.A. Poole-Wilson, and S.D. Anker.2012. Ursodeoxycholic acid in patients with chronic heart failure a double-blind, randomized, placebo-controlled, crossover trial. *Journal of the American College of Cardiology* 59:585–592.

Weist, B.J.D., M. Schmueck, P. Reinke, and N. Babel. 2012. Control and abatement of polyomavirus BK—it's not the CD8+ but multifunctional and cytolytic CD4+ T cells. *American Journal of Transplantation* 12:213 (May 2012).

Chinese University of Hong Kong (CUHK)

Site Address:	Shatin, New Territories, Hong Kong (Meeting took place in Beijing) http://www.cuhk.edu.hk/english/index.html
Date Visited:	November 13, 2011
WTEC Attendees:	S. Palecek (report author), S. Demir, K. Ye, F. Huband
Host(s):	Professor Gang Li School of Biomedical Sciences Dept. of Orthopaedics & Traumatology Tel.: 37636153 gangli@ort.cuhk.edu.hk http://www.sbs.cuhk.edu.hk/TeachingStaffDetails.asp?TE_NAME=LI%20 Gang Professor Gang Xu Depart of Medicine, Chinese University of Hong Kong gangxu@cuhk.edu.hk Professor Zhiyong Zhang The Fourth Military Medical University Tel.: 15291573296 mr.zhiyong@gmail.com

Overview

The School of Biomedical Sciences at the Chinese University of Hong Kong (CUHK) has a Thematic Research Program in Stem Cells and Regeneration. The focus of this program is to understand the role of stem cells in disease and development and to use mesenchymal stem cells isolated from adult tissues in clinical translational research. Prof. Gang Li is a Member and Chief of the Stem Cells and Regeneration Theme, and is a Professor in the Department of Orthopaedics and Traumatology at CUHK. Prof. Gang Xu is an Associate Member of the Stem Cells and Regeneration Theme, and is a Professor in the Department of Medicine & Therapeutics at CUHK. Prof. Zhiyong Zhang is a Professor in the Institute of Orthopaedics and Traumatology at Xijing Hospital, Fourth Military Medical University, Xi'an, China.

Research and Development Activities

Research projects in the Stem Cells and Regeneration Theme include (1) identifying factors and regulatory mechanisms that control MSC proliferation, differentiation, and fate, (2) using MSCs in tissue engineering and regenerative applications including bone-tendon healing, tendon repair, bone fracture healing, and cardiac tissue repair, (3) understanding the role of MSCs in cancer development and using MSCs as gene delivery vectors to treat cancer, (4) cell reprogramming for studying disease and development, and (5) GMP cell manufacturing for cellular therapies.

Prof. Li's lab works on musculoskeletal tissue engineering, with a focus on MSCs. His lab has published on mechanisms of MSC differentiation, MSC recruitment and homing to tumors, the use of MSCs in gene therapy applications, materials for bone tissue engineering, and expansion and GMP processing of MSCs.

Prof. Xu's lab researches diabetes and mechanisms regulating beta cell survival and function.

Prof. Zhang's lab develops culture systems for MSC expansion and differentiation for applications in musculoskeletal tissue engineering.

Translation

The research focus on MSCs by Profs. Li and Zhang has strong clinical translational potential. Culture system development is focused on expanding clinical grade cells while development efforts using these cells in tissue engineering and anti-cancer therapies are under way. Opportunities and financial incentives exist to promote and facilitate translation. Prof. Zhang has commercialized a stem cell bioreactor system.

Sources of Support

The main funding support for Prof. Li and Prof. Xu in Hong Kong are from the Hong Kong government, Research Grant Council, and Innovation and Technology Funding Agency. In addition, Prof. Li has also obtained industrial contract research from Amgen USA and Eli Lilly USA to test novel compounds using well established preclinical animals models in his laboratories.

Prof. Zhong receives funding from China Natural Science Foundation and other grant giving bodies in China. In addition, military hospitals in China have a separate funding mechanism from other academic institutes visited. Investigators at military hospitals are eligible for funding from both civilian and military programs.

Collaboration Possibilities

Discussions of collaborations focused on interests in working with engineers and materials scientists in development of culture systems for stem cells, and in use of these stem cells in tissue engineering applications.

Prof. Li expressed strong interest in collaboration with scientists in the United States on stem cell biology, tissue engineering for musculoskeletal tissue regeneration, and clinical translational research. Hong Kong is a unique place as it is close to China, yet has a western (U.K.) system; the communication with researchers in Hong Kong is very easy. Professors in Chinese University of Hong Kong are now eligible to apply for China National Funding as well through the newly established Chinese University of Hong Kong Shenzhen Research Institute.

Summary and Conclusions

The hosts noted that China provides an excellent environment for tissue engineering. Funding is generally good, and large animal studies and clinical trials are easier to perform in China than in the United States The regulatory environment on clinical trials is changing to be more similar to that in the United States and Europe, however. China is investing heavily in stem cell technology. The hosts indicated that it is difficult to compete with overseas Ph.D. programs in attracting top Ph.D. students, in part because of lower stipends. Hong Kong has a program to recruit foreign students and provides stipends comparable to those at U.S. institutions. There is a strong incentive to publish in high-impact-factor SCI journals, although many Chinese researchers read more papers in Chinese than English because of the language barrier.

Selected References

Fan, R., Z. Kang, L. He, J. Chan, and G. Xu. 2011. Extendin-4 improves blood glucose control in both young and aging normal non-diabetic mice, possible contribution of beta cell independent effects. *PLoS One* 6:e20443.

Green, D.W., G. Li, B. Milthorpe, and B. Ben-Nissan. 2012, Mesenchymal stem cells coated with biomaterials in regenerative medicine. *Materials Today* 5(1–2):626–632.

Ominsky, M.S., C.Y. Li, X.D. Li, H.L. Tan, E. Lee, M. Barrero, F.J. Asuncion, D. Dwyer, C.Y. Han, F. Vlasseros, R. Samadfam, J. Jolette, S.Y. Smith, M. Stolina, D.L. Lacey, W.S. Simonet, C. Paszty, G. Li, and H.Z. Ke. 2011, Inhibition of Sclerostin by monoclonal antibody enhances bone healing and improves bone density and strength of non-fractured bones. *Journal of Bone and Mineral Research* 26(5):1012–1021.

Song, C., J. Xiang, J.Q. Tang, D. Hirst, J.W. Zhou, K.M. Chan, and G. Li. 2011. Thymidine kinase gene modified bone marrow mesenchymal stem cells as vehicles for anti-tumor therapy. *Human Gene Therapy* 22:439–449.

Wang, Y., X. Chen, M. Armstrong, and G. Li. 2007. Survival of xenogeneic bone marrow-derived mesenchymal stem cells in a xeno-transplantation model. *Journal of Orthopaedic Research* 25:926–932.

Xu, L.L., C. Song, M. Ni, F.B. Meng, and G. Li. 2012. Cellular retinol-binding protein 1 (CRBP-1) promotes osteogenic differentiation of mesenchymal stem cells. *International Journal of Biochemistry & Cell Biology* 44:612–619.

Zhang, G., B. Guo, H. Wu, T. Tang, B. Zhang, L. Zheng, Y. He, Z. Yang, X. Pan, H. Chow, K. To, Y. Li, D. Li, X. Wang, Y. Wang, K. Lee, Z. Hou, N. Dong, G. Li, K. Leung, L. Hung, F. He, L. Zhang, and L. Qin. 2012, A delivery system targeting bone formation surfaces to facilitate RNAi-based anabolic therapy. *Nature Medicine* 18(2):307–314.

Fraunhofer Institute for Immunology and Cell Therapy

Site Address:	Perlickstrasse 1
	04103 Leipzig
	Germany
	http://www.izi.fraunhofer.de/ueber-uns.html?&L=1
Date Visited:	February 27, 2012
WTEC Attendees:	T. McDevitt (report author), D. Schaffer, Nicole Moore, H. Sarin
Host(s):	Prof. Dr. Frank Emmrich, Director
	Tel.: +49 341 9725-500
	Frank.Emmrich@izi.fraunhofer.de
	Dr. Thomas Tradler, Head of Business Development
	Tel.: +49 341 35536-9305
	Thomas.Tradler@izi.fraunhofer.de
	Dr. Alexandra Stolzing, Group Leader of Stem Cell Biology and Regeneration
	Tel.: +49 341 35536-3405
	Alexandra.Stolzing@izi.fraunhofer.de

Overview

The Fraunhofer Institute for Immunology and Cell Therapy IZI at Leipzig was founded in 2005 and is a member of the Fraunhofer Life Sciences Alliance (Fraunhofer-Gesellschaft), which consists of 6 institutes and is the youngest of the

Fraunhofer alliances. Ten years ago, the need for biotechnology was noted due to the relative small amount of industry in biotech; now the Life Sciences Alliance has become the most active with regard to start-ups. The mission of the Life Sciences Alliance is to find solutions to specific problems at the interfaces between medicine, life sciences and engineering through partnerships with industry and hospital institutions. The Fraunhofer Institute is the largest organization focused on applied research in all of Europe with an annual budget of 1.8 billion Euro and employs 20,000 people. It has more than 80 research units spread among 60 individual institutes and has research centers in Europe, the United States, Asia, and the Middle East.

Training and Education

The Fraunhofer IZI collaborates in education and training programs with other life science institutions in Leipzig and has graduate students and post-docs among their researchers. They also collaborate with several international companies on specific training courses (i.e., methods, devices, etc.). They offer single to multi-day training that includes both classroom and practical training exercises, often in combination with conferences in the area. For example, the recent World Conference on Regenerative Medicine (http://www.wcrm-leipzig.com/), which is held every 2 years, is organized by the institute and the Translational Center for Regenerative Medicine at Leipzig University.

Research and Development Activities

The Fraunhofer IZI specializes in the area of regenerative medicine and the development of cell therapies and stem cell technologies to generate biologically compatible tissues and organs. The institute consists of four departments representing their core competencies, Cell Engineering, Immunology, Cell Therapy and Diagnostics & New Technologies. Each of the 4 departments functions as an individual business unit with its own operating budget, but they work together in an interdisciplinary manner to develop solutions interfacing medicine, life sciences, and engineering. The Fraunhofer IZI has 169 staff, 89 % of whom are scientific personnel and 70 % of whom are female. In 2010 the overall operating budget was 10 million Euro. The building of the institute was completed in 2008, with funding from the European Union, the Federal Republic of Germany, the Freestate of Saxonia and the city of Leipzig. The first extension to the building was added in 2009, and houses experimental medicine laboratories and a GMP facility. A second building extension that will add nearly 50 % more space is scheduled to be completed in 2012.

The Department of Cell Engineering is focused on GMP manufacturing of cell and tissue products for regenerative medicine applications. The Department of Immunology is focused on the development of immunological products for control of diseases, such as cell therapies and biopharmaceuticals to prevent GvHD, cell and antibody-based therapies, and phage display technologies. The Department of Cell Therapy is exploring new treatment strategies for ischemic diseases, inflammatory diseases, age-related diseases and oncology using mesenchymal and other cell types (Fricke et al. 2009; Stolzing et al. 2008). They have automated the production of human skin with up to 1,000s of units per month that are intended primarily for cosmetic testing purposes; automatization is a historical strength of Germany and the Fraunhofer Institutes. The Department of Diagnostics focuses primarily on the research and development of diagnostic markers and therapeutic targets on ncRNA and miRNA (RNomics).

Translation

The Fraunhofer IZI offers full service packages covering broad parts of value chain development and manufacturing to business partners from all over the world, including Canada, Israel, Australia and the United States, not just Germany and Europe. Their approach is to design specific solutions for individual needs and can go from GLP to GMP to GCP to product. Commercial products based upon Fraunhofer IZI research include autologous dermal skin equivalent grafts produced from as few as 20 hair follicles (EpiDex); this automatization process is now being moved out to a separate company. Patents on virus-free, mRNA reprogramming methods have been filed. A recent paper on this work is the work of Arnold et al. (2012).

Sources of Support

More than 70 % of the funding for the Fraunhofer Life Sciences Alliance is derived from contracts with industry and from publicly financed research projects; about 30 % of funding comes from the government, which is contributed by the German and Länder governments in the form of base funding.

Collaboration Possibilities

Local Collaborations

The Fraunhofer IZI has close ties to the other research institutions located in Leipzig. BIO CITY Leipzig, which houses, for instance, the Biotechnological-Biomedical Center (BBZ) of the University of Leipzig, is nearby, as is the Translational Center for Regenerative Medicine (TRM), which is one of four big regenerative medicine centers in Germany that are funded by BMBF (Federal Ministry of Education and Research).

The Faculty of Veterinary Medicine at the University of Leipzig, one of only five such faculties in all of Germany, is located across the street from the institute and often collaborates with Fraunhofer IZI researchers on the development of large animal models. Internal calls within the Fraunhofer Institutes are intended to encourage investigators from different sites to collaborate.

International Collaborations

The largest German-funded collaborative project with joint funding from BMBF and the California Institute for Regenerative Medicine is in Leipzig where they are working with a large-animal stroke model. Although this relationship has gone well, one problem that can be encountered with multinational funding initiatives currently is that since they are reviewed independently, one can review well in one system and the other may not, which can then stifle the collaboration; scientific cooperation is not a problem but the logistical mechanisms to facilitate international collaborations could be improved.

Summary and Conclusions

The Fraunhofer Institute for Immunology and Cell Therapy IZI at Leipzig provides an interdisciplinary and translational approach to regenerative cell therapies. A number of active collaborations with local and foreign entities have been established to leverage the strengths of the institute research activities. A clear emphasis on translational work permeates all of the activities of the institute.

References

Arnold. A., Y.M. Naaldijk, C. Fabian, H. Wirth, H. Binder, G. Nikkhah, L. Armstrong, and A. Stolzing. 2012. Reprogramming of human Huntington fibroblasts using mRNA. *ISRN Cell Biology* 2012:Article ID 124878, 12 pp., doi:10.5402/2012/124878.

Fricke, S., M. Ackermann, A. Stolzing, C. Schimmelpfennig, N. Hilger, J. Jahns, G. Hildebrandt, F. Emmrich, P. Ruschpler, C. Pösel, M. Kamprad, and U. Sack. 2009. Allogeneic non-adherent bone marrow cells facilitate hematopoietic recovery but do not lead to allogeneic engraftment. *PLoS One* 4(7):e6157.

Stolzing, A., E. Jones, D. McGonagle, and A. Scutt. 2008. Age related changes in human bone marrow-derived mesenchymal stem cells: consequences for cell therapies. *Mech. Ageing Dev.* 129(3):163–173, doi:10.1016/j.mad.2007.12.002.

Fudan University, Zhongsan Hospital

Site Address:	180 Fenglin Road, Shanghai, China Tel:86-21-64041990 http://www.zs-hospital.sh.cn/e/index.asp
Date Visited:	November 17, 2011
WTEC Attendees:	R.M. Nerem (report author), S. Demir, N. Moore, S. Palecek, P. Zandstra, K. Ye, F. Huband
Host(s):	Professor Junbo Ge Director of the Department of Cardiology and Co-Chairman of the Shanghai Institute of Cardiovascular Diseases. Tel.: 86-21-6404-1990 Fax: 86-21-6422-3006 ge.junbo@zs-hospital.sh.cn Dr. Aijun Sun Associate Professor Tel.: 13641882087 sun.aijun@zs-hospital.sh.cn Dr. Shuning Zhang Tel.: 15921766132 zhang.shuning@zs-hospital.sh.cn

Overview

At Fudan University located within Zhongshan Hospital is the Shanghai Institute of Cardiovascular Disease and the Stem Cell and Tissue Engineering Center. Under the direction of Dr. Junbo Ge, the major interests include mechanisms of atherosclerosis, the early diagnosis of coronary heart disease, and the use of bone marrow-derived cells for cardiac repair therapies. Dr. Ge is an accomplished cardiologist who is the editor-in-chief of the *Chinese Journal of Circulation Research* and has more than 200 publications in international journals. In total Dr. Ge's group has more than 50 researchers. It also should be noted that Dr. Aijun Sun, who is an Associate Professor working with Dr. Ge, was of considerable help in making the WTEC visit a productive one.

Research and Development Activities

The major topic of discussion was the use of bone marrow-derived cells in cardiac repair clinical therapies. Two significant clinical trials were discussed that also have been reported in the literature. The one published in the journal *Heart* (Ge et al. 2006) reports on the efficacy of transcatheter transplantation/delivery of bone marrow stem cells in the treatment of acute myocardial infarction (MI). The study included 20 patients who were admitted within 24 h after an acute MI. From 1 week to 6 months after the acute MI there was an increase in left ventricular ejection fraction of up to 8 %. This study demonstrated the practicality of this type of clinical therapy and the results not only showed improved cardiac function, but also increased myocardial perfusion.

A second study (Yao et al. 2009) investigated the repeated administration of bone marrow mononuclear cells as a therapy in patients with a large myocardial infarction. Thirty nine patients were studied. The cells were administered both at 3–7 days and at 3 months, and the results obtained compared with a single infusion of cells. The increase in the left ventricle ejection fraction as evaluated after 12 months by magnetic resonance imaging (MRI) was significantly greater in the patients receiving the repeated administration of cells compared to those patients receiving a single infusion. Myocardial infarct size as derived by MRI also was decreased significantly in those patients receiving repeat administration of cells as compared to the single infusion patient group. The data from this preliminary study thus suggests that repeated bone marrow mononuclear cell administration is safe and might be a feasible approach for patients with large acute MI.

It should be noted that in general the harvested cell population is not expanded before therapy. Also, they are investigating other cell types including snMSCs, i.e., the single non-hematopoietic MSC subpopulation (CD133 + CD344).

Also of interest to this WTEC assessment of stem cell engineering was the use of a magnetic particle technique for tracking MSCs in the pig heart. This technique was developed in collaboration with the Department of Physics at Shanghai University. In this technique cells are labeled using a ferumoxide injectable solution with Resovist, a type of superparamagnetic iron oxide. Only a few percent of the cells were found to be detectable after a few weeks.

Translation

The above cardiac repair clinical studies clearly demonstrate the focus of Dr. Ge's group on the translation of stem cell research into clinical therapies. It should be noted that for both animal and clinical studies approval is granted by the hospital; however, for multi-center clinical trials permission must be sought from the government.

Sources of Support

MOST, Shanghai University, the Shanghai City government.

Collaboration Possibilities

There are some possibilities here; however, with the exception of the magnetic particle tracking technique, there appeared to be little involvement of engineers or even the engineering approach in the studies in Dr. Ge's group.

Summary and Conclusions

The research group of Dr. Ge appears to be aggressively advancing the use of bone marrow-derived cells in cardiac clinical therapies. In total more than 200 patients have been involved in the clinical studies conducted by this group.

Selected References

Ge, J., Y. Li, J. Qian, J. Shi, Q. Wang, Y. Niu, B. Fan, X. Liu, S. Zhang, A. Sun, and Y. Zou. 2006. Efficacy of emergent transcatheter transplantation of stem cells for treatment of acute myocardial infarction. *Heart* 92:1764–1767, doi:10.1136/hrt.2005.085431.
   Yao, K., R. Huang, A. Sun, J. Qian, X. Liu, L. Ge, Y. Zhang, S. Zhang, Y. Niu, Q. Wang, Y. Zou, and J. Ge 2009. Repeated autologous bone marrow mononuclear cell therapy in patients with large myocardial infarction. *European Journal of Heart Failure* 11:691–698.

Institute for Medical Informatics and Biometry (IMB), Dresden University of Technology (TUD)

Site Address:	Blasewitzer Strasse 86, D-01307 Dresden, Germany http://tu-dresden.de/die_tu_dresden/fakultaeten/medizinische_ fakultaet/inst/imb
Date Visited:	February 27, 2012
WTEC Attendees:	D. Schaffer (report author), T. McDevitt, N. Moore, H. Sarin

Host(s):	Prof. Ingo Roeder, Head of the Institute for Medical Informatics
	    and Biometry
	Tel.: +49 (0)351 458 6060
	Fax: +49 (0)351 458 7222
	ingo.roeder@tu-dresden.de
	
	Prof. Dr. Lars Kaderali, Chair for Statistical Bioinformatics
	Tel.: +49 (0)351 458 6060
	Fax: +49 (0)351 458 7222
	lars.kaderali@tu-dresden.de
	
	Dr. Ingmar Glauche, Junior Group Leader for Theoretical
	    Stem Cell Biology
	Tel.: +49 (0)351 458 6051
	Fax: +49 (0)351 458 7222
	ingmar.glauche@tu-dresden.de

Overview

The Institute for Medical Informatics and Biometry (IMB), part of the Medical Faculty Carl Gustav Carus at the Dresden University of Technology (TUD), is an inter-disciplinary institute whose research activities span across medicine, biology, mathematics, statistics, and bioinformatics. The IMB, by utilizing theoretical methods and computer-assisted approaches, supports the planning, implementation, data analysis and interpretation of basic and clinical research projects at its medical facility and the other institutes of the university such as the Coordination Centre for Clinical Trials (KKS) Dresden. Its research activities include: (1) modeling and systems biology (theoretical stem cell biology, disease and treatment models, and image analysis and reconstruction of cellular development); (2) biometry and statistical bioinformatics (classical biometric approaches in the planning and execution of clinical trials, statistics of dynamic processes and structures, genetic statistics, and integrative analysis of molecular and high-dimensional data); and (3) health service research and epidemiology (quality management and evaluation of health care projects, development and supplementation of clinical practice guidelines, and the development and maintenance of clinical-epidemiological registers for chronic diseases).

In 2009, TUD began an association of 14 cultural and research institutions called the DRESDEN-concept; one of the university's major activities under the concept has been to develop a common technology platform with its partners in a single online database. Since 2006, a partial funding for the research at the university is available through the German federal government's Initiative for Excellence (i.e., funding of the Cluster of Excellence "Center for Regenerative Therapies Dresden (CRTD)" and the "Dresden International Graduate School for Biomedicine and Bioengineering (DIGS-BB)").

Research and Development Activities

The IMB has three research areas related to stem cells: (1) Medical Systems Biology and Mathematical Modeling, (2) Medical Bioinformatics and Biometry, and (3) Bioimage Informatics.

In the first area, there are several projects, including theoretical stem cell biology, mechanisms of aging, host-pathogen interactions and immune responses, and analysis of the development and treatment of cancer. For example, they have modeled chronic myelogenous leukemia (CML), a homogeneous disease involving the Bcr-Abl fusion associated with the Philadelphia chromosomal translocation, as a competition between normal and leukemic cells for limited resources, in this case niche locations. Leukemic cells have an advantage, and they have used modeling to hypothesize which parameters may underlie this advantage, enabled by comparison to data on clinical progression. Two parameters could explain the data: the activation rate of cells from dormancy into a proliferative phase, and the deactivation into a quiet state within in niche (Roeder et al. 2006). The behavior of cells in this model is illustrated in Fig. B.4. Normal (blue) and leukemic (gray) stem cells are regularly activated from their bone

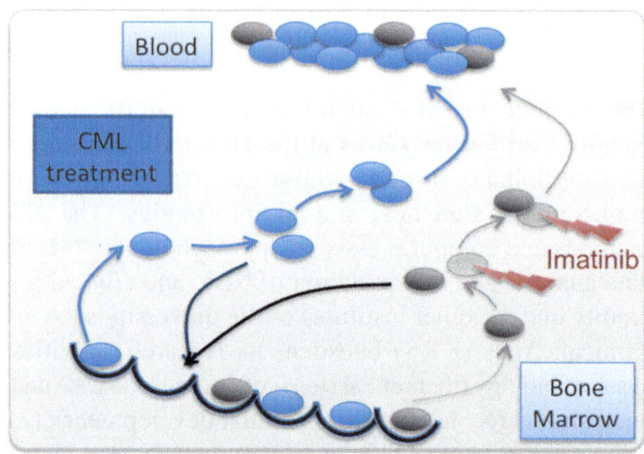

Fig. B.4 Leukemia cell model of stem cell activation used to investigate the role of tyrosine kinase inhibitors. Normal cells are shown in *blue*; leukemic cells in *gray* (From Glauche et al. 2012)

marrow niches (bottom) and subsequently divide. For the maintenance of a balance between quiescent and activated cells, some cells return to the niches and self-renew while others undergo further proliferation and differentiation, and contribute to peripheral blood. Tyrosine kinase inhibitors (TKIs) preferentially target activated leukemic cells, thus leading to a significant reduction of tumor load. Given that leukemic stem cells are less likely to be activated under TKI treatment, a residual pool of leukemic cells persists over long time scales. Furthermore, the model makes predictions of why drugs like imatinib (Gleevac) are only partially effective at eradicating the malignancy, as they apparently affect only proliferating cells, and a dynamic equilibrium of dividing and non-dividing cells that reside in the niche is able to lead to tumor progression once drug treatment is ceased. Furthermore, the modeling makes predictions of the duration of treatment needed to eliminate cancer cells.

In the Medical Bioinformatics and Biometry area, there are three main topics: (1) core biostatistics of high-throughput data, in particular the analysis of genome-wide microscopy based screens of RNAi data, and well as next generation sequencing data analysis, (2) statistical analysis of cellular genealogies and network structures, and (3) network inference and machine learning. Examples include RNAi screens of host factors involved in viral infection (e.g., hepatitis C), DNA sequencing of bacterial samples from lungs of cystic fibrosis patients, and sequencing the mutational spectrum in cancer cells.

In the Bioimage Informatics area, the investigators are developing image based analysis of multicellular systems in space and time. As one example, they have a cell lineage tracing project in which they construct lineage trees of single cells dividing and migrating in culture, with a focus on investigating how the microenvironment impacts hematopoietic stem cell function (Scherf et al. 2012, Fig. B.5).

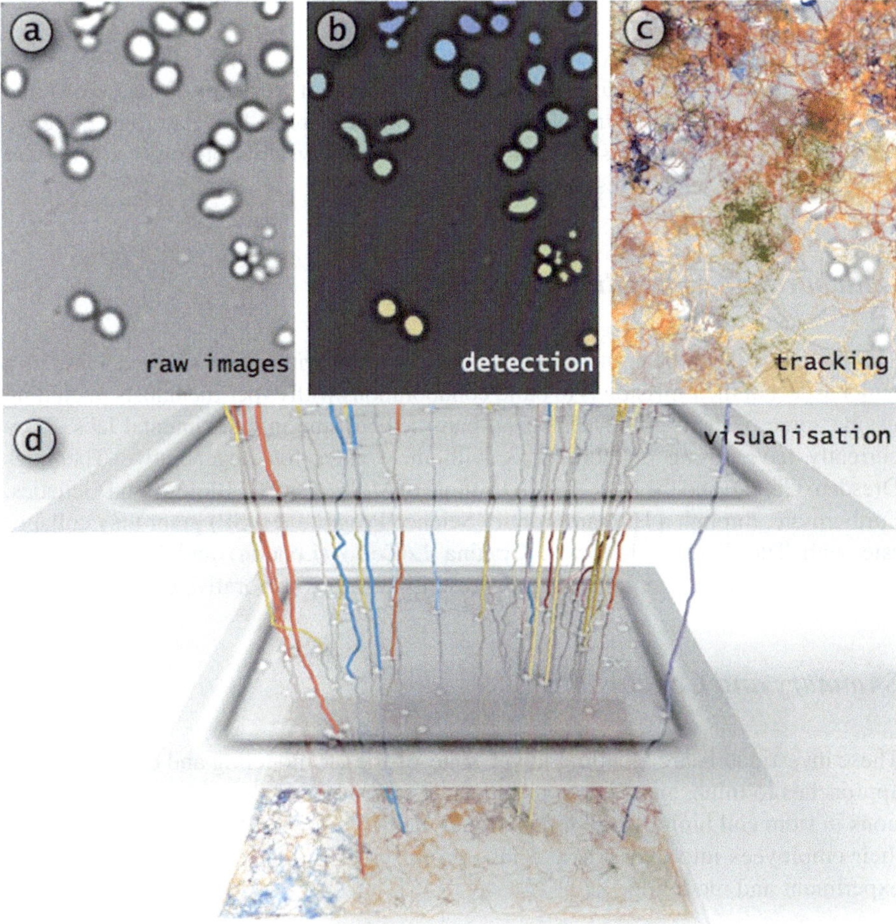

Fig. B.5 Sequence of detection steps for the identification of cellular motion (cell tracking, (**a**)–(**d**)). A spatio-temporal summary is provided in visualization (**d**) (Courtesy of N. Scherf, IMB)

Translation

Several project areas have a strong translational component, in particular the work with leukemia. In addition, the active collaborations of these mathematicians, physicists, and systems biologists with biologists and clinicians are impressive.

Sources of Support

The institute receives funding from the DFG (German Research Foundation) and the BMBF (Federal Ministry of Education and Research). For example, the eBio initiative of the BMBF has the goal of building up systems biology as a regular subject in all life science disciplines.

Collaboration Possibilities

The IMB is a theoretical and computational institute without on-site wet labs, so a major aspect of their mission is to build collaborations with experimentalists, including having some of their employees spend part of their time in experimental labs. They currently have strong collaborations with the Center for Regenerative Therapies Dresden (CRTD) and the Max Planck Institute of Molecular Cell Biology and Genetics. Furthermore, through a Human Frontier Science Program (HFSP) grant they collaborate with Tilo Pompe (Leipzig), Cristina LoCelso (London) and Peter Zandstra (Toronto). Therefore, in general IMB presents strong collaborative opportunities.

Summary and Conclusions

These investigators are blending novel, state-of-the-art modeling and computational approaches to mine, analyze, and synthesize experimental data for several applications in stem cell biology and translational medicine. In addition, the integration of their employees into experimental labs promises to further enable the melding of experiment and modeling.

Selected References

Glauche, I., K. Horn, M. Horn, L. Thielecke, M.A.G. Essers, A. Trumpp, and I. Roeder. 2012. Therapy of chronic myeloid leukaemia can benefit from the activation of stem cells: simulation studies of different treatment combinations. *British Journal of Cancer* Published online April 26, 2012, doi:10.1038/bjc.2012.142.

Roeder, I., M. Horn, I. Glauche, A. Hochhaus, M.C. Mueller, and M. Loeffler. 2006. Dynamic modeling of imatinib-treated chronic myeloid leukemia: functional insights and clinical implications. *Nature Medicine* 12:1181–1184, doi:10.1038/nm1487.

Scherf, N., K. Franke, I. Glauche, I. Kurth, M. Bornhauser, C. Werner, T. Pompe, and I. Roeder. 2012. On the symmetry of siblings: automated single-cell tracking to quantify the behavior of hematopoietic stem cells in a biomimetic setup. *Experimental Hematology* 40:119–130.e9, doi:10.1016/j.exphem.2011.10.009.

Institute for Stem Cell Therapy and Exploration of Monogenic Diseases (I-STEM)

Site Address:	AFM: Genopole Campus 1 5 rue Henri Desbrueres, 91030 Evry Cedex, France http://istem.eu/ewb_pages/e/english.php
Date Visited:	March 1, 2012
WTEC Attendees:	P. Zandstra (report author), T. McDevitt, D. Schaffer, N. Moore, H. Sarin
Host(s):	Dr. Yacine Laâbi Group Leader of Biotechnology of Stem Cells Tel.: +33 1 69 90 85 17 ylaabi@istem.fr Dr. Fulvio Mavilio Scientific Director of Genethon fmavilio@genethon.fr Dr. Pauline Poydenot Group Leader of High-Throughput Screening ppoydenot@istem.fr Dr. Mathilde Girard Group Leader of Pathological iPS Modeling mgirard@istem.fr Dr. Emmanuel Galène Head of the GMP Production at Genethon egalene@genethon.fr

Overview

Institute for Stem Cell Therapy and Exploration of Monogenic Diseases (I-STEM), is a leading French research and translation center dedicated to the development of treatments for monogenic diseases, with a particular focus on harnessing the potential of stem cells for substitutive and regenerative therapies. A second focus of I-Stem is the modeling of monogenic diseases using preimplantation embryo diagnosis-derived embryonic stem cells, and patient-derived induced pluripotent stem cells. It is anticipated that models based on these cells will enable fundamental investigations into disease mechanisms, and be useful as tools for screening compound libraries in order to discover new potential drugs.

I-STEM consists of basic biological research laboratories as well as technological platforms for the development and application of stem cell-based therapies. I-STEM is organized into approximately 12 teams, ranging from core platforms such as the Biotechnology of Stem Cells, Stem Cell Genomics and High-Throughput Screening, to fundamental research teams focused on diseases such as Retinopathies, Motor Neuron Disease, and Neuro- or Muscular Degenerative Disease. I-STEM has a staff of about 85 (35 Ph.D.s) and world class equipment (including for automated high-throughput screening, automated culture systems, cell line cryopreservation and storage, and reprogramming) and facilities.

Research and Development Activities

I-Stem has a major research focus on the use of pluripotent stem cells to model disease. In 2005 I-STEM was the first lab authorized by French Medical Agency to use PGD-derived ESC, and later on iPSCs. To support the use of these cells to model disease I-STEM has developed significant expertise in PSC differentiation protocols, with a particular strength and focus on neural and neural crest lineage differentiation. I-STEM also has activities on mesenchymal progenitor cell culture and the development of screening assays based on these cells.

Dr. Yacine Laâbi is team leader of the Stem Cell Biotechnologies group. This group undertakes three main activities at I-STEM: biobanking of hESC lines, cell culture automation of iPSCs and their progeny, and genomic engineering of hiP-SCs. This group collaborates closely with industry, including with Cellectis, a French tools and technology company (see http://www.cellectis-bioresearch.com/) on iPSC engineering. The strategy in these collaborations is one of open innovation—to develop new technologies, which they then make accessible to industry partners. Examples of technologies under development include PSC production and scale-up; PSC training programs, and access to equipment and resources. In the scale-up area, Dr. Laâbi is using the CompacT SelecT (http://www.tapbiosystems.com/) automated cell production platform with the goal of automating culture for expansion and differentiation of human PSC. In the cell banking area his group has the ambitious goal of generating and banking, in close collaboration with Cellectis, iPSCs covering all 5,000 monogenic disorders (plus sibling controls) at low passage, in traceable (cryotube barcoding), N_2 vapor storage. The bank is operating under strict guidelines set up by the Agence de la Biomédecine (authorizes French teams to work on hESCs and import them), and complies with international banking standards. Thus far his group has banked 38 hESC lines, representing 14 monogenic diseases (e.g., HD, Steinert, fascio-scapulo-humeral dystrophy, SCA7, Marfan syndrome). These lines are available to the scientific community through the European hESC Registry (http://www.hescreg.eu/). Students in this team are primarily from the biology department at the University of Evry-Val d'Essonne. Cell process engineering jobs are typically filled by cell biology Ph.D.s who learn the production side because the University of Evry-Val d'Essonne lacks an engineering school.

Dr. Pauline Poydenot leads the High-Throughput Screening facility. Dr. Poydenot is an engineer, and emphasized that bioengineering was not well integrated into bio-based research or biotechnology efforts in Europe, in part because of the lack of emphasis on bio-related activities in engineering schools. Her platform is well equipped (e.g., Bravo-Benchcel (Agilent) for picking, Biocel 1800 (Agilent) for compound management, Thermo, Biotek for High Content Screening (HCS) campaigns, AnalystGT (Molecular Device), and the Arrayscan (Cellomics) for analysis of PSC responses to drug libraries. A significant effort in this platform is focused on the development of robust and predictive assays for disease relevant cell types. The platform has direct access to screening libraries including the commercially available Prestwick, LOPAC and Sigma libraries. Dr. Poydenot's group also has access to a "Chem-X Infinity" library of 9,864 unique compounds bought from a small French chemical company. Once "hits" are found these are further investigated for fundamental mechanisms with other teams at I-STEM. No structure function chemistry is available at I-STEM and this would have to be pursued in collaboration with industry in the current model. Assays and screens undertaken thus far include myotonic dystrophy 1 (HCS image based assays on nuclear complex formation), Huntington's disease (reporter gene assay for transcriptional activation of REST in neural stem cells (NSCs)), and Lesch-Nyham (HTS viability in selective medium). This group is very open to industrial collaboration, and recently completed a 200k molecule screen for proliferation in NSCs with Roche (www.roche.com).

Dr. Fulvio Mavilio is the Scientific Director of Genethon. Genethon aims to be the European center for enabling gene therapy for rare genetic disorders. Dr. Mavilio, a molecular biologist, is working to create international clinical trial networks for gene therapy around Genethon. The institute should be particularly attractive to international partners as it will soon open what will be the world's biggest plant for producing large volumes of clinical-grade viral vectors—used to transfer therapeutic genes into the cells of patients. This so-called Genethon Bioprod manufacturing plant, represents a ~€28.5 million investment, featuring 5,000 m^2 of facilities, 4 production suites for vectors and cells, and anticipating >10 adeno-attenuated virus and >10 lenti batches per year to support phase I/II clinical investigation. The development of closer and more tangible scientific and translational connections between I-STEM and Genethon represents a wonderful opportunity.

Dr. Mathilde Girard is the leader of the Pathological iPSC Modeling Group. The iPSC pathological modeling team was created to develop two axes of iPSC research: (1) the optimization and standardization of reprogramming and (2) systems for quality control of iPSC manufacture. The Girard group is focusing on tool development for screening on pathological models, including metabolic diseases, enzymatic defects, and mitochondrial pathologies, in which markers of the pathology can be detected by biochemical processes and adapted to high-throughput screening. The proof of concept of this strategy is being developed for Lesch-Nyhan syndrome and on the Friedreich syndrome, for which iPSCs are currently being derived and characterized. QC approaches for iPSCs include imaging based outputs (Chan et al. 2009), and developing fully defined surfaces and media for iPSC derivation and culture. The Girard group is developing in close collaboration with the biotech company Cellectis a bank of GMP-grade, haplotyped iPSC, lines for therapy. It is also is involved in a number of national and international collaborations including with Luc Douay group for the generation of RBCs.

Translation

I-STEM has a close relationship with the Genethon (http://www.genethon.fr/en/about-us/our-mission/), a nonprofit biotherapy R&D organization, and together I-STEM and Genethon cover fundamental and translational aspects of both stem cells and gene therapy vector biology and manufacturing. Reflective of the large investment in this area, Genethon is building a world leading gene therapy/viral manufacturing center to support stem cell and gene theory trials across the world.

Sources of Support

I-Stem is supported by the combination of public, private and philanthropic entities. These include the French Muscular Disease Association (AFM), the French Government (through INSERM, the national institute of health and medical research) the University Evry-Val d'Essonne (founded in 1991) and Genopole, a multi-sector funded biocluster focused on genomics, genetics and biotech. I-STEM receives administrative, financial and logistic support from the Centre pour l'Etude des Cellules Souches (CECS), which is funded by the AFM. I-STEM also receives funding support from the European Union Framework Programs through specific investigator-driven projects and through partnerships with industry, including Roche (CH).

Genethon, like I-STEM, is funded primarily through the AFM, and was a major partner in the mapping of the human genome in the early 1990s (Chumakov et al. 1992).

Selected References

Abbott, A. 2012. French institute prepares for gene-therapy push; Genethon relaunches itself as a force for translational medicine. *Nature* 481:423–424, doi:10.1038/481423a.

Chan, E.M., S. Ratanasirintrawoot, I.H. Park, P.D. Manos, Y.H. Loh, H. Huo, J.D. Miller, O. Hartung, J. Rho, T.A. Ince, G.Q. Daley, and T.M. Schlaeger. 2009. Live cell imaging distinguishes bona fide human iPS cells from partially reprogrammed cells. *Nat. Biotechnol.* 27(11):1033–1037.

Chumakov, I., P. Rigault, S. Guillou, P. Ougen, A. Billaut, G. Guasconi, P. Gervy, I. LeGall, P. Soularue, L. Grinas, L. Bougueleret, C. Bellanné-Chantelot, B. Lacroix, E. Barillot, P. Gesnouin, S. Pook, G. Vaysseix, G. Frelat. A. Schmitz, J.-L. Sambucy, A. Bosch, X. Estivill, J. Weissenbach, A. Vignal, H. Riethman, D. Cox, D. Patterson, K. Gardiner, M. Hattori, Y. Sakaki, H. Ichikawa, M. Ohki, D. Le Paslier, R. Heilig, S. Antonarakis, and D. Cohen. 1992. Continuum of overlapping clones spanning the entire human chromosome 21q. *Nature* 359:380–387, doi:10.1038/359380a0.

Côme, J., X. Nissan, L. Aubry, J. Tournois, M. Girard, A.L. Perrier, M. Peschanski, M. Cailleret. 2008. Improvement of culture conditions of human embryoid bodies using a controlled perfused and dialyzed bioreactor system. *Tissue Eng, Part C Methods* 14(4):289–298.

Guenou, H., X. Nissan, F. Larcher, J. Feteira, G. Lemaitre, M. Saidani, M. Del Rio, C.C. Barrault, F.X. Bernard, M. Peschanski, C. Baldeschi, and G. Waksman. 2009. Human embryonic stem-cell derivatives for full reconstruction of the pluristratified epidermis: a preclinical study. *Lancet* 374(9703):1745–1753.

Lefort, N., A.L. Perrier, Y. Laâbi, C. Varela, and M. Peschanski. 2009. Human embryonic stem cells and genomic instability. *Regen. Med.* 4(6):899–909.

Marteyn, A., Y. Maury, M.M. Gauthier, C. Lecuyer, R. Vernet, J.A. Denis, G. Pietu, M. Peschanski, and C. Martinat. 2011. Mutant human embryonic stem cells reveal neurite and synapse formation defects in type 1 myotonic dystrophy. *Cell Stem Cell* 8(4):434–44, Epub 2011 Mar 31.

Nissan, X., L. Larribere, M. Saidani, I. Hurbain, C. Delevoye, J. Feteira, G. Lemaitre, M. Peschanski, and C. Baldeschi. 2011. Functional melanocytes derived from human pluripotent stem cells engraft into pluristratified epidermis. *Proc. Natl. Acad. Sci. USA* 108(36):14861–14866, Epub 2011 Aug 19. (Erratum in *Proc. Natl. Acad. Sci. USA* 108(43):17856.)

Nissan, X., S. Blondel, and M. Peschanski. 2011. *In vitro* pathological modelling using patient-specific induced pluripotent stem cells: the case of progeria. *Biochem. Soc. Trans.* 39(6):1775–1779.

Tropel, P., J. Tournois, J. Côme, C. Varela, C. Moutou, P. Fragner, M. Cailleret, Y. Laâbi, M. Peschanski, S. Viville. 2010. High-efficiency derivation of human embryonic stem cell lines following pre-implantation genetic diagnosis. *In vitro Cell. Dev. Biol. Anim.* 46(3–4):376–85.

Institute of Biochemistry and Cell Biology, Shanghai Institutes for Biological Sciences

Site Address:	Chinese Academy of Sciences 320 Yue-yang Road, Shanghai 200031, China http://www.sibcb.ac.cn/eindex.asp
Date Visited:	November 17, 2011
WTEC Attendees:	P. Zandstra (report author), S. Demir, N. Moore, R.M. Nerem, S. Palecek, K. Ye, F. Huband
Host(s):	Professor Gang Wang Tel.: 86-21-5492-1083 Fax: 86-21-5492-1085 gwang22@sibs.ac.cn Professor Jinsong Li 86-21-5491-1422 Fax: 86-21-5491-1426 jsli@sibs.ac.cn Professor Guoliang Xu Tel.: 86-021-54921332 glxu@sibs.ac.cn

Overview

Shanghai Institute of Biochemistry and Cell Biology (SIBCB) is the largest institute of the Shanghai Institutes for Biological Sciences (SIBS), Chinese Academy of Sciences (CAS). It was established in 2000 through the merger of Shanghai Institute of Biochemistry (founded in 1958) and Shanghai Institute of Cell Biology (founded in 1950). Both of the former institutions had contributed to scientific advances over the last century including the total synthesis of crystalline bovine insulin, the total synthesis of yeast alanine tRNA, the artificial propagation of domestic freshwater fish, and artificial monogenesis of amphibian oocytes.

Research and Development Activities

The Epigenetics and Stem Cell Biology research clusters at the SIBCB are world-class. The institute has an increasing emphasis on publishing papers in the highest tier journals and having a significant impact internationally. As indicated in the

Selected References, that emphasis is working well, with stem cell related papers in *Molecular Cell, Science, Nature,* and *Cell Stem Cell* in the last 12 months alone. The SIBCB is primarily a fundamental biology institute and interactions with bioengineering are rare. During our visit to the SIBCB we met with Dr. Gang Wang, Li Jinsong and Guoliang Xu.

Dr. Gang Wang received his Ph.D. in Molecular and Cellular Biology from Tulane University in 1998. From 1999 to 2005, he was a postdoctoral fellow and then an Assistant Researcher in the Molecular Biology Institute at the University of California, Los Angeles. He was recruited to the Shanghai Institute of Biochemistry and Cell Biology by the Chinese Academy of Sciences "Hundred Talent Program" (see below) in 2006. Dr. Wang's research is in the area of molecular developmental biology, focusing on the mediator complex. This is a large complex of proteins (and perhaps other molecules) important for integrating signaling, transcription, and diverse biological processes. Dr. Wang is particularly interested in understanding the biological function of this complex and how they are regulated by developmental signaling pathways.

Dr. Jinsong Li received his Ph.D. in 2002 from the Institute of Zoology, Chinese Academy of Sciences and from 2002 to 2007 was a postdoctoral fellow at Rockefeller University. Dr. Li's research is focused on understanding the role of genetic and epigenetic alterations in induced pluripotent stem cell (iPSC) targeted and nuclear transfer (NT) based cell reprogramming. Dr. Li (with Dr. Xu, see below) recently identified a key role for the Tet3 DNA dioxygenase in epigenetic reprogramming by oocytes. This finding is important as it helps us understand differences between the iPSC and NT technologies.

Dr. Guoliang Xu received his Ph.D. in 1993 from the Max Planck Institute (MPI) for Molecular Genetics, Berlin, Germany. Between 1993 and 2001 he undertook postdoctoral fellowships in Germany and the United States, including at Columbia University. Dr. Xu's research is focused on mechanisms of the formation of genomic methylation patterns, especially DNA methyltransferases (Dnmts) in development and disease.

Dr. Naihe Jing received his Ph.D. in 1988 from Shanghai Institute of Biochemistry, Chinese Academy of Sciences and he was a postdoctoral fellow at the Institute of Physical and Chemical Research (RIKEN), Japan from 1989 to 1991. Dr. Jing's research is focused on neural development and neural stem cells, especially bone morphogenetic protein (BMP) signaling and its cross talk with other signaling pathways during CNS development.

Dr. Lijian Hui obtained his Ph.D. degree on cell biology from Shanghai Institute of Biochemistry and Cell Biology (SIBCB) in 2003. He had his postdoctoral training on mouse genetics at the Institute of Molecular Pathology, Vienna, Austria. After moving back to SIBCB as an independent principal investigator at the end of 2008, Dr. Hui continues to study the molecular alterations underlying cell transformation during liver cancer development. In addition, he has expanded his research interests to cell fate conversion. His lab lately demonstrated that fibroblasts can be converted into functional hepatocyte-like cells by overexpression of three transcription factors and inactivation of p19Arf.

Dr. Ping Hu received her Ph.D. in 2003 from the joint graduate program for State University of New York, Stony Brook and Cold Spring Harbor Laboratory, in the United States. From 2004 to 2010, she worked first as a postdoctoral fellow, then as a scientist at the University of California, Berkeley/Howard Hughes Medical Institute. She was recruited to SIBCB as a principal investigator in 2010. Dr. Hu's research focuses mainly on elucidation of the mechanism governing muscle stem cell fate determination, with an emphasis on transcription regulation networks in muscle stem cells and during the process of myogenic lineage commitment.

Dr. Yi Arial Zeng received her Ph.D. in 2005 from Simon Fraser University in Canada and from 2005 to 2010 was a postdoctoral fellow at Stanford University. Dr. Zeng's research focuses on understanding the molecular mechanism of how self-renewal is maintained in adult mammary stem cells and their interaction between the niches. She identified Wnts as the self-renewal factor for mammary stem cell self-renewal and established a mammary stem cell long-term culture and expansion system *in vitro*.

Dr. Ling-Ling Chen received her Ph.D. in 2009 from the University of Connecticut and from 2009 to 2011 was a postdoctoral fellow and assistant professor at the University of Connecticut Stem Cell Institute. Dr. Chen's research specialty is understanding the regulatory function of long, non-coding RNAs that are involved in nuclear architecture and the renewal of human embryonic stem cells.

Other members of the SIBCB who are involved in stem cell research can be found at http://www.sibcb.ac.cn/ep2-1-4.asp.

Building on the success of the "Signal Transduction" International Partnership project, SIBCB set up its Junior PI Mentor System in 2009, the first of its kind among Chinese research institutions. Since then, SIBCB has invited 18 renowned overseas Chinese scientists as mentors, to provide academic mentorship to the Institute's junior PIs and to promote academic exchange between junior PIs and the international academic community.

Admission to the training programs at the SIBCB is very competitive, with <5 % of applicants getting admission to the Ph.D. program. There are plans to hire 30 new faculty members in the next few years. The Hundred Talents program (similar to Thousand Young Talents program but specific to CAS) is an important tool to recruit scientists back from overseas/year to CAS. The program incentives include salary and start-up funds.

Translation

The SIBCB is involved in technology transfer to pharma companies. From 2009 to 2011, SIBCB scientists filed 102 patent applications (including 14 international patent applications), and were granted 54 patents (including 5 international patents). IP transfer is mainly via licensing. For example, in 2010 "Detection and modulation of Slit and Robo mediated angiogenesis and uses thereof," an SIBCB researchers' invention, was successfully transferred to Sanofi-Aventis with a contract totaling $60 million. Recently, the SIBCB's new finding that the Oct4-Vp16 fusion protein improves reprogramming efficiency has been licensed to Novartis.

Sources of Support

Institute revenue in 2011 is $28 million, with 60 % from competitive research grants and 40 % from basic running grants.

Collaboration Possibilities

To strengthen scientific exchanges, SIBCB has been actively promoting research collaborations in China and abroad. It has established more than 22 partnerships with international institutions, universities, industries, and organizations including the Max Planck Society of Germany, Asia-Pacific International Molecular Biology Network (A-IMBN), RIKEN of Japan, and the University of Toronto (Canada). The SIBCB has been collaborating with Nature Publishing Group in issuing the peer-reviewed international journal *Cell Research* (IF 9.5), which is considered a highly respected journal in the molecular cell biology field.

Selected References

Chen, L.L. and G.G. Carmichael. 2012. Nuclear editing of mRNA 3'-UTRs. *Curr. Top. Microbiol. Immunol.* 353:111–21.

Gu, T., F. Guo, H. Yang, H. Wu, G. Xu, W. Liu, Z. Xie, L. Shi, X. He, S. Jin, K. Iqbal, Y. Geno Shi, Z. Deng, P.E. Szabó, G.P. Pfeifer, J. Li, and G. Xu. 2011. The role of Tet3 DNA dioxygenase in epigenetic reprogramming by oocytes. *Nature* 477:606–610.

He, Y., B. Li, Z. Li, P. Liu, Y. Wang, Q. Tang, J. Ding, Y. Jia, Z. Chen, L. Li, Y. Sun, X. Li, Q. Dai, C. Song, K. Zhang, C. He, and G. Xu. 2011. Tet-mediated formation of 5-carboxylcytosine and its excision by TDG in mammalian DNA. Science 333:1303–1307.

Huang, P., Z. He, S. Ji, H. Sun, S. Xiang, C. Liu, Y. Hu, X. Wang, and L. Hui. 2011. Induction of functional hepatocyte-like cells from mouse fibroblasts by defined factors. *Nature* 475:386–389.

Huang, Y., W. Li, X. Yao, Q. Lin, J. Yin, Y. Liang, M. Heiner, B. Tian, J. Hui, and G. Wang. 2012. Mediator complex regulates alternative mRNA processing via the Med23 subunit. *Mol. Cell.* 45:459–469.

Lin, J., L. Shi, M. Zhang, H. Yang, Y. Qin, J. Zhang, D. Gong, X. Zhang, D. Li, and J. Li. 2011. Defects in trophoblast cell lineage account for the impaired *in vivo* development of cloned embryos generated by somatic nuclear transfer. *Cell Stem Cell* 8:371–375.

Xie, Z., Y. Chen, Z. Li, G. Bai, Y. Zhu, R. Yan, F. Tan, Y.-G. Chen, F. Guillemot, L. Li, and N. Jing. 2011. Smad6 promotes neuronal differentiation in the intermediate zone of the dorsal neural tube by inhibition of the Wnt/β-catenin pathway. *Proc. Natl. Acad. Sci. USA* 108:12119–12124.

Yang, L., M.O. Duff, B.R. Graveley, G.G. Carmichael, and L.L. Chen. 2011. Genomewide characterization of non-polyadenylated RNAs. *Genome Biol.* 12:R16.

Institute of Biophysics, Chinese Academy of Sciences

Site Address:	No.8 Beichendong Road, Chaoyang District Beijing, 100101 China http://english.ibp.cas.cn/
Date Visited:	November 13, 2011
WTEC Attendees:	S. Palecek (report author), S. Demir, K. Ye, F. Huband
Host(s):	Professor Yue Ma Tel.: 010-64888818 yuema@ibp.ac.cn http://sourcedb.cas.cn/sourcedb_ibp_cas/en/eibpexport/200904/ t20090403_45243.html Professor Guohong Li Tel.: 86-10-64888795 Mobile: 13651340251 Fax: 86-10-64856269 liguohong@sun5.ibp.ac.cn

Overview

The Institute of Biophysics, Chinese Academy of Sciences, was established in 1958 and its research focus is on molecular and cellular life sciences. Two National Key Laboratories, The National Laboratory of Biomacromolecules and The National Laboratory of Brain and Cognitive Sciences, are located at the Institute of Biophysics. Dr. Yue Ma is a Professor and Principal Investigator at the Institute

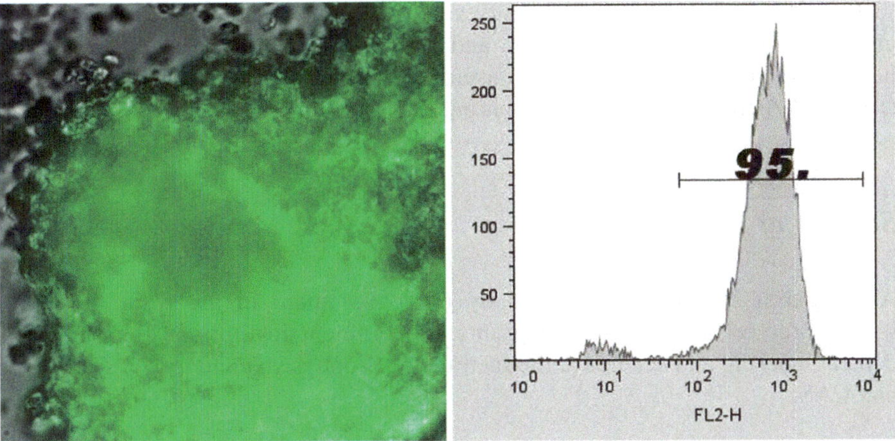

Fig. B.6 Promoter-based drug selection of cardiomyocytes differentiated from human embryonic stem cells (Courtesy of the Institute of Biophysics, Chinese Academy of Sciences)

of Biophysics. He obtained his Ph.D. from the University of Massachusetts, performed postdoctoral research in the laboratory of James Thomson at the University of Wisconsin-Madison, and worked as a researcher at the University of Washington prior to joining the Institute of Biophysics in 2006. Dr. Ma has extensive experience with culturing and differentiating human embryonic stem cells. Dr. Guohong Li is also a Professor and Principle Investigator at the Institute of Biophysics. He obtained his Ph.D. at the Max Planck Institute for Cell Biology and performed postdoctoral research in the United States prior to joining the Institute of Biophysics in 2009. Dr. Li and Dr. Ma are members of the Hundred Talents program of the Chinese Academy of Sciences.

Research and Development Activities

The Institute of Biophysics focuses on basic research. Prof. Ma researches mechanisms of cardiac differentiation of human embryonic stem cells (hESCs), including specification of atrial vs. ventricular cardiomyocytes. The goal of Dr. Ma's lab is to obtain cells suitable for drug screening applications and cellular therapies/cardiac tissue regeneration. Research activities associated with this goal include scaling hESC culture and differentiation of cardiomyocytes, elucidating regulatory mechanisms of cardiac differentiation, and cardiac tissue engineering (Fig. B.6). Dr. Ma's main accomplishment to date is identifying how retinoic acid signaling regulates atrial vs. ventricular cardiomyocyte specification.

Prof. Li's lab studies epigenetic events of hESC differentiation, including changes to chromatin organization and nuclear structure, the establishment of heterochromatin lining the nuclear envelope, and the associated functions of these structural changes in cell differentiation.

Sources of Support

All research funding is provided by the government in the form of multi-investigator or multi-institution grants. Government institutions funding stem cell engineering and science at the Institute of Biophysics include MOST, NSFC, and CAS.

Collaboration Possibilities

The Institute of Biophysics focuses on basic research. Thus, substantial opportunities for collaboration exist in translating stem cell research at the institute to commercial or clinical applications. In particular, development of hESC-derived cells for drug screening or heart tissue engineering could be performed in conjunction with Chinese or international collaborators.

Summary and Conclusions

Grand challenges in stem cell engineering include efficient protocols for generating functional cell types of interest. The work performed in Dr. Ma's lab facilitates generation of specialized cardiomyocyte subtypes, which will enable use of these cells in drug screening/testing and cardiac tissue engineering. Dr. Ma and Dr. Li expressed substantial interest in collaborations with engineers and clinical investigators, especially in hESC culture and differentiation scale-up and animal or clinical evaluation of tissues engineered from hESCs.

Selected References

Zhang, Q., J. Jiang, P. Han, Q. Yuan, J. Zhang, X. Zhang, Y. Xu, H. Cao, Q. Meng, L. Chen, T. Tian, X. Wang, P. Li, J. Hescheler, G. Ji, and Y. Ma, Y. 2011. Direct differentiation of atrial and ventricular myocytes from human embryonic stem cells by alternating retinoid signals. *Cell Research* 21:579–587.

Institute of Zoology, Chinese Academy of Sciences

Site Address:	C107-1, 1 Beichen West Road, Chaoyang District Beijing 100101 http://english.ioz.cas.cn/
Date Visited:	November 15, 2011
WTEC Attendees:	S. Palecek (report author), S. Demir, N. Moore, R.M. Nerem, P. Zandstra, K. Ye, F. Huband
Host(s):	Professor Baoyang Hu Key Lab. of Stem Cell & Developmental Biology Tel.: 86-10-64806251 byhu@ioz.ac.cn www.ioz.ac.cn Professor Jianwei Jiao Tel/Fax: +86-10-64806335 jwjiao@ioz.ac.cn Professor Chunsheng Han Tel/Fax: +86-10-64807105 hancs@ioz.ac.cn http://www.rpb.ioz.ac.cn/index_en.asp Professor Wan-Zhu Jin Tel/Fax: +86-10-64806302 jinw@ioz.ac.cn

Overview

The Institute of Zoology (IOZ), Chinese Academy of Sciences (CAS) is focused on addressing basic questions in biodiversity, ecology, agricultural biology, human health, and reproductive biology. Prof. Baoyang Hu is a Professor of Stem Cells and Regenerative Biology. He obtained his Ph.D. at Fudan University and performed postdoctoral research at the University of Wisconsin-Madison before joining IOZ in 2011 as part of the Hundred Talents program of CAS. Prof. Jianwei Jiao received his Ph.D. from Peking University and performed postdoctoral research at Harvard. He joined IOZ in 2011 as part of the Hundred Talents program. Prof. Chunsheng Han has been a Professor of Bioinformatics at IOZ since 2003. Prior to joining the IOZ he obtained his Ph.D. at the University of Missouri and worked as a bioinformatics scientist at Lexicon Genetics and Pharmaceuticals, Inc. Prof. Wan-Zhu Jin is the Group Leader of the Wild Animal Nutrition and Reproduction group in the

Key Laboratory of Animal Ecology and Conservation Biology. He performed postdoctoral research at the RIKEN Institute and Harvard Medical School prior to joining IOZ in 2010.

Research and Development Activities

The stem cell research at IOZ focuses on basic questions in stem cell biology that will enable clinical translation of stem cell therapies, including neural stem cell transplantation. Dr. Hu has worked on human iPSC differentiation to motoneurons and oligodendrocytes and now focuses on directing iPSCs to pan-neural lineages. He is interested in how mechanical properties of the culture substrate affect neural differentiation and is developing promoter-reporter biosensors to assess cell status. Dr. Hu's lab compares iPSC-derived cells and tissues with native tissues *in vivo*, with the goal of generating physiologically relevant cells and tissues from human stem cells. Dr. Jiao's lab is investigating effects of aging on neural stem cells, including how the stem cell niche affects the number and properties of neural stem cells. He utilizes mouse models of the neural stem cell niche and gene knockouts to dissect mechanisms regulating neural stem cell number and function.

Dr. Han's research program focuses on germ line stem cells, including murine spermatozoa stem cells. He is reprogramming spermatozoa stem cells to pluripotent stem cells. Dr. Jin's lab studies white-to-brown adipose tissue transdifferentiation using transgenic mice and cell culture systems.

The IOZ maintains one of the China stem cell banks. This bank has disease-specific iPSC lines. Dr. Jiao runs a workshop to train researchers in culturing stem cells.

Principle investigators at the CAS can train Ph.D. students from universities around China. The quality of postdocs available to CAS researchers is very good. Most students and postdocs who train in the stem cell field at IOZ are interested in academic careers.

Translation

The IOZ is focused on basic research questions with potential clinical impact. CAS has a patent office to protect intellectual property arising from research at IOZ, and will provide assistance in starting a company.

Sources of Support

The majority of research funding is from competitive applications to MOST. The funding climate in this area is strong.

Collaboration Possibilities

The hosts expressed an interest and potential benefit in collaborating with engineers since there are no engineering principal investigators at IOZ. A collaboration with a biomaterials expert at the Institute of Developmental Biology has been established. The IOZ stem cell community holds meetings and seminars with researchers at Tsinghua University and Peking University. Additional collaborations on stem cell applications and clinical translation would be beneficial.

Summary and Conclusions

There is substantial basic stem cell research expertise at IOZ that has the potential to help drive stem cell translation in China. The mission of the IOZ stem cell community would benefit from interactions with engineering and clinical translation groups in China and around the world. The stem cell community at IOZ is relatively new. Availability of funding and stability of regulation of stem cells has permitted IOZ to recruit top stem cell scientists.

Selected References

Meng, F., S. Chen, Q. Miao, K. Zhou, Q. Lao, X. Zhang, W. Guo, and J. Jiao. 2011. Induction of fibroblasts to neurons through adenoviral gene delivery. *Cell Research* (in press).

Shi, Y.Q., S.Y. Liao, X.J. Zhuang, and C.S. Han. 2011. Mouse Fem1b interacts with and induces ubiquitin-mediated degradation of Ankrd37. *Gene* 485:153–159.

Karolinska Institute and Karolinska University Hospitals

Site Address:	Meeting with Karolinska Institute (KI) Investigators at the Hotel Radisson Blu Sky City SE-190 45 Stockholm-Arlanda, Sweden http://ki.se/ki/jsp/polopoly.jsp?d=9292&l=en http://ki.se/ki/jsp/polopoly.jsp?d=39655&l=en
Date Visited:	March 2, 2012
WTEC Attendees:	T. McDevitt (report author), P. Zandstra, D. Schaffer, N. Moore, H. Sarin

(continued)

Host(s):

Prof. Dr. Matthias Lohr
Department of Clinical Science, Intervention and Technology
  (CLINTEC)
Karolinska Institute, Karolinska University Hospital
Huddinge, SE-144 86
Stockholm, Sweden
Tel.: +46 08-585 895 91
matthias.lohr@ki.se

Prof. Ana Teixeira
Department of Cell and Molecular Biology (CMB), C5
Karolinska Institute
von Eulers vag 3 SE-171 77
Stockholm, Sweden
Tel.: +46 08-524 879 79
Ana.Teixeira@ki.se

Overview

The Karolinska Institute (KI) has research campuses at Solna and Huddinge, which are also affiliated with Karolinska University Hospitals. The Department of Clinical Science, Intervention, and Technology (CLINTEC) at the Karolinska University Hospital in Huddinge has divisions that represent all organ systems. The research at CLINTEC is translational and clinical, being conducted mainly in collaboration with clinics and associated laboratories at the Karolinska University Hospital. The KI conducts its research activities in partnership with Stockholm University, the Royal Institute of Technology (KTH), and the Swedish School of Sport and Health Sciences (GIH). The KI is also home to the Advanced Center for Translational Regenerative Medicine (ACTREG) created in 2011, which is a consortium of scientists whose purpose is to investigate the mechanisms of tissue formation and repair directed by Dr. Paolo Macchiarini (Director) and Dr. Christer Sylvén (Deputy Director). This group performed the world's first transplantation of a stem cell-based tissue engineered organ, the trachea, into a patient, and has continued to improve this method and apply the approach to other organs of more complex architecture. Research at KI is financed primarily by the European Union's Framework Programmes 6 and 7 (FP6, FP7) as well as by the European Union's Public Health Programme. The KI is a participating partner in several of the projects under the FP7 Innovative Medicines Initiative (IMI), which is public-private combined venture with the pharmaceutical industry.

Research and Development Activities

Prof. Dr. Lohr discussed a recent effort of the Karolinska University Hospital in which a Chief Officer of Innovation and Development has recently been appointed and is emphasizing innovation in translational medicine, with the twin topics of individualized medicine and regenerative medicine (including stem cells). The funding for this effort is currently coming from the county, but they are also applying for national government funding. In addition, they are integrating people from technology transfer, business development, and institutional review boards. They are also discussing reimbursement mechanisms. In other words, this effort is increasingly becoming top-down. He emphasizes that there are few boundaries between departments at the KI, and in addition that they have strong collaborations with other institutions. For example, they collaborate with engineers at the KTH on the development of materials and nanoparticles facilitate regenerative medicine and stem cell research. Furthermore, they collaborate with Tokyo Women's Medical University on cell sheets and are interested in scaling up this process to clinical scale.

Prof. Teixeira has joined KI relatively recently. She received her Ph.D. with Prof. Paul Neely at the University of Wisconsin-Madison and conducted a postdoctoral fellowship working on neural stem cells with Prof. Ola Hermanson of the KI. Her current research integrates nano- and microtechnologies in stem cell research, central nervous system development, and cancer. As a recent example, she synthesized a redox sensitive polymer poly(3,4-ethylenedioxythiophene) (PEDOT) and complexed it with heparin (Herland et al. 2011). In the reduced state, the material is neutral, and the heparin presents exposed heparin-binding growth factors such as FGF-2 or FGF-8. However, oxidation results in a positively charged material that more tightly associates with the heparin and as a result masks and reduces the bioavailability of the growth factors. In one application, she used embryonic NSCs from E15 rat brain, which are FGF-2 dependent, and found that polymer presenting the FGF aided self-renewal; however, oxidizing the polymer induced the onset of cell differentiation. In other work, she has been exploring the effect of microenvironmental mechanical properties on NSC function.

Translation

The tight interactions between universities and affiliated hospitals are enabling translation, and engineers are playing a role. For example, Prof. Paolo Macchiarini of the KI led the first effort to transplant a tissue engineered trachea, based on a scaffold (developed at University College London) seeded with the patient's bone marrow derived mononuclear cells (Jungebluth et al. 2011). Alternative biomaterial scaffolds are now being investigated.

Sources of Support

The investigators stated that funding for stem cells has to date primarily been due to bottom-up initiatives, but top-down programs have recently emerged. One is Science for Life Laboratory (www.scilifelab.se), a collaborative venture that is funded by two strategic grants from the Swedish government and involves KTH, KI, UI, and Stockholm University. This initiative provides technological platforms and a scientific community. In addition, the investigators state that like the rest of Europe, investigator-initiated clinical trials are problematic for funding. The Swedish government is examining this challenge, and the next research bill may include funding for this.

Collaboration Possibilities

It is very clear that there is extensive collaboration among the Karolinska Institute, Uppsala University, the Royal Institute of Technology, the University of Stockholm, and Lund University that brings together expertise in biology, engineering, and medicine. Additional opportunities for collaborations in engineering areas such as mathematical modeling and materials science (to interact with their existing strength in this area) may exist.

Summary and Conclusions

The Karolinska Institute has a strong tradition of stem cell biology and translational research (especially in neuronal cell replacement therapy), and they have a growing effort in stem cell engineering. In addition, they are part of a strong network that is melding the country's expertise in stem cell biology, medicine, and to a growing extent stem cell engineering.

Selected References

Herland, A., K.M. Persson, V. Lundin, M. Fahlman, M. Berggren, E.W. Jager, and A.I. Teixeira. 2011. Electrochemical control of growth factor presentation to steer neural stem cell differentiation. *Angew. Chem. Int. Ed. Engl.* 50:12529–12533, doi:10.1002/anie.201103728.

Jungebluth, P., E. Alici, S. Baiguera, K. Le Blanc, P. Blomberg, B. Bozóky, C. Crowley, O. Einarsson, K.-H. Grinnemo, T. Gudbjartsson, S. Le Guyader, G. Henriksson, O. Hermanson, J.E. Juto, B. Leidner, T. Lilja, J. Liskal, T. Luedde, V. Lundin, G. Moll, B. Nilsson, C. Roderburg, S. Strömblad, T. Sutlu, A.I. Teixeira, E. Watz, A. Seifalian, and P. Macchiarini. 2011. Tracheobronchial transplantation with a stem-cell-seeded bioartificial nanocomposite: a proof-of-concept study. *Lancet* 378:1997–2004, doi:10.1016/S0140-6736(11)61715-7.

Keio University, Yagami Campus

Site Address:	Miyata Laboratory, Biophysical & Tissue Engineering
	3-14-1 Hiyoshi, Kohoku-ku, Yokohama 223-8522
	http://www.keio.ac.jp/index-en.html
Date Visited:	November 17, 2011
WTEC Attendees:	T. McDevitt (report author), J. Loring, D. Schaffer, N. Kuhn, H. Ali, T. Satoh
Host(s):	Asst. Professor Shogo Miyata
	Tel.: 81-45-566-1827
	Fax: 81-45-566-1827
	miyata@mech.keio.ac.jp
	http://www.miyata.mech.keio.ac.jp/member.html

Overview

The focus of the Miyata laboratory is on the development of microscale technologies in combination with dielectrophoresis principles to be able to sort cells from heterogeneous populations and assemble cells onto microbeads in a controlled manner (Fig. B.7). Dielectrophoresis (DEP) induces a net force on a polarizable particle (such as cells) in a spatially nonuniform electric field; the strength and direction of DEP forces depend on the electrical properties of the fluid and the particle. In the case of cells, electrical properties of plasma membranes can differ according to cell type or phenotype. In Japan, it is rare for someone in a mechanical engineering department, like Dr. Miyata, to be conducting tissue engineering research.

Research and Development Activities

In the first line of research presented, Dr. Miyata presented work on how the application of DEP was being used to separate different types of cell populations from one another. DEP operates on the principle that a net force can be imposed on a polarizable particle introduced into a spatially nonuniform electric field; the strength and direction of the force are dependent on the electrical properties of the particle as well as the surrounding fluid. The inherent electrical properties of cell membranes vary according to different cell types and phenotypic changes. In the case of stem cell cultures, this could be used to separate feeder cells from stem

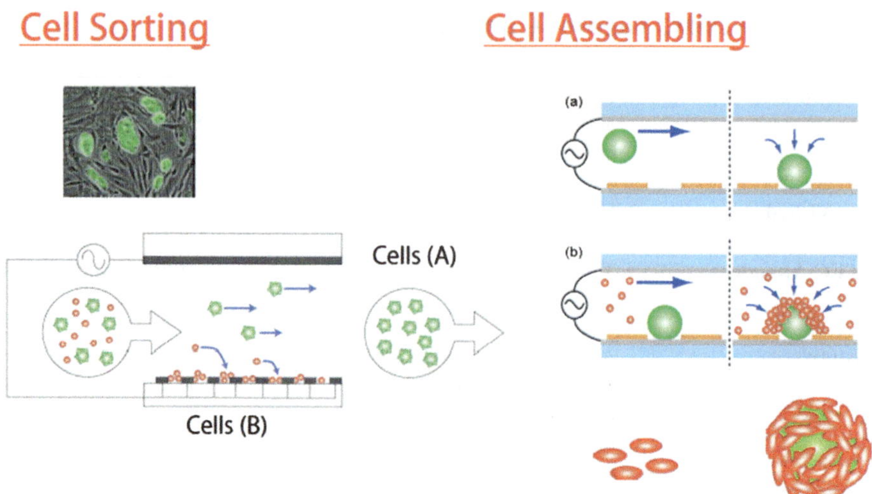

Fig. B.7 Overview of microscale technologies. Dielectrophoresis principles can be used in combination with microfluidics for cell sorting and in combination with microfluidics and microparticles for assembling multicellular constructs (Courtesy of S. Miyata, Keio University)

cells and also to potentially distinguish and purify or enrich specific differentiated cell populations. A DEP cell sorting device has been generated by creating a microfabricated cavity with insulated film within a transparent conducting glass environment and layers separated by a silicone rubber gasket. Cells are introduced in a low-conductivity buffer and AC voltages of varying frequency are applied; at low frequencies, cells move between electrodes (negative DEP), whereas at high frequencies they move toward electrodes (positive DEP). Currently, proof-of-principle studies have been carried out with differentiated (primary isolated) and dedifferentiated (cultured) chondrocytes.

As a second application of DEP, cell adhesion was being directed to microbeads as a way to direct multicellular assembly. The microbeads (50–80 μm diameter) are made of collagen-alginate; collagen is included to support cell adhesion. The microbeads are individually attracted to a pattern of microelectrodes (50 μm diameter) created within a microfabricated device using similar materials as those in the DEP cell sorting technology. The microbeads are attracted to the electrodes first before flowing cells through the chamber and allowing them to accumulate on the beads in a controlled manner. Cells can then be subsequently cultured on the microbeads for at least up to 7 days, during which time they grow to completely cover the bead and exhibit high viability, based on calcein staining. This technology is envisioned as a useful platform to generate multilayered cell constructs that could be used in high-throughput drug screening assays, such as for skin cells.

Translation

At an annual conference at Keio University, Dr. Miyata presented his research and was approached by a company to begin collaborating with him on the cell separation technology. Dr. Miyata has applied for Japanese and U.S. patents and the company has licensed the patents from the university. The university's technology transfer office handles all of the logistics in working with the company.

Collaboration Possibilities

The Keio University medical school is very famous and one of the four main places in Japan for stem cell research, but the medical school campus is far from Dr. Miyata's lab, making collaborations more difficult. Dr. Miyata is planning to work with pluripotent stem cells and intends to get them from the RIKEN cell bank. Traditionally, there are high walls between the medical school and engineering environments, due largely to historical cultural differences, which limit collaborations. Dr. Miyata commented how he was surprised to find both M.D.s and engineers attending conferences together held in the United States. Ten years ago, the Japanese government encouraged collaborations between physicians and engineers with funding from JST or the Ministry of Education, Culture, Sports, Science & Technology (MEXT), which made some improvements.

Summary and Conclusions

Dr. Miyata is engineering strategies to both isolate single cells from heterogeneous populations and construct multilayered cell constructs using principles of DEP in microfabricated devices. Although not doing much work yet with stem cells, both of these technologies have clear applications in stem cell engineering. His primary interest in expanding current work to include stem cells is for cell chip technologies for drug testing applications.

Selected References

Miyata, S., and Y. Sugimoto. Control of cellular organization around collagen beads using dielectrophoresis. *Intel. Autom. Soft Comp.*, in press.

Kyoto University-CiRA (Center for iPS Cell Research and Application)

Site Address:	53 Shogoin-Kawaracho, Sakyo-ku, Kyoto, 606-8507, Japan http://www.cira.kyoto-u.ac.jp/e/
Date Visited:	November 15, 2011
WTEC Attendees:	T. McDevitt, J. Loring (report author), D. Schaffer, L. Nagahara, N. Kuhn, H. Ali, M. Imaizumi
Host(s):	Dr. Takafumi Kimura Tel.: 81-75-366-7051 Fax: +81-75-366-7070 kimura-g@cira.kyoto-u.ac.jp http://www.cira.kyoto-u.ac.jp/e/research/kimura_summary.html Dr. Akitsu Hotta Tel.: 81-75-366-7051 Fax: +81-75-366-7070 hotta-g@cira.kyoto-u.ac.jp http://www.cira.kyoto-u.ac.jp/e/research/hotta_summary.html Saki Tamura Tel.: 81-75-366-7005 saki.tamura@cira.kyoto-u.ac.jp

Overview

CiRA is a well-funded new institute that was founded to take advantage of the intellectual property surrounding induced pluripotent stem cells (iPSCs) developed by Dr. Shinya Yamanaka and his colleagues at Kyoto University. It has several hundred employees and Dr. Yamanaka is the overall director. The center has multiple projects centered on iPSCs, including the development of a haplotype-matched collection of iPSCs that will match the majority of the Japanese population. These cells will be used to develop cell replacement therapies that minimize the need for immunosuppression.

CiRA is a new Institute focused on iPSC technology. Dr. Shinya Yamanaka is the Director of the Institute. The mission of CiRA is to be Japan's core institute for iPS cell research and applications and to promote basic, preclinical, and clinical studies and to train young scientists.

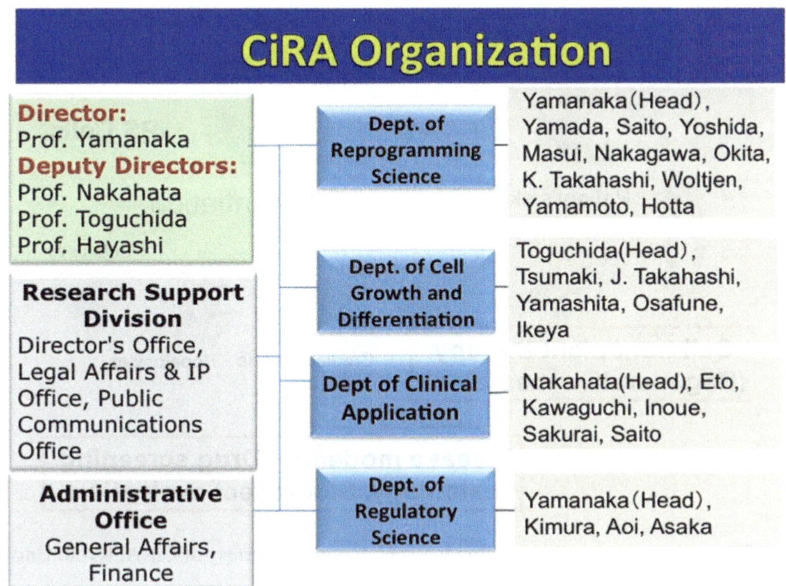

Fig. B.8 Leading researchers and general organization of CiRA (Courtesy of CiRA, Kyoto University)

There are four departments (Fig. B.8), each with a different, complementary mission.

- Department of Reprogramming Science: development of improved methods for producing safe effective reprogramming methods
- Department of Cell Growth and Differentiation: development of methods for inducing cells toward specific fates. Conduct preclinical studies
- Department of Clinical Application: study of disease etiology and mechanisms of pathology
- Department of Regulatory Science: study issues in regulation concerning iPS cells; oversees operations of the GMP facility (Facility for iPS Cell Therapy: FiT)

They have an animal facility that includes mice, rats, dogs, and monkeys.

Research and Development Activities

Regulatory Sciences (T. Kimura) and Reprogramming Sciences (A. Hotta)

Skin punch biopsies are the usual method for obtaining adult somatic cells (Fig. B.9). For cell therapy they plan to produce a bank of GMP-compliant iPSCs that have sufficient haplotype representation that they can be used for allogenic transplants.

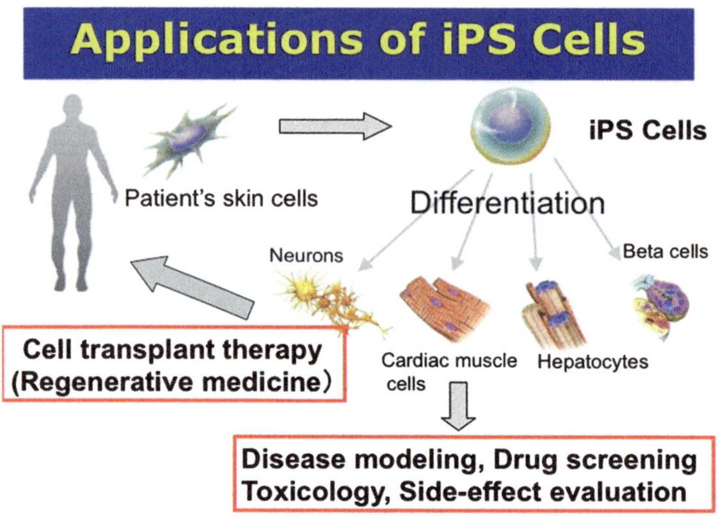

Fig. B.9 Generation of iPS cells for individualized therapy (Courtesy of CiRA, Kyoto University)

Autologous transplants are impractical for most purposes, since it takes 2 months to produce iPSCs, and then 1–2 more months to generate progenitor cells for therapy.

Specific Project: iPSC Therapy for Hemophilia (A. Hotta)

There are 4,000–5,000 hemophilia patients in Japan who have a genetic mutation in clotting factor VIII. The plan is to use iPSCs for gene therapy, using HLA-matched iPSC-derived cells to be factor VIII producing factories in the patient (Fig. B.10). Endogenous factor VIII is produced in the liver but they don't necessarily need to differentiate the iPSCs into "liver-like" cells. Any non-proliferating differentiated cell type that can produce factor VIII would suffice.

New vectors are being tested to improve transgene expression. One is the piggy-Bac (transposon) vector, which has a much higher capacity than retroviral/lentiviral vectors. To prevent silencing of the transgenes, they are using an "insulator" (D4Z4), which allows transgene activity up to 80 days posttransfection whereas without the insulator control, they see a decrease by 30 days posttransfection.

Translation

CiRA is improving methods for using iPSCs for multiple purposes. Their departmental structure (Reprogramming, Differentiation, Clinical, Regulatory) is directed toward clinical applications, including cell therapy.

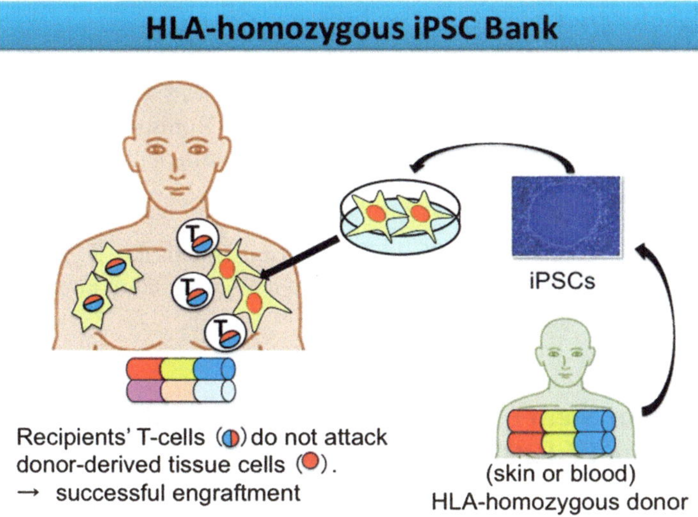

Fig. B.10 Use of HLA-matched iPS cells to support factor VIII production (Courtesy of CiRA, Kyoto University)

Sources of Support

Japan has invested a great deal in iPSC technology, and the main recipient is the new CiRA institute. In 3 years it has grown to 300 members. Dr. Yamanaka is responsible for allocating the funding for individual researchers at CiRA. Young scientists receive start-up funds from CiRA, and are encouraged to apply for external funds in addition to what they receive at CiRA. A company, iPS Academia Japan, has been established to handle licensing of intellectual property. There is considerable Government oversight. For example, there are two face-to-face meetings per year for Akitsu Hotta (10 M yen/year for 3 years). In general, government officials visit 2–3 times per month from different departments to learn the research overview of CiRA and to take a tour of the facilities.

Summary and Conclusions

CiRA is a new government-sponsored institute dedicated to iPSC technology and application, and is associated with Kyoto University. Their departmental structure (Reprogramming, Differentiation, Clinical, Regulatory) is directed toward clinical applications, including cell therapy. With 300 members, this is the largest dedicated stem cell institute in Japan. The institute covers all of the major areas of research and clinical application for iPSC technology. There is considerable government oversight and effort invested in commercializing the developments of the institute.

Selected References

Okita, K., Y. Matsumura, Y. Sato, A. Okada, A. Morizane, S. Okamoto, H. Hong, M. Nakagawa, K. Tanabe, K.I. Tezuka, T. Shibata, T. Kunisada, M. Takahashi, J. Takahashi, H. Saji, and S. Yamanaka. 2011. A more efficient method to generate integration-free human iPS cells. *Nature Methods* 8:409–412, doi:10.1038/nmeth.1591.

Takahashi, K., K. Tanabe, M. Ohnuki, M. Narita, T. Ichisaka, K. Tomoda, and S. Yamanaka. 2007. Induction of pluripotent stem cells from adult human fibroblasts by defined factors. *Cell* 131:861–872.

Laboratory of Stem Cell Bioengineering (LSCB), École Polytechnique Fédérale de Lausanne (EPFL)

Site Address:	Station 15, Building AI 1109 CH-1015 Lausanne, Switzerland http://www.epfl.ch/
Date Visited:	February 28, 2012
WTEC Attendees:	T. McDevitt (report author), D. Schaffer, N. Moore, H. Sarin
Host(s):	Prof. Matthias Lutolf, Laboratory of Stem Cell Engineering Tel.: +41 21 69 31876 matthias.lutolf@epfl.ch
	Prof. Sebastian Maerkl, Laboratory of Biological Network Characterization Tel.: +41 21 69 37835, 31161 sebastian.maerkl@epfl.ch
	Dr. Rajwinder Lehal, Radthke Lab. +41 21 69 30774 rajwinder.lehal@epfl.ch
	Prof. Philippe Renaud, Microsystems Laboratory Tel.: +41 21 69 32596, 36797 philippe.renaud@epfl.ch
	Prof. Bart Deplancke, Laboratory of Systems Biology and Genetics Tel.: +41 21 69 31821, 30982 bart.deplancke@epfl.ch
	Yann Barrondon, Laboratory of Stem Cell Dynamics Tel.: +41 21 69 39491, 31633 yann.barrandon@epfl.ch

Overview

The École Polytechnique Fédérale de Lausanne (EPFL) is composed of several departments, which include the School of Life Sciences and the School of Engineering. In addition, there are more than a dozen of strategic transdisciplinary research institutes at the EPFL, of which, the Interfaculty Institute of Bioengineering (IBI) is one. Led by Professor Jeffrey Hubbell, the IBI interfaces the Schools of Life Sciences and Engineering in order to answer questions in the basic life sciences through the application of engineering methodologies. The EPFL IBI Laboratory of Stem Cell Bioengineering (LSCB) is interested in elucidating how microenvironmental signals control the behavior of adult stem cells, particularly those of hematopoietic and neural origins. Towards this goal, the LSCB is developing innovative bioengineering strategies to learn about the *in vivo* stem cell niches and reconstruct them *in vitro*; it is the laboratory's expectation that this knowledge with be translatable to the clinical setting.

Research and Development Activities

Microfluidic High-Throughput Screening

The three focuses of Prof. Sebastian Maerkl's research program are systems biology, synthetic biology, and health. Much of his work involves live cell imaging and high-throughput protein biochemistry. He has developed high-density arrays of valves to enable single-cell analysis such that he can perform multiplex analysis for up to 4 biomarkers with comparable sensitivity to ELISAs and much smaller volumes on 384 samples in less than 10 cm^2 of space, thus reducing the overall cost of analysis by almost 103. This technology has been used to perform adjuvant screening in less than 2 weeks on dendritic cells subjected to more than 500 different conditions. In addition, Prof. Maerkl is interested in quantifying the strength of protein interactions by examining the affinity and kinetics of protein interactions. Again using microvalve technology, repeated cycling of opening and closing combined with washes can be used to look at brief association and dissociation interactions. This has been used, for example, to quantify the affinity of hundreds of transcription factors for regions of DNA and therefore could be used to screen promoter and enhancer sequences simply by flowing across a protein array.

High-Throughput Screening of Compounds Affecting Notch Signaling

Dr. Raj Lehal works in a mouse genetics lab that focuses primarily on self-renewing tissues and cancer, so the group consists mainly of molecular biologists. Dr. Lehal's research focuses on Notch as a target for drug development because of its involvement in so many different cell processes. Currently, gamma-secretase inhibitors are the primary target for most therapies, but they are not specific to just the Notch signaling pathway. Using a luciferase reporter screening system for DL4, they were able to obtain 3 hits from a library of 50,000 snake venom peptides and 5 hits from a library of 17,000 chemical compounds. One of the hits they found and have characterized is "inhibitor 3" (I3), which is a novel inhibitor of the Notch pathway. I3 is 1,000 times cheaper than gamma-secretase inhibitors (GSIs) with an IC50 value of 1 μM, and it does not act like a GSI. I3 does not block nuclear colocalization, but instead inhibits Notch signaling, induces growth arrest of Notch-dependent tumor cell lines and induces differentiation of skeletal myoblasts (C2C12).

BioMEME for 3D Cell Culture and Tissue Engineering

Ms. Anja Kunze, a Ph.D. student working with Prof. Philippe Renaud, presented research focused on the cortical neural cell niche, and attempts to mimic the native layered structure of the cortical anatomy, including the different organization and populations of cells. Her work combines microfluidics with planar or liquid electrodes and mechanical structures to create the *in vitro* engineered tissues. Stacked layers can be oriented alongside one another and coupled with microfluidics to have both laminar flow regions and perfusion channels. Agarose gels with porosity on the order of hundreds of micrometers enable neurite extensions into the gels and hydrostatically driven flow from reservoirs can establish linear gradients and avoid having to connect tubing to the gels to direct flow.

Systems-Based Quantitative Analysis

The work of Prof. Deplancke focuses on "integrative genomics" to decipher gene regulatory networks using quantitative analysis in order to understand the binding of transcription factors (TFs) to the genome in the context of cofactors. Currently, his laboratory has 900 mouse TFs out of an estimated total of 1,500. Most cofactor analysis focuses on coactivators, but they were interested in co-repressors because most knockout mice for repressors are lethal. Their particular focus is on NCoR2/SMRT repressors, which are large proteins that contain largely undefined domains and when knocked down in pre-adipocytes (3T3-L1), differentiation to adipocytes is enhanced and accelerated. The researchers found that a zinc finger protein, KAISO (ZBTB33), which can bind to methylated motifs, mediates the DNA tethering of SMRT and "primes" the site for transcription. Knockdown of KAISO, similar to SMRT, accelerates and enhances adipogenic differentiation, suggesting that it may be required to transiently increase proliferation before differentiation.

Engineering Artificial Niches

The focus of Prof. Matthias Lutolf's laboratory is on the engineering of *in vitro* environments to recapitulate the cross talk of stem cell niches in order to better understand the niche environment. Using a PEG-based chemically cross-linked hydrogel platform, they imprint PEG microwell arrays to capture individual cells in individual wells and in combination with a conventional DNA spotter can create arrays with different biochemical environments and also possibly different mechanic stiffness (via PEG chemistries; Gobaa et al. 2011). The workers in Prof. Lutolf's lab have developed the means to screen 3D stem cell niches with the use of robotic automatic handling to establish complex combinatorial 3D environments containing cells that can then be imaged and analyzed in the microwell format (i.e., conventional plate/array readers). Culturing Oct4-GFP mouse embryonic stem cells (mESCs) in 3D conditions and using BD Pathway for 3D HTP image analysis, they found that soluble factors had the strongest effect on maintaining Oct4 expression, but in the absence of soluble factors, mechanics of the environment played more of a role. The use of systems biology tools to examine the data provides a way to get a relative quantitative analysis of the role of the different factors comprising the niche environment. Dr. Lutolf also presented work on a single-cell microfluidic trap technology that could be used to perform *in vitro* mapping of single hematopoietic stem cells "on a chip" (Kobel et al. 2012). This system, in combination with single-cell PCR analysis and microvalve technologies can examine individual cells in isolation from one another and the daughter cells produced by asymmetric divisions.

Translation

The start-up environment in and around Lake Geneva is particularly strong due to a relatively large amount of venture capital in the area. Venture capitalists actually come to the campus often looking for technologies to commercialize. Many of the Ph.D. students go to work at related start-ups in the area upon the completion of their degrees, and many of the start-ups are doing well.

Sources of Support

Research funding is largely provided by the Swiss government to individual investigators and to collaborative networks of researchers throughout Switzerland. The EPFL IBI LSCB is funded by several sources including the European Science Foundation (ESF), the Swiss National Science Foundation (SNFS), the Swiss Initiative in Systems Biology (www.systemsx.ch), the European Union Framework Programme 7 (FP7), and the Commission for Technology and Innovation (CTI). Additional funding for some projects comes from SystemsX and Nano-tera.ch (www.nano-tera.ch) EPFL investigators were equipped with many state-of-the-art technologies and modern biological equipment. Principal investigators receive an annual budget but also apply for additional grant funding opportunities.

Collaboration Possibilities

Since its inception in the 1990s, the EPFL has welcomed industrial collaboration on campus. Initially, this laboratory space at the heart of EPFL was known as Science Park (PSE); since 2010, Innovation Square has been the venue at EPFL for collaboration between public and private sectors as it houses 12 buildings to provide start-ups and major companies in the fields of information technology, biotechnology and telecommunications with an optimal scientific environment for growth. A number of collaborations were evident among the investigators the WTEC panel met at EPFL. Many of the collaborations were based on the combination of different technologies to create and analyze higher content biological systems. Collaborations with many external investigators from Europe, Asia, and North America were noted.

Summary and Conclusions

Using high-throughput screening approaches and systems analysis techniques, EPFL researchers are leading the development of several new and innovative technologies to assess the effects of extracellular environments and intracellular signaling pathway on cell fate decisions. Various collaborative activities exist between individual EPFL investigators with different sets of scientific and technological expertise to address these questions. Strong research funding from the government enables investigators to boldly pursue challenging problems and a local community of venture capital supports translation efforts of new technologies.

Selected References

Gobaa, S., S. Hoehnel, M. Roccio, A. Negro, S. Kobel, and M.P. Lutolf MP. 2011. Artificial niche microarrays for probing single stem cell fate in high throughput. *Nat. Methods* 8(11):949–955.

Kobel, S.A., O. Burri, A. Griffa, M. Girotra, A. Seitz, and M.P. Lutolf. 2012. Automated analysis of single stem cells in microfluidic traps. *Lab Chip* 12(16):2843–2849.

Leiden University Medical Center

Site Address:	Einthovenweg 20, Zone S1-P
	PO Box 9600, 2300 RC Leiden, The Netherlands
	http://www.lumc.nl/home/?setlanguage=english

(continued)

Date Visited:	March 2, 2012

WTEC Attendees: R.M. Nerem, J. Loring (report author), S. Palecek, L. Nagahara, H. Ali

Host(s):   Professor Christine Mummery
Professor of Developmental Biology and Chair, Dept. of Anatomy
    & Embryology
Tel.: +31-71-526 9307
Fax: +31-71-526 8289
C.L.Mummery@lumc.nl

Dr. Milena Bellin
Researcher, Anatomy & Embryology Department
Tel.: +31-71-526 9382
M.Bellin@lumc.nl

Dr. Stefan Braam
Chief Scientific Officer, Pluriomics
Tel.: +31-71-526 9585
Stefan.braam@pluriomics.com

Dr. Robert Passier
Associate Professor, Anatomy & Embryology Department
Tel.: +31-71-526-9359
P.C.J.J.Passier@lumc.nl

Dr. Christian Freund
Senior Researcher, Anatomy & Embryology Department
Tel.: +31-71-526 9351
c.m.a.h.freund@lumc.nl

Dr. Siebe Spijker
M.D. Ph.D. student, Department of Kidney Disease
Tel.: +31-71-526 2214 (2148)
Fax: +31-71-526 6868
H.S.Spijker@lumc.nl

Dr. Marten Engelse
Coordinator, Beta Islet Isolation Department of Kidney Disease
Tel.: +31-71-526 6855
Fax: +31-71-526 6868
M.A.Engelse@lumc.nl

Dr. Valeria Orlova
Researcher, Department of Molecular Cell Biology
Tel.: +31-71-526 9265
v.orlova@lumc.nl

Overview

The Leiden University Medical Center (Dutch: Leids Universitair Medisch Centrum) or LUMC, is the university hospital affiliated with Leiden University, of which it forms the medical faculty. It resulted from the merger of the Academisch Ziekenhuis Leiden (Leiden Academic Hospital) and the medical faculty of Leiden University in the late 1990s, forming an academic health science center.

The Department of Anatomy and Embryology is located adjacent to the main hospital. The head of the Department is Professor Christine Mummery, who specializes in stem cell research and differentiation to the cardiovascular lineage. Professor Marco de Ruiter is responsible for teaching and research in Clinical Anatomy with e-learning as a special focus.

The department has two main activities:

- Teaching gross anatomy and embryology to medical students and providing specific courses to postgraduate and postdoctoral medical specialists in training.
- Research on development and function of the normal and abnormal cardiovascular system and mechanisms underlying pluripotency in germ cells. Mouse, chicken and stem cells are used as experimental model systems and where appropriate, results are extrapolated to human development and disease.

In addition, with the Department of Molecular Cell Biology, the department is responsible for the human induced pluripotent cell (hiPSC) core facility of the LUMC.

Research and Development Activities

Research proceeds along three main themes:

- Differentiation of pluripotent stem cells to cardiomyocytes and vascular cells
- Generation and characterization of patient-derived induced pluripotent stem cells
- Development of cardiovascular disease models in mouse and humans

Dr. Mummery's group uses human and mouse embryonic stem cells to study directed differentiation to cardiomyocytes, vascular endothelial cells, and vascular smooth muscle cells. Using defined growth conditions and genetic modification of the human cells, they mark specific cell lineages by selectable fluorescent reporter genes to derive pure populations of differentiated cells. This allows each step in the differentiation process to be optimized using growth factors and small molecules so that more than half of the cells in the culture are either beating heart cells, endothelial cells or smooth muscle cells.

They also derive human and mouse induced pluripotent stem (iPSCs) from mutant mice and patients with genetic cardiac and vascular diseases. This is done in part with the LUMC human iPSC core facility, jointly run with the Department of Molecular Cell Biology.

The vascular disease of interest is hereditary hemorrhagic telangiectasia (HHT), caused by mutations in receptors for transforming growth factor β. Mice with deletions in any components of the TGF-β signaling pathway show defective vasculogenesis in the developing yolk sac at mid-gestation.

Patients with HHT have weak walled vessels that hemorrhage easily, resulting in chronic nose bleeds. The group is creating disease models based on iPSCs derived from these patients to study development of the disease and methods to treat it. Working with clinicians who gave HHT patients thalidomide to help reduce their nosebleeds, they discovered through studies in stem cells and mutant mice that weak vessels are stabilized by thalidomide treatment because the association between the two cell types that make up the vessel wall is enhanced. The vessels thus become strengthened.

The cardiac diseases investigated are primarily channelopathies, caused by mutations in ion channels. Cardiomyocytes from human pluripotent stem cells beat spontaneously. Their electrical properties are characterized by the Electrophysiology Facility using microelectrode arrays and patch clamp with microelectrodes. Researchers use homologous recombination (1) to rescue mutant phenotypes and (2) to introduce mutations in control human embryonic stem cells to prove that the mutation of interest causes the *in vitro* disease phenotype. Other cardiac diseases studied include genetic forms of myopathy and hypertrophy.

Translation

The group encompasses both basic and translational research, with a main focus on human iPSC-derived heart cells, but with links to the diabetes program and beta-islet cell transplantation. They collaborate with other groups in the Netherlands through the NIRM program.

A spin-off company, Pluriomics, was founded in 2010 to commercialize technologies developed by the research group. Stefan Braam serves as its CSO and Herman Spolders is CEO. The goals of the company are to develop and optimize:

- Differentiation technology (defined, scalable, reproducible)
- Phenotypic assays (electrophysiology, calcium handling, contractility, toxicity)
- Genetic/disease models

Their short-term projects revolve around developing cardiac safety assays for pharmaceutical developers, noting that 33 % of drug attrition during late stages of clinical development is caused by cardiac toxicity. They aim for an integrated solution for cardiac analysis that can be directly provided to a pharmaceutical company. For development of specialized materials for culture and functional assays, the company works with Philips and other technology development companies (Imec in Belgium). One example is a multiple electrode array (MEA) that allows cardiac cells to contract while electrodes measure the field potentials. They have a contract with Janssen Pharmaceuticals and have achieved interesting early results using beating cardiomyocytes to screen blind compounds to study cardiac safety pharmacology. They are also working with Johnson & Johnson on validation studies.

Sources of Support

In addition to core funding from LUMC, the major source of support for the academic stem cell research is government funding for the Netherlands Institute for Regenerative Medicine. NIRM comprises two formerly existing research consortia in the Netherlands: Stem Cells in Development and Disease (SCDD) and the Dutch Program for Tissue Engineering (DPTE). NIRM is funded from 2010 to 2015; Dr. Mummery is the principal investigator, and has established collaborative projects within Leiden and at other institutions in the Netherlands.

The program's main aims are:

- Identify and characterize the stem cells and biomolecules contributing to normal tissue organization and function
- Characterize the levels at which tissues become diseased or damaged and investigate the mechanisms underlying aberrant tissue function
- Establish and test novel tissue repair and regeneration strategies and methodologies from gained knowledge and insights.

Collaboration Possibilities

NIRM hosts collaborations within the Netherlands, and many of the senior stem cell and tissue engineering researchers in the country are involved. There is continuing effort to obtain funding through the EU to extend the collaborative group to include other European countries. There is no formal collaboration with U.S. scientists.

Summary and Conclusions

This is one of the best groups in the world for cutting edge research in cell-based cardiac regenerative medicine. The Leiden group has exceptional expertise in producing high quality hiPSCs, and has a core center devoted to culture and banking of stem cells. They also have excellent electrophysiological facilities, and through collaborations they have added tissue engineering expertise and experts on specific areas of cardiac and vascular development and disease. The collaborative efforts are supported by a NIRM grant, which provides research funding from 2010 to 2015. The investigators are already looking ahead to support the work beyond the funding period, and spinning off companies to begin to commercialize their discoveries. The commercial applications revolve around development of integrated systems that provide functional assays using cardiomyocytes and related cell types for toxicity and drug development.

References

Braam, S.R. and C.L. Mummery. 2010. Human stem cell models for predictive cardiac safety pharmacology. *Stem Cell Res.* 4(3):155–156.

Braam, S.R., R. Passier, and C.L. Mummery. 2009. Cardiomyocytes from human pluripotent stem cells in regenerative medicine and drug discovery. *Trends Pharmacol. Sci.* 30(10):536–545.

Freund, C., and C.L. Mummery. 2009. Prospects for pluripotent stem cell-derived cardiomyocytes in cardiac cell therapy and as disease models. *J. Cell Biochem.* 107(4):592–599.

Mummery, C. and M.J. Goumans. 2011. Shedding new light on the mechanism underlying stem cell therapy for the heart. *Mol. Ther.* 19(7):1186–1188.

Mummery, C.L., R.P. Davis, and J.E. Krieger. 2010. Challenges in using stem cells for cardiac repair. *Sci. Transl. Med.* 2(27):27ps17, doi:10.1126/scitranslmed.3000558.

Mummery, C.L., and R. Passier. 2011. New perspectives on regeneration of the heart. *Circ. Res.* 109(8):828–829.

Wu, S.M., K.R. Chien, and C. Mummery. 2008. Origins and fates of cardiovascular progenitor cells. *Cell* 132(4):537–543.

Life&Brain Center, Bonn

Site Address:	Sigmund-Freud-Strasse 25, 53127 Bonn, GERMANY http://www.lifeandbrain.de/
Date Visited:	February 27, 2012
WTEC Attendees:	J. Loring (report author), R.M. Nerem, H. Ali
Host(s):	Professor Oliver Brüstle Director: Institute of Reconstructive Neurobiology Tel.: +49 228 6885 500 Fax: +49 228 6885 501 cellomics@lifeandbrain.com

(continued)

Dr. Manal Hadenfeld
Tel.: +49-228-6885 470
Fax: +49-228-6885 471
mhadenfeld@lifeandbrain.com

Dr. Michael Peitz
Group Leader, Reconstructive Neurobiology (University of Bonn)
peitz@uni-bonn.de

Simone Haupt
Head of Bioengineering Group
Tel.: +49 (228) 6885-470
Fax: +49 (228) 6885-471
shaupt@lifeandbrain.com

Dr. Annette Pusch
Scientist, Quality Management
Tel.: +49 228 6885 472
Fax: +49 228 6885 471
apusch@lifeandbrain.com

Prof. Dr. Björn Scheffler
Head of Stem Cell Pathologies Group (University of Bonn)
bscheffler@uni-bonn.de

Jerome Mertens
Junior Scientist, Reconstructive Neurobiology (University of Bonn)
jerome.mertens@uni-bonn.de

Overview

The Life&Brain Center has a unique structure that encompasses basic research and industrial applications in the field of biomedicine. Life&Brain brings together expertise in genomic research, transgenic animals, neurocognition, and stem cell biology to deliver novel products for disease modeling, early diagnosis, compound development, and tissue regeneration (Fig. B.11). The research at the Cellomics platform is focused on the development of novel applications for human pluripotent stem cells. Particular emphasis is put on the controlled differentiation of pluripotent stem cells (embryonic stem cells and induced pluripotent stem cells). These unique cells are characterized by unlimited self-renewal, pluripotency, and amenability to gene targeting. Neural progenitors/long-term neural epithelial stem cells (lt-NES®) derived from human pluripotent stem cells are being developed for models of human disease, for high-throughput screening of pharmacological compounds, and for

Life&Brain GmbH	
Type of organization	life science company
Founded (year)	2002
Concept	to exploit the power of academic research to support the early R&D phase
Mission Statement	discovery and development of novel strategies for the diagnosis and therapy of nervous system disorders
Areas of Activity	pluripotent stem cell-based disease modeling for drug screening Genomics, Transgenics and Neurocognition

Fig. B.11 Summary history, mission, and activities of the Life&Brain Center (Courtesy of Life&Brain Center)

cell replacement in human neurodegenerative disease. In the area of cell therapy, they are trying to overcome the limited regenerative capacity of the brain and spinal cord by using stem cell-derived neural transplants. They are using information from animal studies showing functional integration of transplanted human stem cell-derived neuronal cells to explore development of treatments for human disease. Their scientific strategy aims at translating mechanisms of nervous system development to stem cell biology, and thus to recreate specific neuronal and glial cell types in the laboratory dish.

Research and Development Activities

The Life&Brain Center is an unusual combination of basic and translational research centered on nervous system disorders. The platform Cellomics is directed by Professor Oliver Brüstle. Its main focus is to develop neuronal cell types from pluripotent stem cells and to employ them to model multiple diseases. The overall strategy is to use the basic research results to develop translational applications (Fig. B.12). Automated methods are being developed for the scale-up and scale-out of PSC-derived cell types for drug screening and prospectively for clinical applications. Having academic institutions and a platform for commercial application under one roof, the goal is to have a seamless transition from research to translational studies and eventually to clinical applications. The research is of excellent quality, and the introduction of biotechnology-based approaches at an early stage enables scientists to have the opportunity to follow through on their experimental results from research to industrial application.

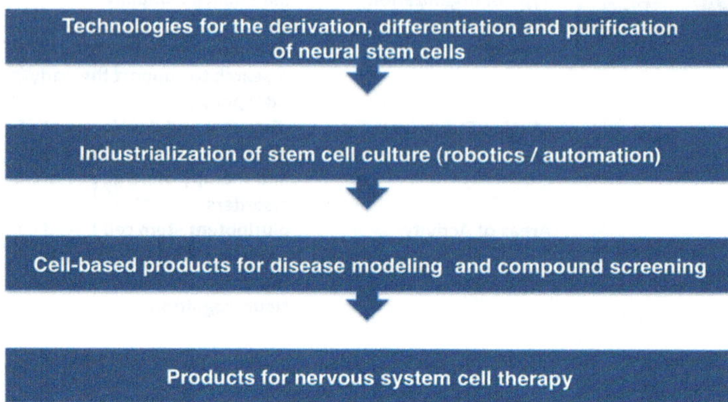

Fig. B.12 The concept of the platform Cellomics is to transfer stem cell-based methods developed in an academic context for the derivation of neural cell types with industrial applications (Courtesy of the Life&Brain Center)

Translation

The mission of Life&Brain Centre is to discover and develop novel strategies for the diagnosis and therapy of nervous system disorders. It is structured to be a center of innovation, and has a unique structure within Germany's translational research landscape. Spread across four R&D platforms covering different topics, more than 300 employees work together under one roof. Novel approaches are typically conceived by academic staff. The biotechnology arm then supports them to further develop their ideas towards commercialization, ideally creating spin-offs within a dedicated incubator segment. The encouragement of translational development within the same organization and infrastructure creates a particularly dynamic atmosphere. Young scientists can explore their entrepreneurial ambitions without abandoning their academic environment while benefiting from flexible support in legal and business-related matters.

Integrating a unique set of expertise in genomics, transgenics, cellomics and neurocognition, Life&Brain Center aims to deliver the next generation of products for disease modeling and prediction, compound development, and tissue regeneration.

Sources of Support

There are multiple sources of support, from government grants to partnerships with industry. The directors are aware of the fragility of funding sources, and are putting considerable effort into obtaining sustainable funding through public-private partnerships.

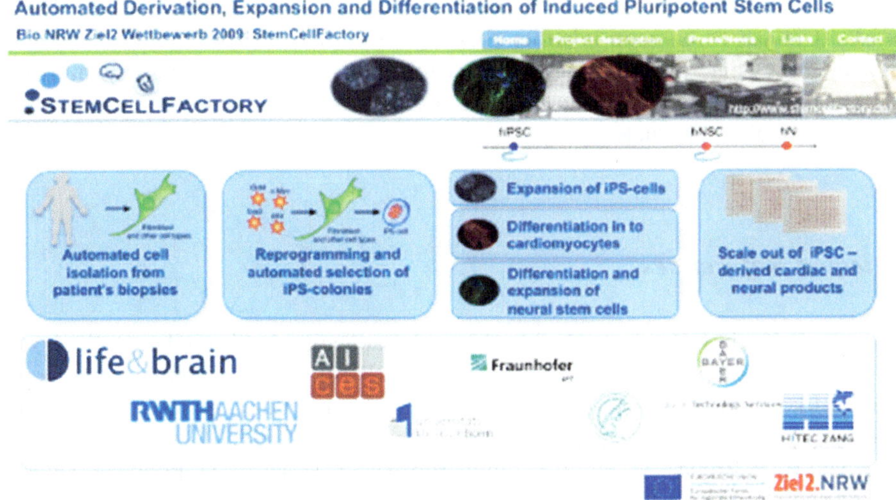

Fig. B.13 StemCellFactory is a collaborative project aimed at developing an automated stem cell production facility. The facility will include automation, standardization, and parallelization of all required cell culture steps, including a comprehensive quality management system (Courtesy of Life&Brain Center)

Collaboration Possibilities

The Life&Brain Centre is very collaborative, and is involved in several European Union projects that include multiple investigators. This group would be an ideal partner for collaboration with U.S. research groups and companies, since they are developing basic technologies and techniques that are of general use in the stem cell field. For instance, Life&Brain is leading, together with RWTH Aachen, the project StemCellFactory, which combines leading forces in stem cell research and engineering technology in North Rhine-Westphalia to build an automated system for the production of patient-specific iPS cells and their neural and cardiac derivatives for drug screening (Fig. B.13). The project is funded by regional funding of the European Commission. Another project is funded by a partnership between German funding agencies and the California Institute for Regenerative Medicine, but extending collaborations across the Atlantic to the United States is still a challenge.

Summary and Conclusions

Life&Brain Center is an excellent example of the effort to move basic research into clinical applications. Its structure is promising, giving investigators the opportunity to help turn their research results into practical products, without taking the risk of moving to a purely biotechnology environment. In theory, this is an excellent mechanism to train researchers in developing commercial applications for their work. It is yet to be seen whether this increases the probability of successful biotechnology enterprise, but it is a worthwhile experiment. Investment by multiple partners in the commercialization of the research is a good strategy for ensuring follow-through to products.

Selected References

Falk, A., P. Koch, J. Kesavan, O. Takashima, J. Ladewig, M. Alexander, O. Wiskow, J. Tailor, M. Trotter, S. Pollard, A. Smith, and O. Brüstle.2012. Capture of neuroepithelial-like stem cells from pluripotent stem cells provides a versatile system for *in vitro* production of human neurons. *PLoS One* 7(1):e29597.

Koch, P., P. Breuer, M. Peitz, J. Jungverdorben, J. Kesavan, D. Poppe, J. Doerr, J. Ladewig, J. Mertens, T. Tüting, P. Hoffmann, T. Klockgether, B.O. Evert, U. Wüllner, and O. Brüstle. 2011. Excitation-induced ataxin-3 aggregation in neurons from patients with Machado-Joseph disease. *Nature* 480(7378):543–546.

Koch, P., T. Opitz, J.A. Steinbeck, J. Ladewig, and O. Brüstle. 2009. A rosette-type, self-renewing human ES cell-derived neural stem cell with potential for *in vitro* instruction and synaptic integration. *Proc. Natl. Acad. Sci. USA* 106(9):3225–3230.

Koch, P., I.Y. Tamboli, J. Mertens, P. Wunderlich, J. Ladewig, K. Stuber, H. Esselmann, J. Wiltfang, O. Brüstle, and J. Walter. 2012. Presenilin-1 L166P mutant human pluripotent stem cell-derived neurons exhibit partial loss of γ-secretase activity in endogenous amyloid-β generation. *Am. J. Pathol.*, Epub ahead of print April 14, 2012.

Ladewig, J., J. Mertens, J. Kesavan, J. Doerr, D. Poppe, F. Glaue, S. Herms, P. Wernet, G. Kögler, F.J. Müller, P. Koch, and O. Brüstle. 2012. Small molecules enable highly efficient neuronal conversion of human fibroblasts. *Nat. Methods*, Epub ahead of print, 8 April 2012, doi:10.1038/nmeth.1972.

Lindvall, O., R.A. Barker, O. Brüstle, O. Isacson, and C.N. Svendsen. 2012. Clinical translation of stem cells in neurodegenerative disorders. *Cell Stem Cell* 10:151–155.

Ming, G.L., O. Brüstle, A. Muotri, L. Studer, M. Wernig, and K.M. Christian. 2011. Cellular reprogramming: recent advances in modeling neurological diseases. *J. Neurosci.* 31:16070–16075.

Oki, K., J. Tatarishvili, J. Woods, P. Koch, S. Wattananit, Y. Mine, E. Monni, D.T. Prietro, H. Ahlenius, J. Ladewig, O. Brüstle, O. Lindvall, and Z. Kokaia. 2012. Human induced pluripotent stem cells form functional neurons and improve recovery after grafting in stroke-damaged brain. *Stem Cells*, Epub ahead of print, doi:10.1002/stem.1104.

Thier, M., P. Worsdorfer, Y.B. Lakes, R. Gorris, S. Herms, T. Opitz, D. Seiferling, T. Quandel, P. Hoffmann, M.M. Nothen, O. Brüstle, and F. Edenhofer. 2012. Direct conversion of fibroblasts into stably expandable neural stem cells. *Cell Stem Cell* 10:473–479.

Lonza Cologne GmbH

Site Address:	Nattermannallee 1, Bocklemünd Mengenich, D-50829 Köln http://www.biocampuscologne.de/index.php?Itemid=144
Date Visited:	March 1, 2012
WTEC Attendees:	S. Palecek (report author), R.M. Nerem, J. Loring, L. Nagahara, H. Ali
Host(s):	Dr. Christian van den Bos Executive Program Manager, LIFT Tel.: 49 221 991 99200 christian.vandenbos@lonza.com

Overview

Lonza is one of the world's leading suppliers to the pharmaceutical, healthcare and life science industries. Its products and services span its customers' needs from research to final product manufacture. Lonza is the global leader in the production and support of active pharmaceutical ingredients both chemically as well as biotechnologically. Lonza has over 11,000 employees and is headquartered in Basel, Switzerland. The Cologne facility, which has 125 employees, is a research and development site focusing on long-term projects and manufactures life science research tools. At the site in Cologne, Lonza develops and commercializes non-viral gene transfer products for primary cells and hard-to-transfect cell lines. These cells are important model systems for both basic and clinical research as they more accurately reflect the condition and behavior of cells within an organism when compared to commonly used laboratory cell lines. With the specially developed Amaxa® Nucleofector® Technology, the functionality of different genes can now be analyzed in biologically relevant cell types. These results are used to more efficiently identify new targets for pharmaceuticals and therapies. Lonza possesses approximately one half of the worldwide market of commercial contract manufacturing of cells, including adult stem cells of various types for cell therapy applications and cells as diagnostic products. The commercial stem cell products Lonza produces include

Cor.At® cardiomyocytes generated from murine ESCs, human adipose-derived stem cells, human mesenchymal stem cells, human hematopoietic progenitors, human neural progenitors, and human bone marrow stromal cells.

Research and Development Activities

Much of Lonza's research and development efforts related to stem cells involve scaling cell culture to transfer processes from research labs to commercial production platforms. For example, cell cubes offer a substantially higher production capacity than T-flasks or Petri dishes. At larger scale, the primary goal is to move to adhesion-independent bioreactor culture (e.g., stirred tanks). While scalable production of cells for protein production is a fairly mature field, it is not trivial to adapt these advances to production of cells as the product. In particular, downstream processing issues are very different when the cells are the product as opposed to the vehicle for producing a protein product. Specific research and development challenges include engineering cells to be anchorage independent while not affecting their desired phenotypes or safety when used *in vivo*, designing equipment that supports large-scale cell production, and expanding primary cells, which exhibit a limited proliferation capacity and may change with time in culture.

Lonza also has research and development interests in general cell manufacturing technologies. Cell manufacturing needs will eclipse current culture technology limits in 3–5 years. Cell quality control remains a major challenge, where it is more difficult to describe and characterize cells than it is to define small molecules or proteins. Characterization methods that directly predict efficacy are needed. Better monitoring and control of the cellular environment to maintain homeostasis are also needed as culture systems grow.

Translation

As a contract manufacturer and supplier of culture platforms, Lonza occupies an important niche in translation of new technologies from research labs to industrial and clinical applications. Lonza has the ability to provide clients with state-of-the-art technologies for stem cell expansion and characterization.

Collaboration Possibilities

Lonza has numerous collaborations with other companies to commercialize stem cell products. As an example, Lonza has collaborated with California Stem Cell, Inc. to commercialize human ESC-derived motor neuron progenitors (MotorPlate™).

Lonza has also had numerous partnerships with academic institutions and labs. These collaborations offer the potential to facilitate development and commercialization arising from both basic and applied research.

Summary and Conclusions

Lonza is an important component of the stem cell research and engineering field in a number of ways. First, the company provides cells and culture platforms for research and development activities and clinical translation. Also, Lonza is at the leading edge of developing new technologies for scaling expansion and characterization of stem cells. In addition, Lonza partners with academic and corporate organizations to commercialize stem cell products. There is substantial uncertainty in predicting future markets for stem cell products. Clear clinical benefits of stem cell therapeutics are needed for the stem cell field to maintain its promising trajectory. With an aging population the demand for successful treatments will grow. A need also exists for clearer and more relevant regulatory standards for cell-based products. The pan-EU mechanism to regulate advanced therapy medical products (ATMPs) provides a model for effective and predictable regulation of stem cell products.

Lund University Biomedical Centre (BMC)

Site Address:	D1, Solvegatan 19/Klinikgatan 30-32
	E-221 84, Lund, Sweden
	http://www.elmat.lth.se/forskning/nanobiotechnology_and_labonachip/
	http://www.med.lu.se/labmedlund/lund_stem_cell_center/research_ groups/mesenchymal_stem_cells_and_cellular_therapies_lab
Date Visited:	March 2, 2012
WTEC Attendees:	P. Zandstra (report author), T. McDevitt, D. Schaffer, N. Moore, H. Sarin
Host(s):	Stefan Scheding, M.D., Associate Professor
	Group Leader of Mesenchymal Stem Cells and Cellular Therapies
	Faculty of Medicine; Stem Cell Center
	Tel.: +46-46-222-3331, 3989
	stefan.scheding@med.lu.se
	Dr. Lars Wallman, Associate Professor
	Department of Electrical Measurements, Faculty of Engineering
	lars.wallman@elmat.lth.se
	Dr. Andreas Lenshof, Post-Doc, Laurell Laboratory
	Department of Electrical Measurements, Faculty of Engineering
	andreas.lenshof@elmat.lth.se

Overview

Lund University is one of the largest and oldest universities in Scandinavia. With eight faculties, the University covers activities in engineering, science, law, social sciences, economics, medicine, humanities, theology, fine arts, music, and drama.

Lund Center for Stem Cell Biology and Cell Therapy is one of six Swedish strategic centers of excellence in life sciences, supported by the Swedish Foundation for Strategic Research. Established in January 2003, the Center focuses on stem cell and developmental biology of the central nervous and blood systems, and development of stem cell and cell replacement therapies in these organ systems as well as research in non-mammalian model systems.

Research and Development Activities

Our visit to Lund included meetings with representatives from the Faculty of Engineering (particularly the Department of Electrical Measurements) and the Faculty of Medicine, Stem Cell Center.

Thomas Laurell is a Professor in Medical and Chemical Microsensors, director of the Nanobiotechnology and Lab-On-A-Chip Group, and co-director of the Clinical Protein Science & Imaging Group. Laurell has a background in engineering with a focus on biomedical technology. Laurell has 20 years experience in the development of lab-on-a-chip based bioanalytical and medical diagnostic technology. He was appointed distinguished professor in the Department of Biomedical Engineering at Dongguk University, Seoul, Korea in 2009, is a member of Royal Swedish Academy of Sciences, the Royal Swedish Academy of Engineering Sciences and the Royal Physiographic Society. Laurell is also the President of the Chemical and Biological Microsystems Society (www.cbmsociety.org). Laurell has published over 130 papers in peer reviewed international journals with more than 3,200 citations. He has filed 25 patent applications. His current research has a focus on nanobiotechnology based diagnostic technologies as well as nanoproteomics where prostate cancer is a major area of interest. Dr. Laurell was not present during the site visit, but did have good representation from his group.

The Laurell group is a world leader in the area of "acoustophoresis," particle migration as a result of sound forces. Particles in suspension exposed to an acoustic standing wave field will be affected by a radiation force. The force will cause the particle to move in the sound field if the acoustic properties of the particle differ from the surrounding medium. The magnitude of the movement depends on many factors, such as the size of the particle, the acoustic pressure amplitude and the frequency of the sound wave. The direction the particle is moved depends on the density and compressibility of the particle as well as the liquid medium.

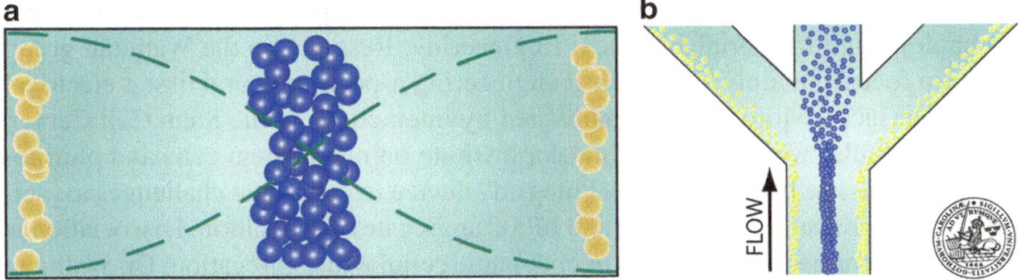

Fig. B.14 Example of an acoustophoresis separation system. (**a**) *Blue* particles aggregate at the pressure nodes. (**b**) Acoustophoresis can be integrated into a microfluidic chip to continuously separate particles (Courtesy of A. Lenshof, Lund University)

Most rigid particles and cells will be affected by the radiation force in such way that they are moved to the pressure node, in the example in Fig. B.14a), present in the center of the channel (blue particles). Liquid elements or air bubbles will move to the anti-nodes at the wall (yellow particles). If the size of the flow channel is in the micro domain, the flow condition is generally laminar, and the particles passing through the standing wave field will be moved to their nodal position and keep that position even after exiting the sound field. By letting the flow channel end in a trifurcation it is possible to separate and/or concentrate the particles from the medium (Fig. B.14b).

The first microfluidic acoustophoresis chips were developed in 2000. Applications of acoustophoresis include; blood washing, automated medium exchange, particle fractionation based on size, and plasmapheresis (separation of blood cells from plasma). A recent exciting application acoustophoresis was to produce clinical grade plasma from undiluted whole blood. The plasma is the subsequently used in a lab-on-a-chip device for detecting prostate-specific antigen (PSA).

Stefan Scheding, M.D. is an Associate Professor and research group leader at the Lund Stem Cell Center, and a senior consultant in the Department of Hematology. Dr. Scheding's research and clinical focus is to improve treatment of patients with hematological malignancies undergoing stem cell transplants. With his group and others at the Lund Stem Cell Center, blood stem cell research is an internationally recognized strength at this institution. Basic research in the Scheding group is focused primarily on the role of mesenchymal stromal cells in modulation of blood stem cell function in the bone marrow niche. In a recent paper his group demonstrated a role for CD146 expression in the localization of blood stem cells in different microenvironments. Dr. Scheding is also interested in the development of better blood processing and isolation technologies during blood stem cell transplantation. It is this interest that catalyzed the relationship with Professor Laurell's group to use acoustophoresis to develop better blood cell harvesting and washing technologies.

Lars Wallman is an Associate Professor in the Department of Measurement Technology and Industrial Electrical Engineering. Research in the Wallman group has been focused mainly on electrical engineering aspects of neural tissue-electrical probe interfaces. More recently, motivated by interactions in the Stem Cell Center, and in particular work at the Karolinska Institute on neural stem cells and pluripotent stem cells, he has developed a "biogrid" device to overcome challenges associated with aggregate (neurosphere and hPSC aggregate) dissociation. Dissociation is typically performed using manual or enzymatic cellular disaggregation; Dr. Wallman had generated a microfabricated cell slicer as an alternative (and apparently more effective) approach. The system can be integrated into a flow cells with a syringe on one side, and cell culture suspension on the other. Application of suction to force cells through a microfabricated grid disrupts the aggregates. The device is manufactured using anisotropic etching.

Andreas Lenshof, the postdoctoral fellow in the laboratory of Dr. Thomas Laurell who led aspects of our visit, has a Ph.D. in Electrical Engineering. Most students in the Laurell laboratory have backgrounds in electrical engineering, physics, and nanotechnology. In contrast, most students and fellows in the Stem Cell Centre have backgrounds in biology or biotechnology (mostly focused on genetic engineering). A master's level Biomedical Engineering program has recently started at Lund; it is anticipated this program will yield opportunities for interdisciplinary training.

Translation

Commercial translation at Lund is active and encouraged. Particularly interesting is the Vinnova Program. Vinnova, the Swedish Governmental Agency for Innovation Systems, invests in research and strengthens Sweden's innovative capacity for competitiveness, sustainable development and growth. Vinnova's efforts range from programs for R&D projects in small companies and at universities to long-term development of strong research and innovation environments that attract R&D investment and expertise from around the world.

Sources of Support

This group has taken advantage of Vinnova funding to start companies including AcouSort (www.acousort.se), which is focused on the commercialization of acoustophoresis systems. Swedish companies involved in human pluripotent cell technology, such as Cellartis (http://www.cellartis.com/), are also involved in industry-academic partnerships.

Collaboration Possibilities

The Lund University Stem Cell Center collaborates with Uppsala University under auspices of the National Initiative on Stem Cells for Regenerative Therapy. Basic research findings of centers are translated to the clinical setting in the clinical medical research centers of the Region Skane, which has led to the developments of the Biomedical Centre (BMC) for research and education in Lund and the Clinical Research Center (CRC) in Malmo. University collaborations with Industry also exist with companies operating at Ideon Science Park.

Summary and Conclusions

Overall, there is a good integration of stem cell biology and engineering at this site.

Selected References

Augustsson P., R. Barnkob, S.T. Wereley, H. Bruus, and T. Laurell. 2011. Automated and temperature-controlled micro-PIV measurements enabling long-term-stable microchannel acoustophoresis characterization. *Lab Chip* 11:4152–4164.

Brune, J.C., A. Tormin, M.C. Johansson, P. Rissler, O. Brosjö, R. Löfvenberg, F. Vult von Steyern, F. Mertens, A. Rydholm, and S. Scheding. 2010. Mesenchymal stromal cells from primary osteosarcoma are non-malignant and strikingly similar to their bone marrow counterparts. *Int. J. Cancer* 129(2):319–330, doi:10.1002/ijc.25697. Epub Dec 1, 2010.

Dykes J., A. Lenshof, I. Åstrand-Grundström, T. Laurell, and S. Scheding. 2011. Efficient removal of platelets from peripheral blood progenitor cell products using a novel micro-chip based acoustophoretic platform. *PLoS One* 6:e23074

Kohler, P., C.E. Linsmeier, J. Thelin, M. Bengtsson, H. Jorntell, M. Garwicz, J. Schouenborg, and L. Wallman. 2009, Flexible multi electrode brain-machine interface for recording in the cerebellum. *Proceedings of the International Conference of the IEEE Engineering in Medicine and Biology Society*, 2009, pp. 536–538.

Lenshof A., C. Magnusson, and T. Laurell. 2012. Acoustofluidics 8: Applications of acoustophoresis in continuous flow Microsystems. *Lab Chip* 12:1210–1233.

Tormin, A., O. Li, J.C. Brune, S. Walsh, B. Schütz, M. Ehinger, N. Ditzel, M. Kassem, and S. Scheding. 2011. CD146 expression on primary non-hematopoietic bone marrow stem cells correlates to *in situ* localization. *Blood* 117(19):5067–5077. Published online before print March 17, 2011, doi:10.1182/blood-2010-08-304287.

Wallman L., E. Åkesson, D. Ceric, P.H. Andersson, K. Day, O. Hovatta, S. Falci, T. Laurell, and E. Sundström. 2011. Biogrid—a microfluidic device for large-scale enzyme-free dissociation of stem cell aggregates. *Lab Chip* 11:3241–3248.

Max Planck Institute for Molecular Biomedicine

Site Address:	Röntgenstraße 20, 48149 Münster, Germany http://www.mpi-muenster.mpg.de/en/
Date Visited:	March 1, 2012
WTEC Attendees:	S. Palecek (report author), R.M. Nerem, J. Loring, H. Ali
Host(s):	Prof. Dr. Hans Schöler, Director, Max Planck Institute for Molecular Biomedicine[1] Tel.: 49 251 70 365 300 Fax: 49 251 70 365 399 office@mpi-muenster.mpg.de Dr. Holm Zaehres Staff Scientist Tel.: 49 251 70 365 360 holm.zaehres@mpi-muenster.mpg.de Dr. Jared Sterneckert, Group leader jsterneckert@mpi-muenster.mpg.de Dr. Boris Greber, Group leader boris.greber@mpi-muenster.mpg.de

Overview

The Max Planck Institute (MPI) for Molecular Biomedicine is one of 80 research institutes of the Max Planck Society for the Advancement of Science (MPG). The institute was founded in 2001 and construction was completed in 2006. Overall, more than 150 scientists from approximately 25 nations work here, among them biologists, physicians, and physicists. The focus of the work is on basic research, the pursuit of new knowledge without focusing on industrial or commercial goals, and serving the public interest. The results of the work can be found each year in numerous scientific publications, e.g., in renowned journals such as *Nature*, *Cell*, and *Cell Stem Cell*.

Basic questions addressed at the Max Planck Institute for Molecular Biomedicine include: How can a complete human being develop from a single fertilized egg cell? How do the cells of an embryo know when and where they should form arteries,

---

[1]Prof. Schöler was not present at the meeting.

nerves or muscles? What tricks do immune cells use to migrate from the blood into infected tissue?

In their various research teams, this institute is seeking answers to these and other questions about important life processes. The teams are subdivided into three permanent departments and several smaller work groups, junior research groups, and research groups, some of which carry out autonomous research projects and some of which are only affiliated with the Institute.

Research and Development Activities

Hans Schöler and his team in the Cell and Developmental Biology department have been key contributors to deciphering the early events in embryonic development. Specifically, they have elucidated molecular mechanisms of pluripotency and reprogramming cells to a pluripotent state. The researchers have shown that Oct4 plays a key regulatory role in pluripotency. Oct4 is only expressed in two types of cells, embryonic stem cells and gametes, precursors of egg and sperm cells. In order to reprogram somatic cells to a pluripotent state, Oct4 expression must be activated.

Researchers in Cell and Developmental Biology at MPI use numerous systems to study development including human embryonic stem cells, induced pluripotent stem cells, adult stem cells, mice, and planaria. Ongoing projects investigate transcriptional regulation of trophectoderm formation and development of pluripotency, maintenance of pluripotency states in embryonic and epiblast stem cells, reprogramming neural stem cells and germ line stem cells, interactions between reprogramming factors, chromatin remodeling during cell reprogramming, differentiation of pluripotent stem cells to germ cells, application of bioinformatic tools to understand stem cell regulation, and disease modeling with induced pluripotent stem cells including generation of dopaminergic neurons from iPSCs.

Research in the Cell and Developmental Biology department is supported by core facilities in MPI including flow cytometry, electron microscopy, confocal microscopy, genomics and proteomics, and an animal facility for mice and zebrafish.

Translation

As a basic research institute, translational research is not performed at MPI. However, intellectual property arising from basic research is patented and licensed through Max Planck Innovation (Munich). Licensing royalties are returned to the institute. Translational research, such as patient-specific disease modeling or drug discovery, is often spun off to biotechnology companies or other R&D institutes. For example, a new Center for Advanced Regenerative Engineering (CARE) is being established in Münster, adjacent to the Max Planck Institute for Molecular Biomedicine, for applied stem cell research and development. CARE will be funded by the German Federal Minster of Education and Research and the North Rhine-Westphalia state.

Sources of Support

Eighty-two percent of the funding comes from the government, of which approximately half is German government funding and half is from the state of North Rhine-Westphalia. The remainder of the funding is from private sources.

Collaboration Possibilities

The MPI for Molecular Biomedicine offers excellent collaborative possibilities for stem cell engineers. They excel at elucidating basic mechanisms of pluripotency, reprogramming, and differentiation. These advances offer the technical basis for surmounting challenges facing stem cell engineers, such as efficient production of stem cells and conversion of these stem cells to differentiated cells of industrial and therapeutic interest.

Dr. Schöler has collaborations with the Max Planck Institute for Surface Chemistry in Stuttgart and the Center for Nanotechnology in Münster. He also has visiting professor appointments at the University of Pennsylvania and the Ulsan National Institute of Science and Technology (UNIST), South Korea.

Summary and Conclusions

The Max Planck Institute for Molecular Biomedicine is a leading research center for mechanisms of pluripotency and reprogramming. While research at this site does not directly involve stem cell engineering, the basic information learned through research here is crucial for improving the ability to control the fates of stem cells. Thus, Dr. Schöler's lab and other groups at this institute will play an important role in the future of stem cell engineering.

Selected References

Greber, B., G. Wu, C. Bernemann, J.Y. Joo, D.W. Han, K. Ko, N. Tapia, D. Sabour, J. Sterneckert, P. Tesar, and H.R. Schöler. 2010. Conserved and divergent roles of FGF signaling in mouse epiblast stem cells and human embryonic stem cells. *Cell Stem Cell* 6:215–226.

Han, D.W., N. Tapia, A. Hermann, K. Hemmer, S. Höing, M.J. Araúzo-Bravo, H. Zaehres, G. Wu, S. Frank, S. Moritz, B. Greber, J.H. Yang, H.T. Lee, J.C. Schwamborn, A. Storch, and H.R. Schöler. 2012. Direct reprogramming of fibroblasts into neural stem cells by defined factors. *Cell Stem Cell* 10(4):465–72.

Han, D.W., N. Tapia, J.Y. Joo, B. Greber, M.J. Araúzo-Bravo, C. Bernemann, K. Ko, G. Wu, M. Stehling, J.T. Do, and H.R. Schöler. 2010. Epiblast stem cell subpopulations represent mouse embryos of distinct pregastrulation stages. *Cell* 143(4):617–627.

Kim, J.B., B. Greber, M.J. Araúzo-Bravo, J. Meyer, K.I. Park, H. Zaehres, and H.R. Schöler. 2009. Direct reprogramming of human neural stem cells by OCT4. *Nature* 461:649–653.

Kim, J.B., H. Zaehres, G. Wu, L. Gentile, K. Ko, V. Sebastiano, M.J. Araúzo-Bravo, D. Ruau, D.W. Han, M. Zenke, and H.R. Schöler. 2008. Pluripotent stem cells induced from adult neural stem cells by reprogramming with two factors. *Nature* 454:646.

Singhal, N., J. Graumann, G. Wu, M.J. Araúzo-Bravo, D.W. Han, B. Greber, L. Gentile, M. Mann, and H.R. Schöler. 2010. Chromatin-remodeling components of the BAF complex facilitate reprogramming. *Cell* 141:943.

Wu, G., L. Gentile, T. Fuchikami, J. Sutter, K. Psathaki, T.C. Esteves, M.J. Araúzo-Bravo, C. Ortmeier, G. Verberk, K. Abe, and H.R. Schöler. 2010. Initiation of trophectoderm lineage specification in mouse embryos is independent of Cdx2. *Development* 137(24):4159–4169.

National Natural Science Foundation of China (NSFC)

Site Address:	Shuangqing Road, Haidian Beijing, China http://www.nsfc.gov.cn/Portal0/default166.htm
Date Visited:	November 14, 2011
WTEC Attendees:	S. Demir (report author), K. Ye
Host(s):	Dr. Zou Liyao Deputy Director Bureau of International Cooperation National Natural Science Foundation of China zouly@nsfc.gov.cn Ms. Liu Xiuping Program Manager, Division of American, Oceania, and East European Programs Bureau Bureau of International Cooperation National Natural Science Foundation of China liuxp@nsfc.gov.cn Dr. Feng Feng Deputy Director, Department of Health Sciences National Natural Science Foundation of China fengf@nsfc.gov.cn

(continued)

Dr. Hujun Jiang
Division of Oncology, Department of Health Sciences
National Natural Science Foundation of China
jianghj@nsfc.gov.cn

Dr. Li Enzhong
Program Director
Department of Health Sciences
National Natural Science Foundation of China

Other Attendees: Dr. Emily Y. Ashworth
Director, NSF China Office
eashwort@nsf.gov

Mr. Sun Bo
Science Program Specialist
NSF China Office

Overview

NSFC funds fundamental research, but does not fund applied research, education and industrial partnerships. The Ministry of Science and Technology (MOST) funds industrial partnerships. Dr. Hujun Jiang stated that research in biomedical engineering, regenerative medicine, biomaterials, sensing, simulations, bioelectric signals, brain-computer interface, equipment, diagnostics, applications of using stem cells, and tissue repair areas have been funded. In 2006, the brain-computer interface was a focus; in 2009 molecular imaging was a focus at a level ten million RMB (These areas are similar to NSF Biomedical Engineering Program and its emphasis areas in the similar year).

Stem cell and regenerative medicine is a national priority. China will invest three billion RMB in R&D during 2011–2015, including:

- 1.2 billion RMB in basic science during 2011–2015, mainly through MOST
- 150 million RMB in clinical science during 2010–2014, mainly through MOST
- 100 million RMB in new drug development during 2010–2014, mainly through the MOST
- 820 million RMB in stem cell and regenerative medicine during 2011–2015, mainly through Chinese Academy of Sciences (CAS)

There are 500 stem cell groups in China, mainly allocated in the following regions:

- Beijing: CAS, Peking university, Tsinghua University, Academy of Military Medical Sciences, MAMS
- Shanghai: CAS, Tongji, Shanghai Jiao Tong, Fudan University

- Guangzhou: CAS, Zhongshan University, Guangzhou Medical College
- Tianjin: Academy of Military Medical Sciences,, Nankai University, Wuhan University, Xi'An Jiaotong University

Sources of Support

The NSFC budget in 1995 was 30 million RMB; the NSFC budget in 2010 was nine billion RMB. Overall in China there is a 20 % annual increase in R&D. In 2002 NSFC received 378 proposals in stem cell research and awarded 59 projects, whereas in 2011 NSFC received 2,365 proposals in stem cell research and awarded 502 projects.

- Single principal investigator (PI) projects are 4 years long, funded at about 600,000 RMB.
- Key projects are three million RMB for 5 years, and are single PI projects.
- General programs are 15 million RMB for 5 years.
- Project plans are 150 million RMB for 7–8 years (2–3 key projects and 7–8 general programs).

Collaboration Possibilities

There is a bilateral stem cell research workshop between United Kingdom and China. Among the future possibilities are the following:

- A China-USA workshop on stem cell engineering, or China-UK-USA workshop.
- Supplementing awards that can start collaborations on stem cell engineering between USA and China (a Dear Colleague Letter).
- Developing a joint solicitation on stem cell engineering (like the biodiversity one).

National Tissue Engineering Center, Shanghai Jiao Tong University School of Medicine

Site Address:	Jiangchuan East Road, Minhang District
	Shanghai
	China
	http://en.sjtu.edu.cn/research/centers-labs/
	national-tissue-engineering-research-center
Date Visited:	November 16, 2011

(continued)

WTEC Attendees:	R.M. Nerem (report author), S. Demir, N. Moore, S. Palecek, P. Zandstra, K. Ye, F. Huband
Host(s):	Prof. Yilin Cao Shanghai 9th People's Hospital Tel.: 86-21-34290613 Fax: 86-21-34292305 yilincao@ntec.org.cn Prof. Wei Liu Shanghai 9th People's Hospital Tel.: 86-21-5351-2182 Fax: 86-21-5307-8128 liuwei_2000@yahoo.com Wen Jie Zhang Shanghai 9th People's Hospital Tel.: 86-21-2327-1699 x 5606 Fax: 86-21-5307-8128 wenjieboshi@yahoo.com Guangdong Zhou Shanghai 9th People's Hospital Tel.: 86-21-2327-1699 x 5606 Fax: 86-21-5307-8128 guangdongzhou@126.com

Overview

The National Tissue Engineering Center (NTEC) is a separate organization and has a well designed complex of buildings for basic research and the translation and commercialization of the technologies developed. This center builds on the Shanghai Tissue Engineering Key Laboratory, which in 2004 became the Tissue Engineering R&D Center, and a year ago led to the establishment of NTEC. Dr. Yilin Cao is a Professor of Plastic Surgery at Shanghai's 9th People's Hospital and he heads NTEC. Professor Cao has an international reputation and is the president-elect of the Asia-Pacific Chapter of the Tissue Engineering and Regenerative Medicine International Society (TERMIS). Dr. Wei Liu also is a Professor of Plastic Surgery and he serves as the Chief Scientific Officer of NTEC. Drs. Cao, Liu, Zhang spent time in the United States prior to returning to Shanghai to provide the leadership that has led to the establishment of NTEC.

Research and Development Activities

Activities of NTEC include tissue constructs and repair, stem cell research, and clinical applications. The stem cell research involves both adult and embryonic stem cells, including bone marrow derived cells, adipose tissue derived cells, epidermal stem cells, dermal MSCs, and hair follicle stem cells. The overall effort is focused on autologous cell approaches.

Clinical studies have focused for the most part on the musculoskeletal area, in particular tissue engineered bone. Applications have included repair of cranial bone defects, humerus cysts, and a thumb deficit. In each of these there have been impressive results. For the bone cranial defect, bone marrow-derived mesenchymal stem cells in a scaffold have been used; without stem cells no repair has been observed in large defects. For bone repair the results from human clinical studies were impressive.

NTEC also is working on tendon tissue engineering and in this project has a bioreactor that incorporates mechanical cues. The animal studies are being conducted with a hen model. The tracheal repair project involves a cartilage tube with muscle wrapped around the tube and epithelialization with native epithelial cells. For skin tissue engineering, the goal is to use cultured and expanded follicle stem cells. Dermal MSCs also are being investigated. Finally, another project is focused on blood vessel development and vascularization.

Translation

The NTEC facility is not totally built out; however, when completed it will include a GMP laboratory for producing scaffolds and a GMP facility for culturing and expanding cells. NTEC looks to the U.S. FDA for the standards necessary for GMP, and the cell culture laboratory is being renovated due to changes in regulations. The NTEC facility also plans to include a clinical trial base in the future, thus acquiring the ability to go from bench top to bedside. NTEC also has an office for quality control. The long-range strategy of NTEC is to create an initial clinical and commercial success, and then use the financial return on such a success to recruit additional talented researchers who can provide the leadership to develop additional areas that can lead to new clinical and commercial successes. It was stated, however, that the translation of products to clinical and commercial success in China is difficult. Furthermore, the Chinese government does not have enough experience with regulatory issues. NTEC did have some interactions with both Genzyme and Organogenesis, but these relationships did not work out.

Sources of Support

Shanghai government, MOST, and NSFC. The funding from the Shanghai government is for three years, after which NTEC is expected to become self sufficient.

Collaboration Possibilities

There clearly are some possibilities for collaborations, in particular in the musculo-skeletal area and with centers in North America interested in translational activities. Both Drs. Cao and Liu have extensive friends and contacts in the United States that could provide the foundation for future collaborations.

Summary and Conclusions

In NTEC there are eight principal investigators, three of whom are engineers, one of whom works on bioreactors. One of the issues discussed was how to integrate and involve engineers in tissue engineering and regenerative medicine. Dr. Wei Liu stated that engineers in China seldom work on stem cells as most engineers either do not have the interest or the biological background. He went on to say that the education and training of engineers in China does not prepare them to work with cells. Thus, from his perspective, ideally cell biologists need to work on the biology and engineers on the engineering aspects; however, he recognized the need to physically locate biologists and engineers together. He also did appear to recognize that education in biomedical engineering at the graduate level might create the right kind of engineer to work in the stem cell area.

Selected References

Chen, F.G., W.J. Zhang, D. Bi, W. Liu, X. Wei, F.F. Chen, L. Zhu, L. Cui, and Y. Cao. 2007. Clonal analysis of nestin- vimentin+ multipotent fibroblasts isolated from human dermis. *J. Cell Sci.* 120(Pt 16):2875–2883.

Liu, W., and Y. Cao. 2007. Application of scaffold materials in tissue reconstruction in immunocompetent mammals: Our experience and future requirements. *Biomaterials* 28(34):5078–86. (Invited review)

Liu, W., L. Cui, and Y.L. Cao. 2003. A closer view of tissue engineering in China—The experience of tissue construction in immunocompetent animals. *Tissue Eng.* 9(Suppl. 1):S17-31.

Liu, W., W. Zhang, and Y. Cao. 2007. Bone and cartilage reconstruction, chapter 57, Part 15. In *Musculoskeletal system. principles of tissue engineering*, 3rd edition, ed. R. Lanza, R. Langer, and J. Vacanti. San Diego: Elsevier Life Sciences.

Zhang, W.J., W. Liu, L. Cui, and Y. Cao. 2007. Tissue engineering of blood vessel. *J. Cell. Mol. Med.* 11(5):945–57.

Zhou, G., W. Liu, L. Cui, X. Wang, T. Liu, and Y. Cao. 2006. Repair of porcine articular osteochondral defects in non-weightbearing areas with autologous bone marrow stromal cells. *Tissue Eng.* 12(11):3209–3221.

Netherlands Initiative for Regenerative Medicine

Site Address:	http://www.nirm.nl/index.html
Date Visited:	February 28, 2012
WTEC Attendees:	R.M. Nerem (report author), J. Loring, S. Palecek, H. Ali
Host(s):	Professor Dr. Ruud A. Bank, Chair in Matrix Biology & Tissue Repair
	University Medical Center Groningen, The Netherlands.
	Tel.: +31 50 361 19 83
	Fax: +31 50 361 99 11
	r.a.bank@med.umcg.nl
	Professor Dr. Ir FPT (Frank) Baaijens, Eindhoven University of Technology
	Biomedical Engineering, Materials Technology
	PO Box 513, GEM-Z 4.117, 5600 MB Eindhoven, The Netherlands.
	Tel.: +31 40 247 4888
	Fax: +31 40 244 7355
	f.p.t.baaijens@tue.nl
	Professor Wouter JA Dhert, M.D., Ph.D.
	Dept. Orthopaedics, UMC Utrecht
	Heidelberglaan 100, 3584 CX Utrecht, The Netherlands
	Tel.: +31 (88) 755 6971
	Fax: +31 (30) 251 0638
	Orthopaedie@umcutrecht.nl
	Professor Dr. Clemens A. Van Blitterswijk, Department of Tissue Regeneration
	MIRA Institute, University of Twente, Faculty Science & Technology
	Zuidhorst ZH143, Drienerlolaan 5, 7522 NB Enschede,
	The Netherlands
	Tel.: +31-(0)53-489-3400
	Fax:+31-(0)53-489-2150
	c.a.vanblitterswijk@utwente.nl

(continued)

Professor Pieter Doevendans, Chief, Department of Cardiology
University Medical Center Utrecht, E 03.511
P.O. Box 85500, 3508 GA, Utrecht, The Netherlands
Tel.: +31 88 75 598 01
Fax: +31 30 25 163 96
p.doevendans@umcutrecht.nl

Professor Willem Fibbe, Professor of Hematology
Leiden University Medical Center, Albinusdreef 2
P.O. Box 9600, 2300 RC Leiden, The Netherlands
Tel.: +31 71 526 3827, 5267
W.E.Fibbe@lumc.nl

Overview

The Netherlands Initiative for Regenerative Medicine (NIRM) grew out of two previous national initiatives, one being the Dutch Program for Tissue Engineering (DPTE) and a similar stem cell initiative that was called Stem Cells in Development and Disease (SCDD). These were funded in 2004 for 5 years. There were only two Dutch investigators who were a part of both of these earlier initiatives. These are Dr. Christine Mummery, now at the Leiden University Medical Campus, and Dr. Elaine Dzierzak who is at the Erasmus Medical Campus.

NIRM was funded in 2009 for 5 years for a total of $81 million. The aims of the NIRM research program may be summarized as follows:

- To identify and characterize the stem cells and biomolecules contributing to normal tissue organization and function
- To characterize the levels at which tissues become diseased or damaged
- To establish and test novel tissue repair and regeneration strategies and methodologies from knowledge and insights gained in aims 1 and 2

The co-leaders of NIRM are Dr. Ruud Bank and Dr. Elaine Dzierzak. Unfortunately, Dr. Dzierzak was not available to meet with the WTEC panel; however, there were a number of investigators participating in the meeting in addition to Dr. Bank. It should be noted that, as a fusion of the earlier tissue engineering and stem cell initiatives, the NIRM has a total of 56 principal investigators. This total is made up of 20 internationally competitive Dutch stem cell research scientists and 36 tissue engineering research scientists. The research program involves 6 academic medical centers, 5 universities, 3 institutes, and 25 commercial and start-up firms. Although it was 2009 when the funding decision was made, there needed to be a business plan approved because of the translational nature of NIRM. This approval was obtained in December 2010, and thus NIRM is only in its second year of the 5-year funding period. NIRM's program is divided into three phases. These are the Building Phase for fundamental research, the Bundle Phase for translational research, and the Benefit Phase for industry involvement.

Research and Development Activities

Over the past 10 years, life sciences research has been a top priority in the Netherlands. The science in the Netherlands is excellent, and there are also real strengths in engineering in the country, particularly in the area of biomaterials. Also, many professors are entrepreneurial. From the University of Twente alone there have been 700 spinoffs in the last 20 years. The venture capital situation in the Netherlands is good. There are 3–4 big venture capitalists and a network of smaller venture capitalists.

Dr. Frank Baaijens, who chaired a committee that created a Dutch view of regenerative medicine, discussed the outcome from this group. Strategically, the focus of the Dutch is on off-the-shelf technology, cost effectiveness, patient-specific implants, and disease models. The last of this list includes organ-on-a-chip technologies. This committee also believed that a part of the strategy must be to identify patients who actually would benefit.

At the site visit meeting, the Dutch investigators who attended each discussed their own research. For Dr. Ruud Bank, the host for this visit with Dutch investigators, his main interest is in tissue engineering and the use of stem cells in wound repair. He has expertise in collagen (connective tissue diseases; fibrosis) and cell/matrix interactions.

His research is focused on BioKid, a kidney bioreactor. This device involves a bioactive coating and epithelial cells on a hemocompatible membrane. The membrane is provided by the University of Twente and the coating from the University of Eindhoven. The cells are an immortalized cell line isolated from the renal proximal tubule and are obtained from Nijmegen.

Dr. Baaijens is full professor in Soft Tissue Biomechanics & Tissue Engineering. In 1985 he joined Philips Research Laboratories in Eindhoven to work on computational mechanics. Since 1990 he has been part-time professor in the Polymer Group of the division of Computational and Experimental Mechanics of the Eindhoven University of Technology, in the area of Computational Rheology. In October 2002 he was appointed full professor in the Department of BioMedical Engineering (division of Biomechanics and Tissue Engineering). From 2003 to 2007 he was Dean of the Department of Biomedical Engineering, and he is currently Scientific Director of the national research program on BioMedical Materials (BMM). His current research focuses on soft tissue biomechanics and tissue engineering. A primary focus of Professor Baaijens involves tissue engineering of a heart valve. The approach is one that uses autologous cells and a minimally invasive, transapical implantation method. Their most recent work involves decellularized valves. When implanted, these valves are fully repopulated with endogenous cells, no thickening occurs after 8 weeks, and there is no retraction of these decellularized valves. In contrast to their earlier tissue engineering studies, this work at the Eindhoven University of Technology is a collaborative one with Dr. Simon Hoerstrup in Zurich.

Dr. Wouter Dhert is the professor of translational research of the musculoskeletal system (2008-present) and director of orthopedic research. His main research focus is on regeneration of tissues of the musculoskeletal system, particularly bone, articular cartilage and intervertebral disc, including the development and use of biomaterials.

Specific applications in man and animal are: spinal surgery and joint reconstruction. Furthermore, specific expertise exists with regards to biomaterials related infections in orthopedics. He discussed regenerative medicine and stem cell research within the Utrecht Medical Center. The main foci are cardiovascular regeneration, the musculo-skeletal area, and stem cell-based therapies. In the orthopedic group, activities include 3D tissue printing, controlled release approaches, and *in vitro* disease models, e.g., for osteogenesis imperfecta and metastatic bone disease. They also are using gene therapy, and Dr. Dhert reported on the use of a gelatin bead delivery of a BMP-2 plasmid in an alginate hydrogel to drive ectopic bone formation. The results were as good as those achieved with the transient transfection approach.

Prof. Dr. Pieter Doevendans is interested in cardiovascular disease. Together with the Hubrecht Institute he tries to unravel the embryonic development of heart and heart muscle cells. Ultimately he hopes this will lead to stem cell therapies for treating cardiovascular disease. Also, he is working on understanding sudden cardiac death. This may be due to heart muscle defects, but also due to failing electrical properties of the heart, which may cause arrhythmias. They have just started a project to identify relatives from sudden cardiac death victims and subsequently screening for cardiac pathology. They expect to identify relatives who may be at risk and can be treated prophylactically.

He has been involved in a number of clinical trials involving cell therapy approaches for cardiac repair. This work has included a variety of different cell types. The new clinical trial employs allogeneic MSCs and intercoronary injection. It should be noted that earlier in his career Dr. Doevendans worked with Dr. Ken Chien on mESCs and currently is collaborating with both Singapore and UCSD on clinical trials with exosomes.

Dr. Willem Fibbe is a professor of hematology at the Leiden University Medical Center. His specific research interest is in stem cell biology, including hematopoietic and mesenchymal stem cells. The work focuses on *in vitro* characterization of human MSCs and the use of murine MSCs in *in vivo* translational animal models. He has been one of the founders of an academic European MSC consortium that has developed a common expansion protocol in order to perform multi-center clinical studies. He discussed the activities of his department and explained the three themes of their research, which are cancer immunotherapy, autoimmunity and transplantation, and regenerative medicine. In the last of these there is a focus on hematopoietic stem cells (HSCs) and includes the HSC niche, homing of these cells, expansion of HSCs, and cord blood transplantation. Dr. Fibbe was involved in the founding of the Dutch cord blood bank in 1997. There now are 4,000 cord bloods banked. He also mentioned the existence of a GMP facility where cells are produced for the clinical trials. At Leiden they lack engineering expertise, but do collaborate with various engineering groups in NIRM.

Professor Clemens van Blitterswijk has an appointment at University of Twente in the Netherlands where, as chair of Tissue Regeneration, he heads a team of Ph.D. students active in the field of tissue engineering. He has cofounded four biomedical companies and held several functions in these organizations. He acted as CEO of IsoTis (a public life sciences company in the Netherlands) from 1996 to 2002. In total he raised over 120 million Euros in funding through equity and/or grants. Resulting from his work 10 implant technologies were brought into clinical evaluation in humans. He made the point that building complex tissues requires building new technologies. He also emphasized that biology is not necessarily predictive, that *in vitro* experiments do not predict *in vivo* results, and animal studies are not necessarily predictive of what will happen in humans. As an example, although the use of MSCs in nude mice led to bone formation, bone was only formed in one of ten patients. He also raised the question "how does one deal with patient variability?" He also described the high-throughput screening of materials. This involves printing a topography of 10,000 samples on a chip, and this work was published in 2011 in the *Proceedings of the National Academy of Sciences of the USA*. Dr. van Blitterswijk also discussed the self-organization of tissues from building blocks of several hundreds of cells to millimeter-size micro units to centimeter-size tissues.

Sources of Support

The funds to support infrastructure development in the country have come from natural gas profits. Science has been part of this; however, the economic crisis has ended this source of funds. Thus, it is difficult to predict what the funding will be after the current 5-year commitment to NIRM is over. It should be noted, however, that there is additional funding from industry and from more local sources.

Collaboration Possibilities

Funding provided through DPTE was the first to motivate investigators in tissue engineering to work together. It changed the mindset of the researchers in this area and helped them recognize the added value of collaboration and the consortium approach. DPTE thus set the stage for the new initiative, i.e., NIRM. The same also can be said for the earlier stem cell initiative. With these two initiatives now joined in NIRM, there is increased interaction and collaboration.

Summary and Conclusions

In the Netherlands the new NIRM initiative represents an exciting new chapter in stem cell research and regenerative medicine for the country. It proactively links the engineering approach with the stem cell community, and there already is evidence of collaborative activities taking place.

Okayama University, Graduate School of Medicine, Dentistry and Pharmaceutical Sciences

Site Address:	5-1 Shikata-cho, Kita-ku, Okayama 700-8558
	(Meeting took place at Gakushi Kaikan, Tokyo)
	http://www.okayama-u.ac.jp/index_e.html
Date Visited:	November 17, 2011
WTEC Attendees:	T. McDevitt, J. Loring (report author), D. Schaffer, N. Kuhn, H. Ali, T. Satoh
Host(s):	Professor Keiji Naruse
	Department of Cardiovascular Physiology
	Tel.: 81-88-235-7112
	Fax: 81-88-235-7430
	knaruse@md.okayama-u.ac.jp
	http://www.okayama-u.ac.jp/user/med/phy2/staff.html

Overview

Professor Naruse focuses on the physiological consequences of mechanical deformations of cell membranes, which he believes underlies multiple disorders, including hypertension, heart failure, and asthma. He designates his areas of research as mechanobiology, mechanophysiology, and mechanomedicine (Fig. B.15). He has designed devices for measuring and controlling mechanical stress. He is entrepreneurial, having founded the company STREX, Inc, that makes devices for controlled mechanical deformation of cells and tissues, and has developed many other technologies, including a novel scaffold material for medical uses and a microfluidic system to improve assisted reproduction methods.

Research and Development Activities

The research activities in this group are diverse, but center on the development and use of instrumentation and materials for studying the effects of mechanical manipulations on cells and tissues. Examples include:

- Studies of mechano-sensitive ion channels
- Studies of single cardiomyocytes using carbon nanofibers to hold and stretch the cells, calculating the force by bending of the carbon fibers (Fig. B.16)
- Effects of dynamic compression and tension on bone and cartilage

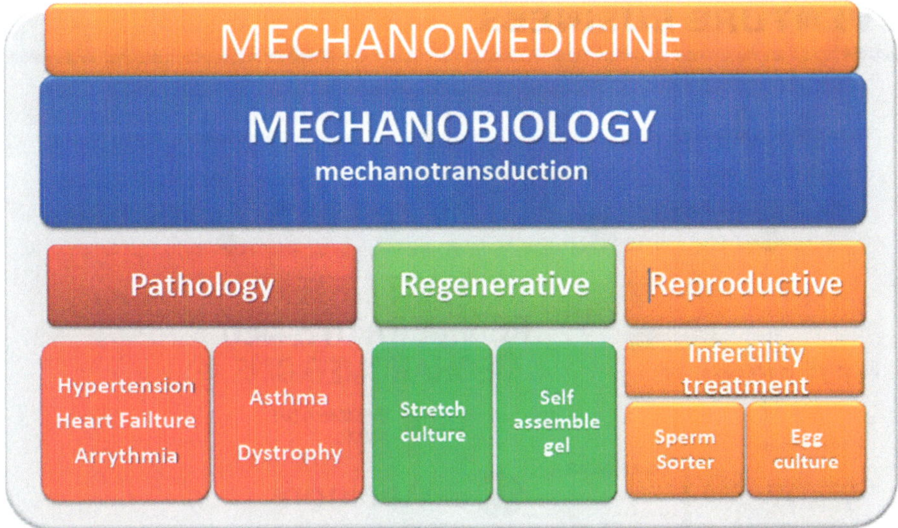

Fig. B.15 Applications of mechanomedical technology (Courtesy of K. Naruse, Okayama University)

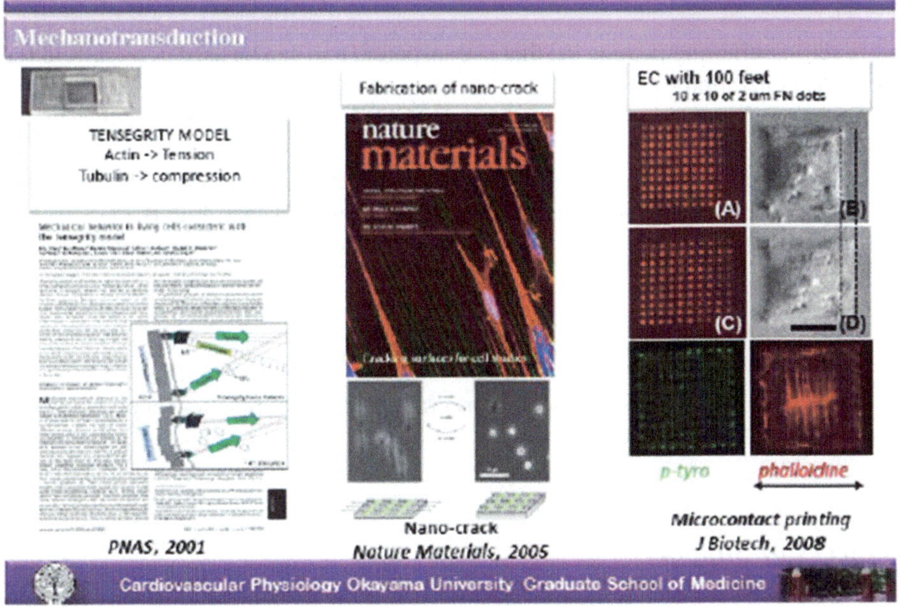

Fig. B.16 Outcomes from application of the carbon nanofiber stretching device (Courtesy of K. Naruse, Okayama University)

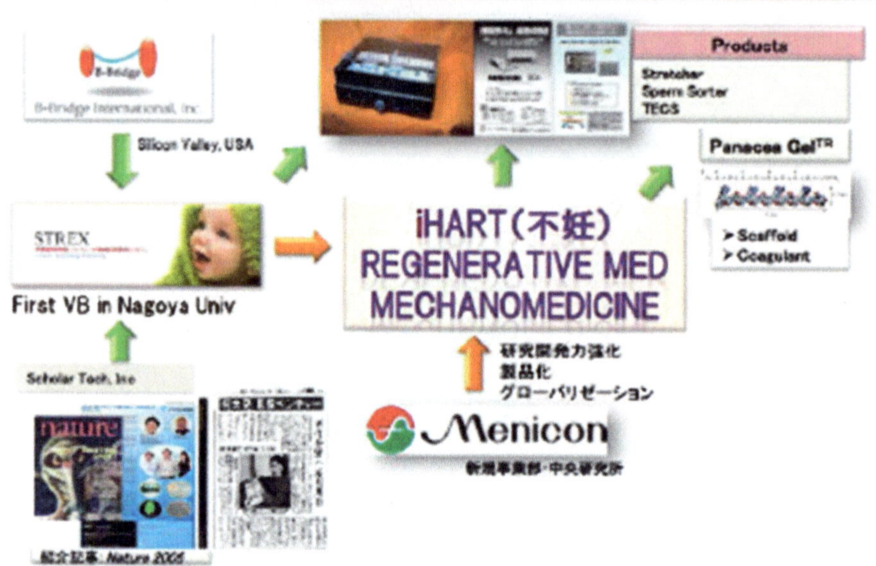

Fig. B.17 Dr. Naruse's commercialization process for iHART (Courtesy of K. Naruse, Okayama University)

- A self-assembling peptide gel that is transparent and can be used as a scaffold for tissue generation
- Fast scanning atomic force microscopy (AFM) for studying stretch-activated ion channels
- Microfluidic systems for improved sperm selection and embryo development (improved human assisted reproduction technology, iHART, Fig. B.17)

Translation

Every aspect of Dr. Naruse's research program is aimed at practical applications of the developed technologies.

Sources of Support

Dr. Naruse is one of the few researchers visited who has venture capital support for his companies.

Collaboration Possibilities

There is interest in collaboration, especially if it provides opportunities to develop new applications for his technology.

Summary and Conclusions

Dr. Naruse is very entrepreneurial, and has involvement in companies that make research products, and recently, products for assisted reproduction, which has a big market in Japan. His venture business, STREX, was the first for Nagoya University. He has connections in Cupertino, CA (B-Bridge) that help to establish his companies.

Selected References

Matsuura, K., M. Takenami, Y. Kuroda, T. Hyakutake, S. Yanase, and K. Naruse. 2012. Screening of sperm velocity by fluid mechanical characteristics of a cyclo-olefin polymer microfluidic sperm-sorting device. *Reprod. Biomed. Online* 24(1):109–115, Epub 16 Sep 2011.

Shimizu, N., K. Yamamoto, S. Obi, S. Kumagaya, T. Masumura, Y. Shimano, K. Naruse, J.K. Yamashita, T. Igarashi, and J. Ando. 2008. Cyclic strain induces mouse embryonic stem cell differentiation into vascular smooth muscle cells by activating PDGF receptor β. *J. Appl. Physiol.* 104:766–772.

Peking University, The College of Life Sciences

Site Address:	5 Yi He Yuan Road, Haidian, Beijing, China, 100871 http://www.bio.pku.edu.cn/news/headline/index.php?styleid=2
Date Visited:	November 14, 2011
WTEC Attendees:	P. Zandstra (report author), S. Demir, N. Moore, S. Palecek, K. Ye, F. Huband
Host(s):	Professor Hongkui Deng Department of Cell Biology and Genetics Tel.: 86-10-62756954 Fax: 86-10-62756954 hongkui_deng@pku.edu.cn

Overview

In 2006 Hongkui Deng was one of two Chinese scientists (among 43 international winners) awarded the Bill & Melinda Gates Foundation Grand Challenges in Global Health Award. Deng won the award for a proposal to use stem cells to create mouse models for testing HIV and hepatitis C vaccines. Dr. Deng is a clear leader, both in China and internationally, in the development of stem cell technologies based on pluripotent cells.

Deng left China in 1989 to enroll in a Ph.D. program in immunology at the University of California in Los Angeles. He worked on HIV during a postdoctoral fellowship with Dan Littman at New York University. He gained significant experience in biotechnology as the research director of the cord blood banking and blood stem cell expansion company ViaCell (later acquired by Perkin Elmer) in Cambridge, Massachusetts.

He returned to Beijing in late 2001, recruited back to his home country when the Chinese Ministry of Education offered him the Cheung Kong Scholar Professorship at Peking University. Hongkui Deng also serves as Chair, of the International Stem Cell Institute (ISSCR) Stem Cell Policy Committee and is a member of the Board of Governors of the ISSCR.

In January, 2012 when China's Ministry of Health announced it was initiating a year-long campaign to more tightly regulate the use of stem cell treatments, including an immediate halt to unapproved stem cell treatments Deng commented "It is a good sign that the Chinese Government is very serious about this issue …. In the near future, an urgent action for the Chinese government is to formulate detailed regulation specific for stem cell-based therapies."

Research and Development Activities

Dr. Deng's research involves the differentiation of human pluripotent stem cells into beta cells to treat diabetes. He has been instrumental in leading the domestic development of stem cell research and has been funded through the National Natural Science Foundation (NSFC), the National "973" and "863" programs (managed by MOST, the Ministry of Science and Technology), and other local and national research programs. Dr. Deng is a member of the Stem Cell Center of Peking University; the Center is directed by Dr. Lisong Li of the Department of Basic Medical Cell Biology.

Research in Peking University has made significant international contributions, including: establishment of the world's first monkey iPS cells, establishment of new molecular adjuvants to accelerate human iPSC generation; significant advances in the directed differentiation of human pluripotent cells to islets; methods to establish nuclear transfer (NT) mouse ES cell lines, including differentiation into islet cells and transplantation into mouse models (demonstrating the therapeutic feasibility of NT strategies technology to treat diabetes).

Links between stem cell biology and bioengineering at Peking University are relatively new and are largely based on new hires in the Department of Biomedical engineering in Peking University's College of Engineering. Peking University Biomedical Engineering department has 14 faculty. Bio-imaging is most mature field; with a few new hires in Biomaterials. The Coulter Foundation sponsors a joint Ph.D. program with Georgia Tech/Emory. In this program students spend at least 1 year on the counterpart's campus and have a coadvisor at the partner institution.

Prof. Deng is chairing an effort to coordinate the many stem cell related activities at Peking University through the creation of a Stem Cell Consortium (to consist of 40–50 faculty in multiple departments). Examples of bioengineering related collaborative projects highlighted during the WTEC panel's visit include:

- Construction of *in vitro* 3D microenvironment to support the phenotype of hepatocytes differentiated from human pluripotent stem cells (hPSCs). This project is part of an interaction with Ying Luo, Ph.D.
- Development of encapsulation technologies to immunoisolate hPSC-derived islets-like cells.
- Interactions between stem cell groups and materials and nanotechnology groups (the latter emerging aggressively in the business area) has been encouraged through a recent workshop to explore the interface between these fields.
- Computational biology and bioinformatics are strong at Peking University; interaction with stem cell researchers, especially in the area of understanding and modeling the molecular events in reprogramming, is under way.
- Peking University has the strongest chemistry program in China (with >100 faculty). Stem cell biology and chemical biology are poised to be significant strengths in the drug development area with increasingly coordinated interactions between these groups. These interactions are strengthened by the advantages in large-scale screening projects that China has in labor costs (e.g., technician salaries typically $300–$400/month).

From a training perspective, Peking University is exploring initiating a stem cell degree program. Current stem cell training is mainly in the developmental biology program. These programs will be important as fundamental/mechanistic studies have been more difficult in China due to the large number of varying quality Ph.D.s produced. Efforts to increase Ph.D. and postdoctoral fellow salaries and to support "home-grown" talent are under way and important to solve this problem. Most instruction at Peking University is in Chinese but teaching in English is encouraged. Concern was raised in a few instances about appropriate mechanisms to recruit the highest quality students (including the ability to pay competitive salaries and mechanisms in grants for longer term staff support).

Translation

The Peking University Stem Cell Consortium works with clinical partners to translate stem cell therapies. Translational grants are typically led by the industry/clinical partner. This is a particular opportunity as clinical research in China is very active and access to larger animal models in not as costly as in North America.

Table B.1 Stem cell research funding mechanisms in China (From Yuan et al. 2012)

Agency	Program	Grant Size[a]	2011 funding level[b]
NSFC	General programs	92.3	1,405
	Key program	452	224.4
	Major program	3,000	6–9
	Total stem cell funding	Varies	223
MOST	863	1,000–5,000	-
	973	3,000–5,000	-
	Stem cell program	4,600	40
	Stem cell initiative for CAS	1,500	150
Other ministries, provinces, cities, major universities, industry	Varies; mainly follow NSFC and/or MOST funding mechanisms	Varies; usually smaller than similar NSFC programs, except for a few special initiatives	Total annual funding estimated to exceed the NSFC total

[a]Unit = thousands, US dollars
[b]Unit = millions, US dollars

Commercialization in the regenerative medicine area has been inconsistent. Part of the problem expressed is that, at least from an engineering perspective, efforts are focused in areas where direct job creation can be traced. The longer term impact of bioengineering and regenerative medicine applications have had more problems attracting industrial partners.

There is some evidence of cofunding provided by foreign pharmaceutical companies (Roche, Johnson & Johnson). Chinese pharmaceutical firms were described by our hosts as being more risk averse and investing in more mature technologies. Cord blood banking is a potential exception.

Increasing efforts at regulation of the stem cell industry and clinical development programs is under way, including the Ministry of Health's current moratorium on clinical trials and a 1-year effort to develop a new and enforceable regulatory framework.

Support for start-up companies appears to be quite region/municipality specific. Shenzhen is one particular area mentioned.

Sources of Support

Data on stem cell funding in China has been published recently by Yuan et al. (2012, Table B.1).

The Thousand Young Talents program (wherein the government offers 1 million yuan (US$146,000) in subsidies to help recruit overseas researchers) is being used to try and bring internationally renowned scientists (ideally full professors with a strong publication record) back to China. A program exists to recruit foreign postdocs with salaries of US$20,000–30,000.

Government funding for interdisciplinary research is provided to the university. One barrier to collaboration between universities is competition (e.g., Peking University vs. Tsinghua). Both Peking University and Tsinghua get substantially more funding from the central government under the 5-year plan than other universities.

The reporting burden on grants in China is high (in-person evaluations). In a 5-year grant, there is usually an evaluation after year 2 that determines whether the funding will continue.

Collaboration Possibilities

Funding exists for international collaborations. Peking University has a Ph.D. student exchange with Georgia Tech and an undergrad student summer exchange with the University of Toronto. MOST has initiated a number of international cofunded collaborations in stem cell research and translation including with the Ontario Ministry of Economic Development and Innovation in Ontario, Canada.

Selected References

Cai. J., C. Yu, Y. Liu, S. Chen, Y. Guo, J. Yong, W. Lu, M. Ding, and H. Deng. 2010. Generation of homogeneous PDX1(+) pancreatic progenitors from human ES cell-derived endoderm cells. *J. Mol. Cell Biol.* 2(1):50–60, Epub Nov. 12, 2009.

Jiang, W., X. Sui, D. Zhang, M. Liu, M. Ding, Y. Shi, and H. Deng. 2011. CD24: a novel surface marker for PDX1-positive pancreatic progenitors derived from human embryonic stem cells. *Stem Cells* 29(4):609–617, doi:10.1002/stem.608.

Li, Y., Q. Zhang, X. Yin, W. Yang, Y. Du, P. Hou, J. Ge, C. Liu, W. Zhang, X. Zhang, Y. Wu, H. Li, K. Liu, C. Wu, Z. Song, Y. Zhao, Y. Shi, and H. Deng. 2011. Generation of iPSCs from mouse fibroblasts with a single gene, Oct4, and small molecules. *Cell Res.* 21(1):196–204, Epub Oct. 19, 2010.

Song, Z., J. Cai, Y. Liu, D. Zhao, J. Yong, S. Duo, X. Song, Y. Guo, Y. Zhao, H. Qin, X. Yin, C. Wu, J. Che, S. Lu, M. Ding, and H. Deng. 2009. Efficient generation of hepatocyte-like cells from human induced pluripotent stem cells. *Cell Res.* 19(11):1233–1242, Epub Sept. 8, 2009.

Wang, C., X. Tang, X. Sun, Z. Miao, Y. Lv, Y. Yang, H. Zhang, P. Zhang, Y. Liu, L. Du, Y. Gao, M. Yin, M. Ding, and H. Deng. 2012. TGF-β inhibition enhances the generation of hematopoietic progenitors from human ES cell-derived hemogenic endothelial cells using a stepwise strategy. *Cell Res.* 22(1):194–207, doi:10.1038/cr.2011.138, Epub Aug. 23, 2011.

Yu, C., Y. Liu, Z. Miao, M. Yin, W. Lu, Y. Lv, M. Ding, and H. Deng. 2010. Retinoic acid enhances the generation of hematopoietic progenitors from human embryonic stem cell-derived hemato-vascular precursors. *Blood* 116(23):4786–4794, Epub Apr. 28, 2010.

Yuan, W., D. Sipp, Z.Z. Wang, H. Deng, D. Pei, Q. Zhou, and Cheng T. 2012. Stem cell science on the rise in China. *Cell Stem Cell* 10(1):12–15, doi:10.1016/j. stem.2011.12.002.

Zhu, F.F., P.B. Zhang, D.H. Zhang, X. Sui, M. Yin, T.T. Xiang, Y. Shi, M.X. Ding, and H. Deng. 2011. Generation of pancreatic insulin-producing cells from rhesus monkey induced pluripotent stem cells. *Diabetologia* 54(9):2325–2336, Epub July 14, 2011.

RIKEN Institute, Kobe

Site Address:	Center for Developmental Biology The Sasai Lab 2-2-3 Minatojima minamimachi Chuo-ku, Kobe, Hyogo 650-0047 JAPAN http://www.kobe.riken.jp/en/
Date Visited:	November 15, 2011
WTEC Attendees:	T. McDevitt (report author), J. Loring, D. Schaffer, L. Nagahara, N. Kuhn, H. Ali, M. Imaizumi
Host(s):	Dr. Yoshiki Sasai Tel.: +81-78-306-1841 Fax: +81-78-306-1854 yoshikisasai@cdb.riken.jp http://www.cdb.riken.jp/sasai/index-e.html Douglas Sipp Tel.: +81-78-306-3043 Fax: +81-78-306-3090 sipp@cdb.riken.jp

Overview

The Center for Developmental Biology at the RIKEN Institute in Kobe, established in 2000, focuses on development and regeneration for the purpose of addressing health challenges posed by an aging society. This national center for developmental biology is physically located within a world-class and growing biomedical research park setting that provides the organizational structure and supportive environment necessary to facilitate local and international collaborations. The primary objective of Dr. Sasai's research is to determine the fundamental principles of developmental biology that regulate the size and shape of organs during morphogenic processes.

The types of studies that Dr. Sasai is performing are very interdisciplinary and as a result he collaborates with various types of other researchers in engineering and the basic sciences.

Research and Development Activities

Dr. Sasai is studying the formation of 3D tissues from stem cells via serum-free floating cultures of mouse ES cells aggregated like embryoid bodies (EBs) and differentiated with growth factor-minimized medium (SFEBq). Almost all ES cells cultured as aggregates in SFEBq media form neuroepithelia, based on Sox1 expression after 7 days. Aggregation of ~3,000 ES cells in low adhesion, multi-well plates can be achieved quickly in 5 h time and although some random differentiation can be observed, it can appear as self-organization of the cells. Using this general approach, Dr. Sasai has examined the multistep morphogenesis of ES cells into optic cup and pituitary gland (hypophysis)-like multicellular structures.

By controlling the starting cell number and soluble media conditions, multiple optic cup structures can be formed at the periphery of a single aggregate and the resulting size of the cultured structures accurately match that of actual mouse optic cups *in vivo* (~200 μm diameter). Using a GFP reporter construct, the initial evagination of the cells followed by invagination can be monitored in real time. If the structures are removed from the rest of the mouse ES cell mass by physical dissection, the optic cup still forms, suggesting that the morphogenic evagination/invagination process is not due simply to hydrostatic pressure from the adjacent regions of the attached cell mass.

Dr. Sasai has gone on to examine the mechanics of the "hinge" region that forms and is responsible for the invagination process. They noted that the neural retina cells flatten immediately prior to invagination due to loss of phosphoMLC2 and that the invagination process could be prevented by treatment with a myosin inhibitor. In additional studies, atomic force microscopy measurements were taken using a 100 μm glass bead affixed to a cantilever to measure the rigidity of the different cell regions. They found that the retinal pigment epithelial (RPE) cells exhibited a higher Young's modulus than the neighboring neural retina (NR) cells, indicating that the cells on the outside were stiffer than the ones on the inside of the invagination structure. The stiffness of the RPE cells could be decreased by treatment with several inhibitors (i.e., blebbistatin and cytochalasin D), so rigidity due to actin-myosin activity alone cannot be responsible for invagination.

Following AFM analysis, ablation studies were performed to determine whether tensile or compression forces were responsible for the hinge activity that is necessary prior to invagination. The hinge region has a wedge shape (single focal point on basal side of cells) compared to the RPE (columnar) and NR (pseudo-stratified) cells on either side of the wedge. Treatment with a mitotic inhibitor (aphidicolin, 5 μM) attenuates invagination over a period of hours. 3D ablation was performed using a multiphoton system and cell debris was observed always to be pushed out to the apical side and the hole closes, indicating that neighboring cells push due to RPE expansion.

Based on these experimental studies, a computational model of the process has been developed with a collaborator (T. Adachi, Kyoto Univ.), who had an appointment at RIKEN and a lab nearby before moving to Kyoto. The model consists of local vertices (representing cells) connected by virtual springs. The equation of motion was solved at the vertices and it was found the vertices move to minimize total potential energy.

Translation

Much of the research currently being conducted is focused on contributing directly towards regenerative medicine of the central nervous system (CNS). The cerebral cortex, cerebellar cortex and neural retina are three major regions of the CNS consisting of multilayered cell structures, and they have a strong translational interest and effort in RPE cells.

Collaboration Possibilities

Compared to the United States, there is generally more separation between engineers and scientists in Japan except in the area of medical engineering, although the community of investigators in Japan is easier to access and networking is more efficient due the relative medium size of the country. RIKEN is making a concerted effort to promote interdisciplinary collaborative studies by housing three theory-based laboratories in the Center for Developmental Biology and an optics support unit, as well as forming a joint program with supercomputer facilities. Thus, many of the collaborations being forged are being driven by needs from the inside and seeking expertise outside of the specific research group and RIKEN Institute. For example, Dr. Sasai has collaborated with Olympus on the development of a new multiphoton 3D imaging instrument capable of imaging whole aggregates with single-cell resolution and less light, making it more sensitive. This system can be used for long-term culture studies.

Summary and Conclusions

The RIKEN Institute and Dr. Sasai's research in particular represent a unique environment where principles of developmental biology are being examined from a mechanistic and quantitative perspective with a long-term goal of applying research to regenerative medicine. The research results attained provide novel examples of how cell-mediated morphogenic processes can form complex spatially organized structures, thereby serving as models for stem cell tissue engineering efforts. The pursuit of challenging interdisciplinary science questions is forcing investigators to address key technological issues, such as complete imaging of 3D multicellular structures, and to collaborate with scientists on development of computational models capable of predicting the cellular phenomena.

Selected References

Eiraku, M., N. Takata, H. Ishibashi, M. Kawada, E. Sakakura, S. Okuda, K. Sekiguchi, T. Adachi, and Y. Sasai. 2011. Self-organizing optic-cup morphogenesis in three-dimensional culture. *Nature* 472:51–56.

Suga, H., T. Kadoshima, M. Minaguchi, M. Ohgushi, M. Soen, T. Nakano, N. Takata, T. Wataya, K. Muguruma, H. Miyoshi, S. Yonemura, Y. Oiso, and Y. Sasai. 2011. Self-formation of functional adenohypophysis in three-dimensional culture, *Nature* 480:57–62.

Shanghai Jiao Tong University, School of Medicine

Site Address:	280S. Chongqing Rd., Shanghai, China 200025 http://en.sjtu.edu.cn/
Date Visited:	November 17, 2011
WTEC Attendees:	R.M. Nerem (report author), S. Palecek, P. Zandstra, S. Demir, N. Moore, K. Ye, F. Huband
Host(s):	Professor Fanyi Zeng Tel./Fax: 86-21-6467-3297 fzeng@sjtu.edu.cn

Overview

Fanyi Zeng is an M.D., Ph.D. who has emerged as a leader in China with several national awards to her credit. This includes the first Young Woman Scientist Award from the Third World Organization of Women in Science and the 10th National Award for Achievement in Science and Technology. She grew up in the genetics laboratories of her parents, and she went to the United States for her university education. Her undergraduate degree is from the University of California, San Diego. After a year as an intern at NIH, she went to the University of Pennsylvania for her M.D. and Ph.D. degrees with her Ph.D. being in developmental biology. She has published in top journals such as *Nature* and *PNAS*. Returning in 2005 to China which always had been her plan, she is at Shanghai Jiaotong University both in the Shanghai Institute for Medical Genetics and the Shanghai Stem Cell Institute. Whereas the Shanghai Institute for Medical Genetics is a real institute, the Shanghai Stem Cell Institute is a virtual organization. Dr. Zeng also is affiliated with the children's hospital. She recently participated in a bilateral meeting between the Chinese Academy of Engineering and the U.S. National Academy of Engineering.

Research and Development Activities

Fanyi Zeng has an interest in technology development, bioinformatics, computational biology, and systems biology. She also is interested in ethical issues as they relate to human stem cell clinical trials and is the representative from China to an international ethics working group. Fanyi Zeng's research interests include animals as bioreactors, transgenic animals, and cloning. She studies birth defects and embryo development. She has used cows as a bioreactor for the production of proteins in milk. She also works on stem cell biology *in vivo*. As an example, she proved the pluripotency of iPS cells in an animal model, and she has worked with human ESCs in a goat model. In the context of stem cell engineering, Fanyi Zeng has a variety of interactions. In her laboratory she has a material scientist. She also collaborates with Dr. Jianhua Quin from Dalian on the use of microfluidic devices. Finally, she also has a collaboration with Dr. Hui Lu in the area of computational biology. She stated that there is a need for engineers who have the flexibility to understand biology and personalized medicine issues.

Translation

Applications of her work include Pharma models, personalized medicine, protein production, and animal models of human disease. She indicated that forming a start-up company in China is very difficult as investors do not understand science; however, she does have patents on cloning, animal bioreactors, and technology chips.

Sources of Support

Funding comes from MOST, NSFC, Ministry of Education, and Ministry of Health for clinical trials. She also has an NIH grant. She indicated that many grants in China have an age limit that may differ for a woman compared to a man. She also has some educational funding.

Collaboration Possibilities

Fanyi Zeng has a number of collaborations as indicated above. She thus appears to be the kind of researcher who readily collaborates. Her interest in engineering is in the development of enabling technologies that will accelerate biological discovery.

Summary and Conclusions

Although Fanyi Zeng returned to China to have an impact on science in her native country and to contribute to the advancement of science there. She recognizes that certain challenges need to be overcome. These challenges include delays in receiving supplies ordered, to the "orthogonal" thinking of her graduate students, and to age and gender discrimination issues that exist. She commented that, although there are a number of Chinese faculty with dual China/United States. appointments, most of them ultimately return to the United States. This is because it is hard to set up a new laboratory and to fulfill obligations at both institutions.

State Key Laboratory of Bioreactor Engineering

Site Address:	East China University of Science and Technology Building 18, Bioengineering School, No. 130, Meilong Road http://www.ecust.edu.cn/s/2/t/31/a/4080/info.jspy
Date Visited:	November 17, 2011
WTEC Attendees:	R.M. Nerem, S. Palecek, P. Zandstra (report author), S. Demir, K. Ye, N. Moore, F. Huband
Host(s):	Professor Wen-Song Tan Tel.: 86-21-6425-09348 Fax: 86-21-6425-3394 wstan@ecust.edu.cn

Overview

East China University of Science and Technology (ECUST) was founded in 1952. ECUST has three campuses. The university has 15 academic schools, including, related to Stem Cell Engineering, the School of Chemical Engineering, School of Bioengineering, School of Chemistry & Molecular Engineering, School of

Pharmacy and the School of Material Science & Engineering. ECUST has more than 20,000 students; of these 7,500 are graduate students (1,000 of these Ph.D.s). It is has one of the top-ranked biochemical engineering programs in China. ECUST ranks third in China in patents and technology transfer and has developed memoranda of understanding and other types of agreements with institutions in more than 20 countries.

The State Key Laboratory of Bioreactor Engineering (SKLBE) was founded in 1989, and formally recognized by the Ministry of Science and Technology (MOST) in 1996. The director of SKLBE is Prof. Jianhe Xu, associate directors are Profs. Wen-Song Tan, Jie Bao, and Changhua Chen. The overall research program at the SKLBE is focused on engineering and fundamental aspects of bioreactor engineering and related disciplines. There are 49 faculty members in SKLBE. Between 2001 and 2010, SKLBE researchers produced around 900 international academic journal papers (SCI cited), 300 patents, and more than 40 academic books or book chapters and supervised 200 Ph.D. and 600 M.S. students in their thesis research projects.

Research and Development Activities

At the SKLBE Professor Wen-Song Tan leads efforts at the integration of bioprocess engineering, principally with stem cell culture and expansion. Professor Tan received his Ph.D. in Chemical Engineering 1993 from ECUST; his dissertation was focused on Hybridoma Cell Suspension Culture Technology and Airlift Bioreactors. His professional career has been primarily at ECUST, although he did do a 1 year sabbatical 1997 at University of Bielefeld, Germany (with a group interested in animal cell culture technology). This visit seemed instrumental in his emerging focus on stem cell bioprocessing.

Areas of research focus related to stem cells in the Tan lab include:

- Studying the influence of the culture environment on growth, biological function, and gene expression of the human hematopoietic stem cells. In particular, there are projects examining the impact of cytokines and oxygen tension on the growth and differentiation of hematopoietic progenitor cells.
- *Ex vivo* cultivation, expansion, and directed differentiation of human mesenchymal stem cells (MSC) in bioreactors. There is a particularly innovative focus on culturing MSC on microcarriers, and directly seeding and integrating these MSC loaded microcarriers into 3D tissues constructs for differentiation and bone development. There are also efforts under way to model hematopoiesis *in vitro* by seeding blood progenitors onto *in vitro* engineered MSC derived bone tissue.

- The group has a number of studies under way in the area of bioreactor operating condition optimization, including the effect of Vitamin C to regulate aggregation, and the role of hydrodynamic and mass transfer phenomena in tissue engineering bioreactors.
- There are efforts to engineer functionalized tissue using dynamic scaffolds, either directly, or in forms that can be integrated into bioreactor-based stem cell growth and differentiation. These efforts are led, in part, by Dr. Zhaoyang Ye, a Senior Scientist in Prof. Tan's group. Dr. Ye did his postdoctoral research with Jennifer Elisseeff (Johns Hopkins), and thus has gained significant experience in human pluripotent cell growth and differentiation and on scaffold functionalization. It is anticipated that Dr. Ye will have a significant impact on the research ongoing in the Tan group.

Although the SKLBE appears adequately funded (especially given the supplements associated with being a Key State Lab, see below), it appears there are challenges in terms of infrastructure and supplies that are faced. The Tan group has undertaken the *de novo* design and manufacture its own bioreactors (for example ones similar to Applikon 3L bioreactors) and the access to reagents (media and cytokines) for larger volume stem cell cultures represents a significant cost.

Translation

As a whole, the SKLBE appears to be involved in a number of protein and vaccine manufacturing projects with pharmaceutical and biocenology companies. It appears that industrial support for stem cell related applications is not as mature, both because the industry receptors are not as prevalent and because the technologies in the lab have, largely, not matured to a clinical implementation stage. Opportunities exist for domestic bioreactor manufacturing, as well as interactions with cord blood expansion, MSC cell therapy, and pharmaceutical drug screening interests.

Sources of Support

The annual total budget for the SKLBE is $6–8 million USD with research funding from the National Basic Research Development Program (973), National High Technology Research and Development Program (863), and National Natural Science Foundation of China (NSFC) representing 20–30 % of total budget. Additional funds come from the institution, Shanghai city and industry partners.

Collaboration Possibilities

Both academic and industrial collaborations are strongly encouraged in SKLBE. The academic committee of SKLBE approved 72 Open Funding grants to domestic and foreign scholars in the past 10 years. SKLBE members have given invited presentations in major international conferences, and scholars are invited to the annual SKLBE forum to share their research findings. A major goal of this international collaboration is to increase the international impact and visibility of the work ongoing at the SKLBE.

One clear missing element in this collaboration strategy is a direct link to stem cell biology and clinical groups, either locally or internationally. This represents a significant growth opportunity. They do obtain primary cells from hospital donors and have discussed some ideas with the National Tissue Engineering Center, located nearby.

Selected References

Chen, M., X. Wang, Z. Ye, Y. Zhang, Y. Zhou, and W.S. Tan WS. 2011. A modular approach to the engineering of a centimeter-sized bone tissue construct with human amniotic mesenchymal stem cells-laden microcarriers. *Biomaterials* 32(30):7532–7542, Epub July 20, 2011.

Chen, T., Y. Zhou, and W.S. Tan. 2009. Influence of lactic acid on the proliferation, metabolism, and differentiation of rabbit mesenchymal stem cells. *Cell Biol. Toxicol.* 25(6):573–586, Epub Jan. 8, 2009.

Fan, J., H. Cai, and W.S. Tan. 2007. Role of the plasma membrane ROS-generating NADPH oxidase in CD34+ progenitor cells preservation by hypoxia. *J. Biotechnol.* 130(4):455–462, Epub May 31, 2007.

Fan, J., H. Cai. Q. Li, Z. Du, and W. Tan. 2012. The effects of ROS-mediating oxygen tension on human CD34(+)CD38(-) cells induced into mature dendritic cells. *J. Biotechnol.* 158(3):104–111, Epub Jan. 15, 2012.

Mei, Y., H. Luo, Q. Tang, Z. Ye, Y. Zhou, and W.S. Tan. 2010. Modulating and modeling aggregation of cell-seeded microcarriers in stirred culture system for macrotissue engineering. *J. Biotechnol.* 150(3):438–446, Epub Oct. 1, 2010.

Wu, W., Z. Ye, Y. Zhou, and W.S. Tan. 2011. AICAR, a small chemical molecule, primes osteogenic differentiation of adult mesenchymal stem cells. *Int. J. Artif. Organs* 34(12):1128–1136, doi:10.5301/ijao.5000007.

Yang, S., H. Cai, H. Jin, and W.S. Tan. 2008. Hematopoietic reconstitution of CD34+ cells grown in static and stirred culture systems in NOD/SCID mice. *Biotechnol. Lett.* 30(1):61–65, Epub Sept. 11, 2007.

Zhang, Y., Y. Zhang, M. Chen, J. Yan, Z. Ye, Y. Zhou, W. Tan, and M. Lang. 2012. Surface properties of amino-functionalized poly(ε-caprolactone) membranes and the improvement of human mesenchymal stem cell behavior. *J. Colloid Interface Sci.* 368(1):64–69, Epub Nov. 18, 2011.

Swiss Center for Regenerative Medicine (SCRM), University Hospital Zurich and University of Zurich

Site Address:	Moussonstrasse 13 CH-8091 Zurich, Switzerland http://www.scrm.uzh.ch/index.html
Date Visited:	February 29, 2012
WTEC Attendees:	T. McDevitt (report author), D. Schaffer, P. Zandstra, N. Moore, H. Sarin
Host(s):	Prof. Dr. Simon Hoerstrup, Head of Regenerative Medicine Program Tel.: +41 (0)44 255 38 01 Fax: +41 (0)44 634 56 08 Simon_Philipp.Hoerstrup@usz.ch

Overview

The Swiss Center for Regenerative Medicine (SCRM) operates under the auspices of the University Hospital Zurich and Zurich Hospital Center for Clinical Research (CRC). The mission of the SCRM is to apply the principles of regenerative medicine for the repair or replacement of diseased tissues with a goal of efficient translation of basic biomedical research to applied regenerative therapies. The SCRM is developing clinically relevant protocols under Good Clinical Practice (GCP) guidelines for the application of cell-based therapies produced using Good Manufacturing Practice (GMP) conditions for cell isolation, expansion and cryopreservation (SCRM Cell and Tissue Biobank). In addition to cell-based therapies, the center specializes in microscale tissue engineering, tissue engineering of autologous living replacement materials, and disease modeling. The Clinical Trials Center (CTC) of the SCRM is a multidisciplinary clinical research unit that provides clinical investigators with the infrastructure and services to conduct high quality patient-oriented research at University Hospital Zurich. The center is funded by several entities, both national (University and Foundation) and international (Netherlands and the European Union Framework Programme 7). The center has collaborations within Switzerland with the Foundation Biobank-Suisse (BBS) and the Blutspende Zurich, and abroad with the Fraunhofer Institute for Biomedical Engineering (IBMT), St. Ingbert, Germany and the VECURA Clinical Research Center, Karolinska University Hospital, Stockholm, Sweden.

The SCRM is led by Dr. Simon Hoerstrup who did postdoctoral research with Dr. John Mayer in Boston before starting his tissue engineering lab in the late 1990s. He was appointed as a professor in biomedical engineering in 2003 and is currently the head of the experimental surgery department, which is rare to find as an independent unit in most universities now.

Research and Development Activities

In 2000, an independent regenerative medicine center was started between the University of Zurich (UZ), the University Hospital of Zurich (UHZ), and ETH Zurich. UZ has excellent basic science, UHZ is the biggest hospital in Switzerland, and ETH is world-renowned for its engineering and has had an increased interest in tissue engineering over the past 10 years. The cell therapies center synergizes the strengths of the three institutions to provide the full pipeline of research to preclinical to human pilot to clinical therapy, including Good Laboratory Practice (GLP) animal studies along the way. A high-quality GMP manufacturing facility was constructed within an existing building on the medical campus to provide the necessary infrastructure to house 5 independent suites that can accommodate up to five clinical trials at a time. Some of the current and anticipated clinical trials being facilitated by the center include:

- an antibody-based approach to block a protein that prevents spinal cord regrowth
- unsorted fractions of bone marrow cells for cardiology repair
- cranio-maxillofacial surgical approach to regrow bone
- an adaptive immunotherapy for hematology and oncology
- clusters of skin as opposed to sheets for pediatric plastic and reconstructive surgery

The regenerative medicine program was brought together by specific topics of interest after first identifying researchers, organizing investigator meetings, and then symposia, first at UHZ and then involving more of the other entities in Zurich. The translational center that houses the GMP facility was the first result of these efforts, but now the research end is being built up with faculty from all three institutions in Zurich.

Dr. Hoerstrup has developed stem cell-based heart valves from a variety of somatic and stem cell sources with a long-term goal of creating a living valve to treat congenital disorders that can grow with an individual during their life-time. As a step in this direction, a clinical study for a tissue engineered artery using amniotic fluid or chorion-derived cells is being pursued based upon encouraging studies in sheep to demonstrate the proof-of-principle for tissue growth. In addition, Dr. Hoerstrup has long-standing collaborations with researchers at Eindhoven in the Netherlands and they have a large EU-funded project on degradable stents for valve deployment in children.

Education and Training

Most of the researchers at UHZ have a dual training background (i.e., M.D./Ph.D.). A new master's program is now being offered and they have one in medicine; this is a new thing in Europe in general. More openness between the M.D. and Ph.D. tracks is being encouraged by allowing passage between UHZ and ETH and this is spawning a new "species" of clinical scientists with more protected time for research. Accelerated clinical training is also being offered for those doing the Ph.D. to reward clinicians in training for pursuing scientific research and this program is actively recruiting and targeting the best people.

The ETH has created a new department of Health Science Technology in order to close the gap between their faculty and UHZ. Truly joint professorships with a 50 % appointment and equal faculty rights in each institution are being created.

Translation

The exit plan strategy from clinical studies is to take the technology transfer out to companies such as Novartis and Roche after good preliminary clinical trials and proof-of-concept have been firmly established. For example, Novartis has a venture fund of $100 million per year, so if something being worked on turns out well, then a company can be started. One example of a successful start-up is a company call "insphero" which produces 3D engineered organoids and functional tissues using a "GravityPLUS" (patent pending) technology. The primary focus is on disease modeling high-throughput platforms for the pharmaceutical industry.

Regulatory Environment

Since they are not part of the EU, Switzerland has their own regulatory agency, but they largely agree with the European Medicines Agency (EMA). The ATMP (Advanced Therapy Medicinal Products) regulation is applied to products that are "living cells" vs. a drug; this includes gene therapy products and somatic cells. Demonstration of the potency and functionality of cell-based products is important, especially the consistency of products. In Switzerland, tissue engineered products are considered transplant products. The Committee for Advanced Therapies (CAT) was established in accordance with regulation 1394/2007 on ATMPs; they serve an advisory (not approval) role.

Sources of Support

As noted above, the center is financially supported by national (University and Foundation) and international funding sources (Netherlands and the European Union Framework Programme 7).

Collaboration Possibilities

A stem cell network in Switzerland was founded as a grass roots effort and initially organized an annual meeting to foster interactions among basic science and basic to applied/translational researchers. It also serves as a bridge to society and provides public education on stem cell research and includes those who know and study educational methods, as well as translation to different languages. Although no money was involved initially, the establishment of a visible and well-connected community could be used to convince the government to provide some funding that individuals can now apply for that is appropriated specifically for stem cell research. The next initiative is to build on this to foster centers of research to further build community.

Summary and Conclusions

The focus of the SCRM is to develop GCP protocols and GMP cell-based therapies for regenerative medicine applications. Center activities combine the expertise and strengths of the 3 primary universities in Zurich in order to provide a complete pipeline from basic to translation to clinical therapies. Successful therapies and technologies are intended to be transferred out to large industry partners for broad implementation.

Selected References

Emmert, M.Y., B. Weber, P. Wolint, L. Behr, S. Sammut, T. Frauenfelder, L. Frese, J. Scherman, C.E. Brokopp, C. Templin, J. Grünenfelder, G. Zünd, V. Falk, and S.P. Hoerstrup. 2012. Stem cell-based transcatheter aortic valve implantation: first experiences in a pre-clinical model. No examples of representative publications? *JACC Cardiovasc. Interv.* 5(8):874–883.

Weber, B., J. Scherman, M.YT. Emmert, J. Gruenenfelder, R. Verbeek, M. Bracher, M. Black, J. Kortsmit, T. Franz, R. Schoenauer, L. Baumgartner, C. Brokopp, I. Agarkova, P. Wolint, G. Zund, V. Falk, P. Zilla, and S.P. Hoerstrup. 2011. Injectable living marrow stromal cell-based autologous tissue engineered heart valves: first experiences with a one-step intervention in primates. *Eur. Heart J.* 32(22):2830–2840.

Tokyo Women's Medical University

Site Address:	Institute of Biomedical Engineering 8-1 Kawadacho Shinjuku, Tokyo 162-8666 Japan http://www.twmu.ac.jp/U/english/index.html
Date Visited:	November 18, 2011
WTEC Attendees:	T. McDevitt (report author), J. Loring, D. Schaffer, N. Kuhn, H. Ali; T. Satoh

(continued)

Host(s):	Professor Teruo Okano
	Tel.: 81-3-5367-9945 Ext.6201
	Fax: 81-3-3359-6046

Professor Mime Egami
Tel.: 81-3-3353-8112 Ext.66213
Fax: 81-3-3359-6046
megami@abmes.twmu.ac.jp
http://www.twmu.ac.jp/ABMES/en

Professor Tatsuya Shimuzu
Tel.: 81-3-5367-9945 Ext.6212
Fax: 81-3-3359-6046
tshimizu@abmes.twmu.ac.jp
http://www.twmu.ac.jp/ABMES/en

Res. Asst. Professor Daisuke Sasaki
Tel.: 81-3-5367-9945 Ext.6223
Fax: 81-3-3358-7428
dsasaki@abmes.twmu.ac.jp

Dr. Manabu Mizutani
Tel.: 81-3-5269-7425
Fax: 81-3-3358-7428
mmizutani@abmes.twmu.ac.jp
Dr. Rie Tanaka
Tel.: 81-3-5269-7425
Fax: 81-3-3358-7428
rtanaka@abmes.twmu.ac.jp

Prof. Masahiro Kino-oka
Laboratory of BioProcess Systems Engineering
Department of Biotechnology
Osaka University
kino-oka@bio.eng.osaka-u.ac.jp

Overview

Tokyo Women's Medical University (TWMU) has the biggest hospital in Japan with many patients who present with difficult diseases that they cannot currently treat. The research institute led by Dr. Okano would like to develop new technologies that can be directly implemented in the hospital setting in order to treat such patients. TWMU and Waseda, which has an engineering school, together formed a 22,000 m² Joint Institute for Advanced Biomedical Sciences ("TWIns") to create a unique interdisciplinary environment. The institute houses a GMP cell processing center and an intelligent alternative operating system to remove tumors during surgery. In Japan, living donors are not allowed, so the need for cell and tissue transplantation therapies is even greater than in countries that do permit organ transplants.

Research and Development Activities

Dr. Okano's personal research program focuses on the use of poly(N-isopropylacrylamide (PNIPAAm) surface coatings to efficiently retrieve cell layers without the use of enzymatic dissociation in order to form cell sheets for tissue engineering applications. They applied for their first patents 20 years ago and began treating patients more than 3 years ago with corneal epithelial constructs; this work was published in the *New England Journal of Medicine* and the clinical study in Japan is ongoing with plans to begin treating patients in France in 2013. In addition, they have treated 10 patients for esophageal cancer with oral mucosal cell sheets, 12 cardiomyopathy patients with myoblast cell sheets, and most recently treated their first patient with a periodontal ligament cell sheet. Last year, the president of the Karolinska Institute visited and has plans to use the cell sheet technology to begin to treat various diseases. Within Dr. Okano's laboratory, medical doctors work with engineers on the further development and applications of cell sheet engineering.

Dr. Tatsuya Shimuzu, a cardiologist, focuses on myocardial tissue repair via the transplantation of cell sheets. They performed animal studies first in rats and then in pigs and found they could obtain significantly greater cell survival with sheets compared to injection of a suspension of the equivalent number of cells. Thicker, electrically interconnected constructs can be obtained simply by stacking individual layers of cardiomyocytes. They have examined a variety of different types of stem and mature cells, including cardiomyocytes, skeletal myoblasts, mesenchymal stem cells (MSCs), cardiac stem cells and cocultures of MSCs and embryonic stem cells with endothelial cells as a means to embed a prevasculature within their constructs. In the future, Dr. Shimuzu would like to use human iPS cells from his collaborator Dr. Matsura as a source for cardiac tissues. They are also working on a tissue factory system that can automatically layer cell sheets as well as monitor culture parameters such as media volume and temperature in order to create ten-layer constructs.

Professor Masahiro Kino-oka is a chemical engineer located at Osaka University who is focused primarily on biomanufacturing technologies for process design and control. He collaborates closely with Drs. Okano, Shimuzu, and other investigators at TWIns on the development of manufacturing platforms for the cell sheet engineering technology. For example, they are developing 3D analysis tools to quantify the distribution of cells within multilayer constructs and the migration of cells over time, analogous to diffusion properties. In a collaborative project with investigators at RIKEN, they are examining the quality of undifferentiated iPS cells and differentiated retinal pigment epithelial (RPE) cells using imaging analysis to study cultures at the individual cell (micro-) to colony (meso-) to entire wells (macro-) simultaneously. The basic research objective is to develop a kinetic model of this multiscale system to understand the extent of de-undifferentiation. They are also working on the design of relatively small (5 mm diameter) automated bioreactors for sub-culture of iPS-derived RPE cells to generate ~10,000 cells for individual retinal surgeries.

Translation

The research being conducted by the team of investigators at TWIns is highly translational, evidenced by the various clinical trials they have been able to initiate for the treatment of different diseases using their cell sheet technology. The infrastructure is critically important to efficient translation of experimental laboratory research to clinical studies. The philosophy of building teams of clinicians, scientists, and engineers clearly permeates all of the lines of research being pursued.

Sources of Support

TWIns is very well-funded by several major sources, including three integrated projects that altogether account for $150 million total. These include the 3rd cabinet promotion fund for Science and Technology (2006–2016) to support the Cell Sheet Tissue Engineering Center (CSTEC), a Global COE (2009–2014) for multidisciplinary education and research center for regenerative medicine (MECREM), and FIRST (Funding Program for World-Leading Innovative R&D on Science and Technology) funding (2010–) to develop the Cell Sheet-Based Tissue & Organ Factory (CSTOF). They benefit from a matching fund program in which 50 % comes from the government and 50 % from industry. The JST funded Dr. Kinooka's collaborative project with RIKEN through a new S-innovation award.

Collaboration Possibilities

In addition to the plethora of local collaborations among investigators, TWIns has a large and growing number of international collaborations for cell sheet engineering. Collaborating institutions include the Karolinska Institute, KorTERMCore (Seoul National), the Wake Forest Institute of Regenerative Medicine, and the University of Utah, which is interested in using cell sheets as a novel platform for predictive and preclinical investigations, as well as the Universities of Alberta, Pittsburgh, Michigan, Rome and Harvard.

Summary and Conclusions

TMWU and TWIns is a very unique and powerful environment to efficiently and effectively translate promising biomedical technologies into clinical practice within the largest hospital in Japan. The long-term vision provided by the leadership and strong collaborations that exist among clinical investigator and basic scientists and engineers have yielded a platform technologies with broad applications for regenerative medicine therapies. Significant funding and strong external collaborations with partners in Japan and internationally enable the technologies being developed at TWMU to be globally implemented.

Selected References

Haraguchi, Y., T. Shimizu, T. Sasagawa, H. Sekine, K. Sakaguchi, T. Kikuchi, W. Sekine, S. Sekiya, M. Yamato, M. Umezu, and T. Okano. 2012. Fabrication of functional three-dimensional tissues by stacking cell sheets *in vitro. Nat. Protoc.* 7(5):850-858.

Kawamura, M., S. Miyagawa, K. Miki, A. Saito, S. Fukushima, T. Higuchi, T. Kawamura, T. Kuratani, T. Daimon, T. Shimizu, T. Okano, and Y. Sawa. 2012. Feasibility, safety, and therapeutic efficacy of human induced pluripotent stem cell-derived cardiomyocyte sheets in a porcine ischemic cardiomyopathy model. *Circulation* 126(11 Suppl. 1):S29-S37, doi:10.1161/CIRCULATIONAHA.111.084343.

Tongji University School of Medicine

Site Address:	Tongji University School of Medicine 1239 Siping Road, Medical Building, Room 505, Shanghai, 200092, China http://med.tongji.edu.cn/oldtj/english/index.htm
Date Visited:	November 16, 2011
WTEC Attendees:	R.M. Nerem, S. Palecek, P. Zandstra (report author), S. Demir, K. Ye, N. Moore, F. Huband
Host(s):	Professor Xiaoqing Zhang Department of Regenerative Medicine and Stem Cell Research Center Tel.: 86-21-65985003 xqzhang@tongji.edu.cn http://med.tongji.edu.cn/shownews.asp?id=2489

(continued)

Professor Guoping Fan
Department of Regenerative Medicine and Stem Cell Research Center
Tel.: 86-21-65985616
guopingfan@gmail.com

Overview

The Tongji University School of Medicine (TUSM) has four departments and one institute:

- Department of Regenerative Medicine
- Department of Pathology and Pathophysiology
- Department of Immunology and Microbiology
- Department of Nursing,
- Institute of Biomedical Engineering and Nanoscience (under reorganization)

TUSM also has five affiliated hospitals:

- Tongji Hospital
- The 10th People's Hospital of Shanghai
- Shanghai East Hospital
- Shanghai Pulmonary Hospital
- The First Maternity and Infant Care Hospital of Shanghai

TUSM offers undergraduate and graduate programs and awards bachelor, master's and doctoral degrees (both M.D. and Ph.D.). TUSM has a faculty of 801 teachers, of whom 210 are professors or associate professors.

Research and Development Activities

The Stem Cell Research Center at Tongji University School of Medicine (TjSCRC) was established in the summer of 2009 under the strong support of Dr. Gang Pei, the President of Tongji University. The center includes one of the four national stem cell banks under the leadership of Dr. Guo-Tong Xu, the Dean of Tongji University School of Medicine. As of 2010, TjSCRC has recruited more than a dozen of principal investigators (PIs), including Drs. Yi E. Sun (Chair)/Yuping Luo, Guo-Tong Xu, Xiaoqing Zhang, Jun Xu, Siguang Li, Guoping Fan/Zhigang Xue, Weidong Zhu, Yunfu Sun, Jialin Zheng and Zhengliang Gao. The center is expected to accommodate 30 PIs and 400–500 people, about half of whom will be clinicians and physician scientists. The mission of the center is to perform excellent basic and translational stem cell research using human stem cells in combination with modern biomedical approaches including next-generation sequencing, genomic, epigenetic, and proteomic profiling, as well as functional physiology and imaging technologies (including optogenetics and fMRI), to model human diseases, to enable stem cell-based drug screenings, and to explore cell replacement/transplantation-related translational approaches. Currently, research of the center is focused primarily on neurological diseases, including retinal degeneration diseases.

Dr. Xiaoqing Zhang is the Assistant Dean of the School of Medicine and a professor in the Department of Regenerative Medicine. This department is the academic home for investigators of the TjSCRC outlined above. Dr. Zhang received his Ph.D. in the State Key Laboratory of Neurobiology and Pharmacology Research Center, Shanghai Medical College, Fudan University, Shanghai, China. He did further research training between 2005 and 2009 at the Waisman Center, University of Wisconsin-Madison. In 2009–2010 he was an Assistant Scientist, in charge of the iPS core facility in Waisman Center. His research is in the area of human embryonic stem cells (hESCs) and human induced pluripotent stem cells (hiPSCs). In particular, his work focuses on the development of neural differentiation systems. Projects under way in his lab include understanding the role of the transcription factor Pax6 in human neuroectoderm differentiation; neural disease modeling using iPSC; and drug screening using iPSC derived neural cells.

Dr. Guoping Fan, a member of the TjSCRC, is also a professor in the Department of Human Genetics at UCLA. He has obtained a visiting professorship at the Tongji University School of Medicine and spends several weeks per year in China. He earned his Ph.D. degree in Neuroscience at Case Western Reserve University. After conducting his postdoctoral training at the Whitehead Institute for Biomedical Research/MIT, he joined UCLA faculty in early 2001. His current lab at UCLA focuses on addressing epigenetic mechanisms underlying self-renewal and differentiation of pluripotent stem cells, neural development, and adult brain function. In particular he utilizes molecular and genetic approaches to investigate how DNA cytosine methylation and its associated components, which include methyl-CpG binding proteins and histone modification enzymes, regulate gene expression, cell differentiation and reprogramming, and neural plasticity in mammalian systems. At TUSM, he teamed up with Dr. Zhigang Xue's lab to focus on translational studies such as stem cell-based therapy of eye disorders.

Stem cell bioengineering research is less developed than the basic stem cell areas, however exciting new interactions, especially with the Institute for Advanced Materials and Nanobiomedicine, are emerging. This institute will join Tongji University School of Medicine in the near future. The institute includes tissue engineering researcher Dr. Xuejun Wen, who teamed up with Dr. Peng Zhao's lab to focus on the design of novel biomaterials and scaffolds. Interactions between the stem cell and biomaterials researchers are growing quickly and can be a significant strength of this group. Prof. Xuejun Wen holds a visiting professorship at the Tongji University. He is also the Hansjörg Wyss Endowed Chair and Professor in the Department of Bioengineering at Clemson.

Example areas of research that have a bioengineering component include:

- Injectable hydrogels for stroke and spinal cord injury
- Cartilage tissue engineering
- Synthetic substrates for hPSC culture

Translation

The perspective was expressed that industry in China is more focused on cord blood stem cells than any other type. Local and international pharmaceutical companies are investing heavily in high-throughput screening facilities and projects in China. One advantage in the commercialization train is the ability to move to larger animal models quickly; for example a new hydrogel developed is moving to a monkey model.

Sources of Support

Support comes from various sources, including Tongji University intramural start-up funds, the Ministry of Science and Technology of China (973 programs), and grants funded by the National Natural Science Foundation of China and the Shanghai municipal government.

Summary and Conclusions

Given the tremendous support through Chinese government and Shanghai municipal government, stem cell research develops quickly at Tongji University. By intense collaboration between researchers in basic science and materials engineering with clinicians, Tongji University is paving a way to promote translational research of stem cells and regenerative medicine.

Selected References

Beachley, V., and X. Wen. 2009. Fabrication of nanofiber reinforced protein structures for tissue engineering. *Mater. Sci. Eng. C Mater. Biol. Appl.* 29(8):2448–2453.

Beachley, V., and X. Wen. 2010. Polymer nanofibrous structures: Fabrication, biofunctionalization, and cell interactions. *Prog. Polym. Sci.* 35(7):868–892.

Hu, B.-Y., J.P. Weick, J. Yu, L.-X. Ma, X.-Q. Zhang, J.A. Thomson, S.C. Zhang. 2010. Neural differentiation of human induced pluripotent stem cells follows developmental principles but with variable potency. *Proc. Natl. Acad. Sci. USA* 107(9):4335–4340.

Li, X.-J., and X.-Q. Zhang (co-first author), M.A. Johnson, Z.-B. Wang, T. Lavaute, and S.C. Zhang. 2009. Coordination of sonic hedgehog and Wnt signaling determines ventral and dorsal telencephalic neuron types from human embryonic stem cells. *Development* 136(23): 4055–4063.

Ren, J., P. Zhao, T. Ren, S. Gu, K. Pan. 2008. Poly (D.L.-lactide)/nano-hydroxyapatite composite scaffolds for bone tissue engineering and biocompatibility evaluation. *J. Mater. Sci. Mater. Med.* 19(3):1075–1082, Epub Aug. 15, 2007.

Wang, A., K. Huang, Y. Shen, Z. Xue, C. Cai, S. Horvath, and G. Fan. 2011. Functional modules distinguish human induced pluripotent stem cells from embryonic stem cells. *Stem Cells Dev.* 20(11):1937–1950, Epub June 20, 2011.

Whatley, B.R., J. Kuo, C. Shuai, B.J. Damon, and X. Wen. 2011. Fabrication of a biomimetic elastic intervertebral disk scaffold using additive manufacturing. *Biofabrication* 3(1):015004, Epub Feb. 22, 2011.

Zhang, X.-Q., T.C. Huang, J. Chen, T.M. Pankratz, J.-J. Xi, J. Li, Y. Yang, M.T. LaVaute, X.-J. Li, M. Ayala, I.G. Bondarenko, Z.-W. Du, Y. Jin, G.T. Golos, and S.-C. Zhang. 2010. Pax6 is a human neuroectoderm cell fate determinant. *Cell Stem Cell* 7(1):90–100.

Tsinghua University, School of Medicine

Site Address:	School of Life Sciences
	Tsinghua University
	Beijing, China 00084
	http://www.tsinghua.edu.cn/publish/meden/3349/index.html
Date Visited:	November 15, 2011
WTEC Attendees:	R.M. Nerem (report author), S. Palecek, P. Zandstra, S. Demir, K. Ye, N. Moore, F. Huband
Host(s):	Professor Wei Guo
	Center for Stem Cell Biology and Regenerative Medicine
	Tel.: 86-10-62782975
	weiguo@mail.tsinghua.edu.cn
	http://www.tsinghua.edu.cn/publish/meden/6975/2010/201012161817355 18709779/20101216181735518709779_.html
	Professor Jie Na
	Center for Stem Cell Biology and Regenerative Medicine
	Tel.: 86-10-62781094
	Fax: 86-10-62772741
	jie.na@mail.tsinghua.edu.cn
	Professor Yanan Du
	Department of Biomedical Engineering
	Tel.: 86-10-62781691
	Fax: 86-10-62781545
	duyanan@mail.tsinghua.edu.cn

Overview

At Tsinghua University the WTEC team met with three young investigators: Professor Wei Guo, Dr. Jie Na, and Professor Yanan Du. All three are part of the Stem Cell Center, which is located in the Department of Basic Medical Science. This center was started a year ago at the time they returned to China in 2010. Professor Guo Wei and Dr. Jie Na are in the School of Life Sciences at Tsinghua University. Dr. Du is in the Department of Biomedical Engineering which is in the School of Medicine. The Department of Biomedical Engineering has 24 faculty, having doubled in size over the past 2 years. The main growth has been in cell and tissue based research. Professor Wei Guo gave an overview of the School of Medicine and the School of Life Sciences. In total there are 85 full professors and 45 associate professors in this school. The plan is to expand over the next 5 years to a total of 175 in 2013 and 250 in 2016. Professor Guo also indicated that science and technology in China is growing at 20 % annually. He indicated that there are pockets of research excellence and that in the government funding system the quality of peer review varies by agency. Also, because of limitations on salary, it is hard to get talented postdoctoral researchers. Finally, in the context of the translation of discovery science into therapies, the Chinese government appears to recognize the need for the regulation of stem cell therapies.

Research and Development Activities

Professor Guo was trained at M.D. Anderson in Houston and at UCLA. He is interested in hematopoiesis and in the self-renewal and loss of fetal and adult human stem cells. He also investigates cancer stem cells compared to normal stem cells (Fig. B.18). He uses English in the laboratory and "forces" his students to use English as much as possible.

Dr. Jie Na did her Ph.D. in the United States and then spent a number of years in the United Kingdom at Cambridge and Sheffield Universities. She is interested in signaling pathways that regulate stem cell fate decisions and in the regulation of cell division in early embryos and in embryonic stem cells. She also is interested in chromosomal instability in human ESCs and in the development of an RNA-based technology for cell programming. Professor Yanan Du worked with Dr. Hanry Yu for his Ph.D. at National University of Singapore in the area of liver tissue engineering. He then did postdoctoral research in the United States before returning to China. He has an interest in microscale technologies including three-dimensional patterning. He is quite new to the stem cell field. His laboratory has just started and focuses on the differentiation of embryonic stem cells into hepatocytes and uses perfusion culture and both 2D and 3D platforms with mechanical and chemical cues. In addition to Dr. Du, who is in the School of Medicine and closely linked to

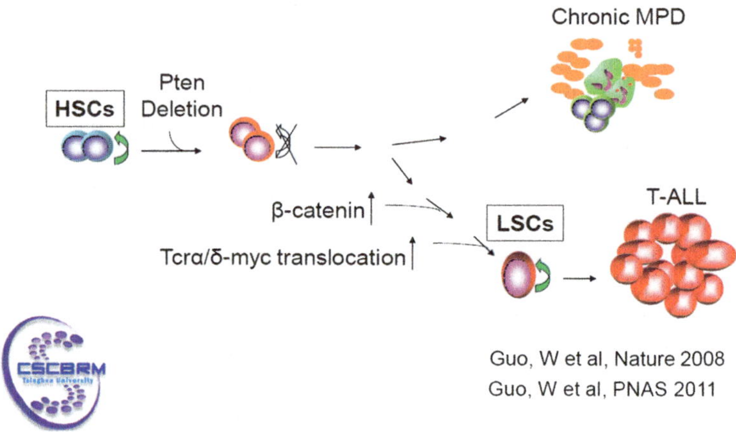

Fig. B.18 Multiple "Hits" result in leukemia stem cell (*LSC*) formation (Courtesy of Prof. W. Guo)

the School of Life Sciences, there are four other faculty applying engineering approaches to the field of stem cells. These are as follows:

- Fu-zai Cui, Department of Material Sciences and Engineering, working on the development of biomaterials designed to regulate stem cell fate, especially differentiation down the neuronal lineage
- Guoqiang Chen, School of Life Sciences, also using biomaterials to control mesenchymal stem cell fate
- Guo-An Luo, Department of Chemistry, applying traditional Chinese medicine to control the differentiation of human embryonic stem cells to the cardiac lineage
- Wei Sun, Department of Mechanical Engineering, using biomechanical approaches to regulate stem cell fate

Although Drs. Cui and Sun are in engineering departments, there appears to be no mechanism to promote interaction between engineering and the life sciences and medicine. Although there is strong basic stem cell biology, the interactions and collaborations between engineers and biologists are still under development.

Translation

This new Stem Cell Center is focused on basic, discovery research. The goal is 15 principal investigators in stem cell biology, however, collaborations with clinicians also is a focus. It was noted that there is an ongoing stem cell therapy clinical trial using cells from aborted embryos in the treatment of spinal cord injury. The military FDA is providing oversight to this trial. It also was noted that regulations for clinical trials in China are becoming more stringent.

Sources of Support

Research funding from MOST and NSFC. The Chinese Department of Education provided the initial start-up funding of the center.

Collaboration Possibilities

All three investigators with whom the WTEC team met have strong ties to the United States and this provides the foundation for collaborations in the future.

Summary and Conclusions

Tsinghua University is the MIT of China. The university is No. 1 in SCI publications among Chinese universities, second to the United States overall; however, SCI publications are doubling every few years. Tsinghua University is the top feeder school to Ph.D. programs in the United States in science and engineering. In addition, China is offering incentives for Chinese researchers to return to the country, and two of the Stem Cell Center faculty are based in the United States, with appointments and laboratories also at Tsinghua University.

Selected References

Guo, W., S. Suzanne, J.Y. Chen, B. Valamehr, S. Mosessian, H. Shi, N.H. Dang, C. Garcia, M.F. Theodoro, M. Varella-Garcia, and H. Wu. 2011. Suppression of leukemia development caused by PTEN loss. *Proc. Natl. Acad. Sci. USA* 108:1409–1414.

Guo, W., J.L. Lasky, C.J. Chang, S. Mosessian, X. Lewis, Y. Xiao, J.E. Yeh, J.Y. Chen, L.M. Iruela-Arispe, M. Varella-Garcia, and H. Wu. 2008. Multi-genetic events collaboratively contribute to Pten null leukemia stem cell formation. *Nature* 453:529–533.

University of Tokyo, Hongo Campus, Department of Biomedical Engineering

Site Address:	Department of Biomedical Engineering
	Graduate School of Medicine
	7-3-1, Hongo, Bunkyo-ku,
	Tokyo, 113-0033 Japan
	http://www.c.u-tokyo.ac.jp/eng_site/about/history.html
Date Visited:	November 14, 2011
WTEC Attendees:	T. McDevitt, J. Loring (report author), D. Schaffer, L. Nagahara, N. Kuhn, H. Ali, M. Imaizumi
Host(s):	Kimiko Yamamoto, M.D., Ph.D
	Tel.: 81-3-5841-3564
	Fax: 81-3-5841-3589
	k-yamamoto@umin.ac.jp
	http://bme-sysphysiol.m.u-tokyo.ac.jp/

Overview

Dr. Yamamoto has a small basic research laboratory working primarily on the biomechanics of blood flow and cellular response to shear stress in the vascular system. She has designed devices to control flow across cultures of endothelial precursor cells, with the goal of understanding the effects of fluid shear on differentiation and signal transduction in endothelial cells. The major areas of investigation are:

- Endothelial cell responses to shear stress
- Shear stress-mediated regulation of endothelial gene expression
- Shear stress-induced cell differentiation
- Shear stress signal transduction in endothelial cells

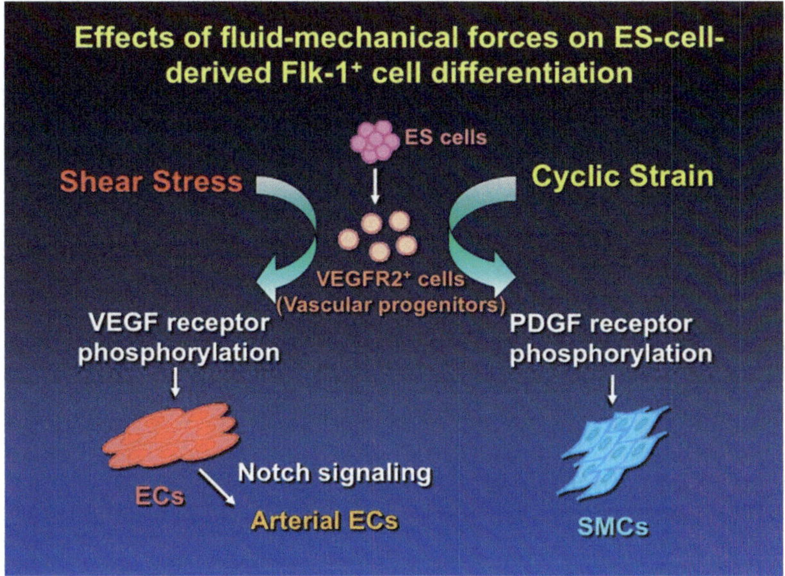

Fig. B.19 Effects of fluid-mechanical forces on ES-cell-derived Flk-1* cell differentiation (Courtesy of Dr. K. Yamamoto)

Research and Development Activities

Dr. Yamamoto has designed devices for exposing cells to shear stress to induce vascular differentiation of mouse embryonic stem cells (ES cells, Fig. B.19). She starts with an ES cell-derived vascular precursor population that expresses VEGF receptor 2, and looks for the effects of shear stress on their subsequent development. She finds that flow increases the expression of markers of endothelium, suggesting that this may be a way to promote vascular differentiation of ES cells. She also sees that tube-like structures form in cultures subjected to shear stress. Using cyclic strain instead of shear stress, she can induce expression of smooth muscle actin.

Translation

There are no immediate plans for translation; the work is early-stage and limited to mouse ES cells. Interaction with Kyoto University would be necessary to obtain human pluripotent stem cells.

Sources of Support

Dr. Yamamoto's funding has been from principal investigator (PI)-initiated grants from the Ministry of Education and JST. There is no formal mechanism for joint funding for collaborators.

Collaboration Possibilities

The devices are cleverly designed and there may be options for collaboration with investigators using human cell types.

Summary and Conclusions

This is a small laboratory with a single PI who is building and testing innovative devices for controlling differentiation of stem cells into vascular tissue. There are promising early results, but not yet any testing with human cells.

Selected References

Ishibazawa, A., T. Nagaoka, T. Takahashi, K. Yamamoto, A. Kamiya, J. Ando, and A. Yoshida. 2011. Effects of shear stress on the gene expressions of endothelial nitric oxide synthase, endothelin-1, and thrombomodulin in human retinal microvascular endothelial cells. *Invest. Ophthalmol. Vis. Sci.* 52:8496–8504.

Yamamoto, K., and J. Ando. 2011. New molecular mechanisms for cardiovascular disease: blood flow sensing mechanism in vascular endothelial cells. *J. Pharmacol. Sci.* 116:323–331.

University of Tokyo, Hongo Campus, Laboratory of Cell Growth and Differentiation

Site Address:	Laboratory of Cell Growth and Differentiation, 3rd floor, Life Sciences Research Bldg, Institute of Molecular and Cellular Biosciences 1-1-1 Yayoi, Bunkyo-ku Tokyo 113-0032, Japan http://www.c.u-tokyo.ac.jp/eng_site/about/history.html
Date Visited:	November 16, 2011

(continued)

WTEC Attendees:	T. McDevitt, J. Loring, D. Schaffer (report author), N. Kuhn, H. Ali, M. Imaizumi
Host(s):	Professor Atsushi Miyajima Tel.: +81-3-5841-7884 Fax: 81-3-5841-8475 miyajima@iam.u-tokyo.ac.jp http://www.iam.u-tokyo.ac.jp/cytokine/index.html

Fig. B.20 WTEC panel members at Professor Miyajima's laboratory

Overview

Professor Miyajima works on developing functional liver cells from stem cells, both tissue stem cells and induced pluripotent stem cells. In general, the Institute of Molecular and Cellular Biosciences has 20 principal investigators (PIs) and some others who work on neural, cardiac, and cancer stem cells. All work has a strong developmental biology approach.

Research and Development Activities

Functional hepatocytes can be useful for numerous applications, including cell/ gene therapy, an artificial liver, drug development, or basic investigation of disease mechanisms. Primary cells are currently available in limited amounts, and

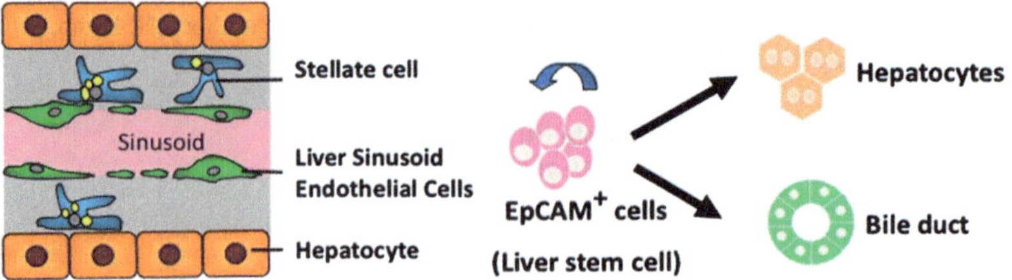

Fig. B.21 Basic unit of liver and liver stem cells (Courtesy of A. Miyajima, University of Tokyo)

stem cells represent a promising means to create hepatocytes at scale (Tanaka 2011; Okabe 2009). However, even if cells could be isolated or differentiated, liver function is lost, likely because differentiated hepatocytes probably need other cell types and more complex structures that mimic native liver tissue. Professor Miyajima's goals are to develop better means to isolate, differentiate, and culture the stem cells.

For isolation, he conducts gene expression analysis to identify new markers, generates antibodies against these markers, and utilizes the antibodies for prospective cell isolation. He has isolated and differentiated hepatoblasts (from E8.5) into hepatocytes and bile duct cholangiocytes (in 3D in a Matrigel+collagen mixture, Fig. B.21). He has also isolated mesenchymal cells from liver that can become portal fibroblasts and stellate precursors. He has also identified markers for endothelial cell precursors. Furthermore, his group is attempting to combine these various cell types toward making a functional liver. An alternative is to generate these various cells from embryonic stem or iPS cells. He is also developing a culture system to create insulin-producing pancreatic islets from iPS cells.

Translation

Prof. Miyajima stated that in general it is challenging to translate work to the private sector in Japan. Major pharmaceutical companies have not yet exhibited a strong interest in the stem cell field. Approximately 10 years ago, the government had some limited funding programs to encourage professors to start venture companies. Prof. Miyajima and his colleagues formed a company focused on tumor-associated antigens.

Sources of Support

The lab has been funded by MEXT and JST.

Collaboration Possibilities

Prof. Miyajima is in general open to collaboration with U.S. investigators, including tissue engineers who work with scaffolds and matrices, medical doctors involved in deriving patient-specific iPS cells, and potentially investigators working with hES cells (as there are considerable restrictions on such research in Japan).

Summary and Conclusions

Prof. Miyajima has a strong program in basic hepatic stem cell biology with applications toward the clinic. His general approach of using stem cells to make the individual cellular components, then reassemble them into an organ, can potentially interface well with biomaterials scientists. In addition, the lab has a strong record and is well-resourced, and there is strong potential for collaboration with both tissue engineers and medical researchers.

Selected References

Okabe, M., Y. Tsukahara, M. Tanaka, K. Suzuki, S. Saito, Y. Kamiya, T. Tsujimura, K. Nakamura, A. Miyajima. 2009. Potential hepatic stem cells reside in EpCAM+ cells of normal and injured mouse liver. *Development* 136:1951–1960.

Saito, H., M. Takeuchi, K. Chida, and A. Miyajima. 2011. Generation of glucose-responsive functional islets with a three-dimensional structure from mouse fetal pancreatic cells and iPS cells *in vitro*. *PLoS One* 6:e28209.

Tanaka, M., T. Itoh, N. Tanimizu, and A. Miyajima. 2011. Liver stem/progenitor cells: their characteristics and regulatory mechanisms. *J. Biochem.* 149:231–239.

University of Tokyo, Komaba Campus, Research Center for Advanced Science and Technology

Site Address:	Room#507 Bldg. 4 4-6-1 Komaba, Meguro-ku, Tokyo 153-8904 JAPAN http://www.c.u-tokyo.ac.jp/eng_site/about/history.html
Date Visited:	November 17, 2011
WTEC Attendees:	T. McDevitt, J. Loring, D. Schaffer (report author), N. Kuhn, H. Ali, T. Satoh

(continued)

Host(s):	Professor Koji Ikuta
	Tel.: 81-3-5452-5160
	Fax: 81-3-5452-5161
	ikuta@rcast.u-tokyo.ac.jp
	http://www.micro.rcast.u-tokyo.ac.jp/index_e.htm

Overview

Professor Ikuta was previously at Nagoya University and recently moved to the University of Tokyo. He has a background in materials science, biophysics, and robotics. His research program focuses on lab-on-a-chip analysis, biomedical microelectromechanical systems (MEMS), and medical robotics. The overall goal is to develop novel technologies in materials science and microfabrication to address problems in biology and medicine.

Research and Development Activities

Professor Ikuta works on several project areas, and each is built on the idea of translating progress in solid state materials science to problems in biology. This entails moving from 2D to 3D processing, from silicon to polymers, and from dry to wet chemistry. The first problem area applies the concept of integrated circuits to problems related to biochemical reactions and biomedical diagnostics/analysis. He has created a modular, layered system in which each layer has a modular function (e.g., reactor, concentrator, mixer, pump) as well as fluidics, and these modules can be stacked within a frame to build a device with complex functionality.

In other work, he is creating technologies for photopolymerization (for example using multiphoton radiation for precise spatial control) to fabricate structures in three dimensions (Ikuta et al. 2004; Yamada et al. 2008). These include multimaterial component systems. In one example (Fig. B.22), he created a microscale polymeric spiral structure. Magnetic particles can be embedded in such a structure, which provides a means to induce the material to spin. This has applications both for locomotion or pumping. In another example, he can build multimaterial structures by photopolymerization of one component, rinsing, adding another monomer solution, and further polymerization to build up the structure.

He has also microfabricated small devices that can be actuated by optical trapping. Fig. B.23 shows a pincer device in which optical trapping at several positions (which appear as the halos) can enable locomotion or pinching of objects. This can be utilized to deliver force to biological samples including individual cells, and can even be used to gather mechanical information about cells (e.g., force-displacement curves). Furthermore, Professor Ikuta has an interest in interfacing devices in general, such as "microfingers" for surgical application with robotic control. He has even moved a "nanorobot" inside the cell, as a means to gather intracellular mechanical information.

Fig. B.22 Microscale
polymeric spiral (Courtesy
of K. Ikuta Laboratory)

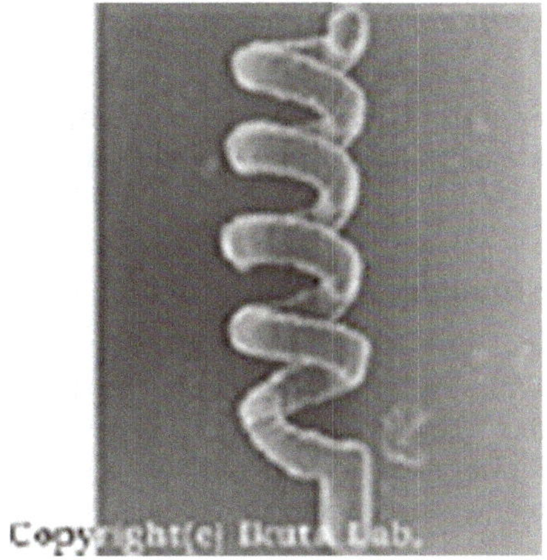

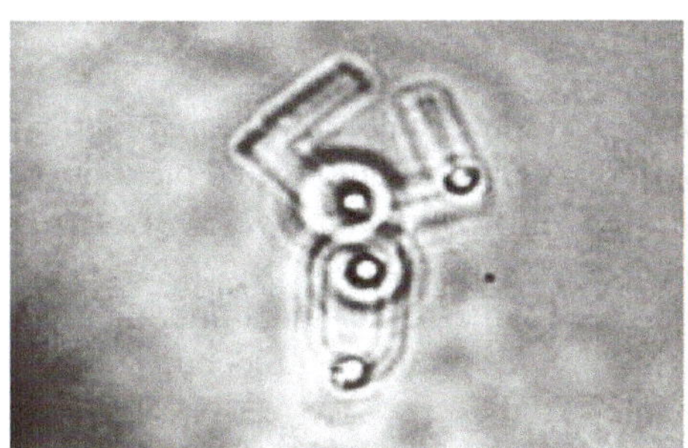
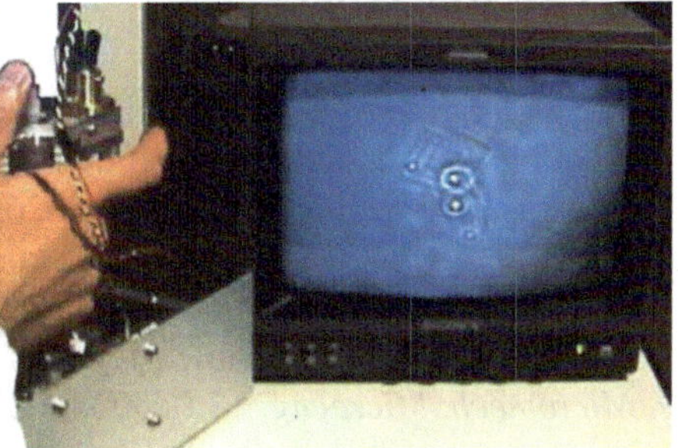

Fig. B.23 Microscale pincer (*left*) and control station (*right*) (Courtesy of K. Ikuta Laboratory)

Translation

Professor Ikuta did not express a strong interest in translating his work into the private sector, for example through the formation of a venture or start-up company. In general, the venture capital community in Japan is apparently relatively small, and the large companies are typically either instrumentation companies (e.g., Hitachi) or pharmaceutical companies that do not strongly support biotechnology or biomedical research.

Sources of Support

The lab is supported by the Japan Science and Technology Agency (JST, a 5 year Core Research for Evolutional Science and Technology (CREST), and by the Japan Society for the Promotion of Science (JSPS).

Collaboration Possibilities

Professor Ikuta leads a highly innovative research program with many opportunities for collaboration with biologists and clinicians.

Summary and Conclusions

While Professor Ikuta has not broadly applied his work to study stem cells to date, the suite of technologies he is developing has strong potential for small-scale biological manipulation and analysis. The potential for collaboration with stem cell biology and engineering groups is very strong.

Selected References

Ikuta. K., A. Yamada, and F. Niikura. 2004. Real three-dimensional microfabrication for biodegradable polymers: demonstration of high-resolution and biocompatibility for implantable microdevices. *26th Annual International Conference of the IEEE Engineering in Medicine and Biology Society (IEMBS '04)*, pp. 2679–2682.
   Yamada, A., F. Niikura, and K. Ikuta. 2008. A three-dimensional microfabrication system for biodegradable polymers with high resolution and biocompatibility. *J. Micromech. Microeng.* 18:025035–025043.

University of Tokyo, Komaba II Campus, Institute of Industrial Science

Site Address:	Institute of Industrial Science (IIS), 4-6-1 Komaba, Meguro-ku Tokyo 153-8505, Japan http://www.c.u-tokyo.ac.jp/eng_site/index.html
Date Visited:	November 18, 2011
WTEC Attendees:	T. McDevitt (report author), J. Loring, D. Schaffer, N. Kuhn, H. Ali, T. Satoh
Host(s):	Professor Yasuyuki Sakai Tel.: 81-3-5452-6352 Fax: 81-3-5452-6352 sakaiyas@iis.u-tokyo.ac.jp http://www.bioeng.t.u-tokyo.ac.jp/english/faculty/members/sakai.html (Dept) http://envchem.iis.u-tokyo.ac.jp/sakai/english/index.html (Lab)

Fig. B.24 Schematic of two-chambered bioreactor. A low volume chamber fabricated in PDMS and containing cells sits above a large volume perfused chamber. The two chambers are separated by a porous membrane (Courtesy of Y. Sakai, University of Tokyo)

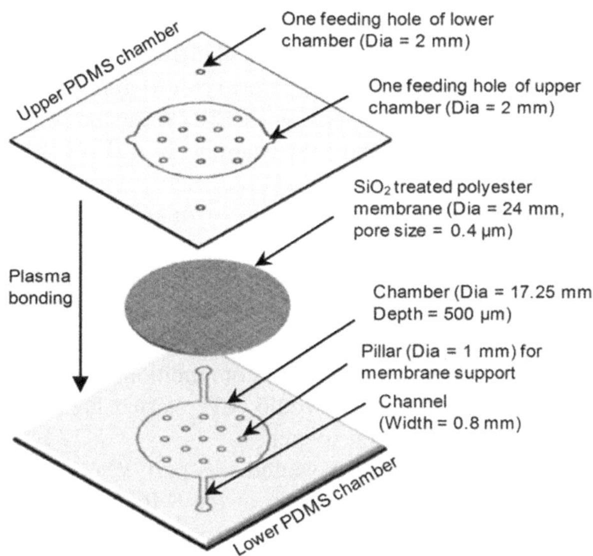

Overview

Dr. Sakai's research group is interested in stem/progenitor cell engineering for applications in organ engineering and cell-based assays. Much of his research is particularly focused on the liver because of the prevalence of hepatitis (B and C) among the Japanese due to reuse of syringes to administer vaccinations nearly 20 years earlier. A significant focus of Dr. Sakai's research program is on the application of microfluidic bioreactor systems for the development of cell-based assays.

Research and Development Activities

In one line of research, the objective was to examine the effects of diffusional signaling of endogenously produced morphogens secreted by ESCs, analogous to self-regulation of embryonic patterning of germ lineages. For these studies, bioreactor chambers were separated by a permeable membrane (Fig. B.24) and the cells were cultured in the presence or absence of leukemia inhibitory factor (LIF). The lower chamber was intended to supply nutrients and the top chamber was used to actually grow the cells. Their published results found better preservation of pluripotency using the microfluidic device configuration compared to conventional culture settings (Chowdhury et al. 2010). They found that this was due to mES cells secreting low amounts of BMP4, which is not relevant in large-scale culture (i.e., 6-well plate), but accumulates to much higher concentrations in the low volume microcultures they are studying. In macroscale cultures, neuronal differentiation is observed, but in microscale volumes, mesodermal differentiation is prevalent, due to the accumulation of BMP4, in this case secreted by differentiated cells cultured with serum-free, KSR medium (Chowdhury et al. 2012). During the course of the studies, mathematical modeling was performed to examine ligand secretion, binding, and trafficking. Although many microfluidics researchers focus on flow and delivery of exogenous factors, few people focus on use of microfluidics to examine effects of endogenously secreted factors. Combining modeling with experimental research remains rare among most faculty. In the instance of Dr. Sakai's work, modeling was initiated after experimental work and interesting findings were attained, but measuring the "parameters of interest" is difficult (nearly impossible) at present. In retrospect, Dr. Sakai noted that he wished that computational modeling had been started earlier in the project because they would have started measuring certain factors and taken them into consideration sooner.

The second project presented was part of a collaboration with Prof. Kino-Oka (Osaka University) that is part of the S Innovation project, which is an industry-academia collaborative research program. The project is focused on the system development of process and quality controls for the differentiation of hiPS toward retinal pigment epithelial (RPE) cells. The industry partner, Shimadzu Corporation, is responsible for making the microscale devices. The Japanese government is supportive of RPE therapy development; one advantage compared to other therapeutic targets is the relatively small number of cells needed to treat individual patients. Dr. Sakai's contribution to the project is the large-scale propagation of ES/iPS cells in suspension culture using hydrogel microencapsulation. The advantages of hydrogen microencapsulation are avoiding direct exposure to shear forces, prevent uncontrolled aggregate development and potentially make better use of autocrine/paracrine factors. They compared three different types of alginate capsule under LIF+ conditions: (1) solid alginate, (2) solid alginate coated with PLL (MWCO = 20,000), and (3) hollow alginate created by EDTA treatment following PLL coating (Fig. B.25). They observed expression of pluripotent transcription factors (Sox2, Oct4, Nanog) in all cases and the coated microcapsules promoted significant increases in the expression of the stem markers compared to unencapsulated aggregates, indicating better pluripotency preservation in suspension culture format.

Translation

Translational activities include the large-scale expansion of differentiated cells from hiPS, and the system development of process and quality controls concerning the differentiation of hiPS cells toward RPE cells, in collaboration with Shimadzu Corporation and Osaka University. Shimadzu is making special microdevices to control the differentiation of the cells. Commercialization plans are to transfer the technology to a large company; the technology is currently being evaluated and negotiations are under way.

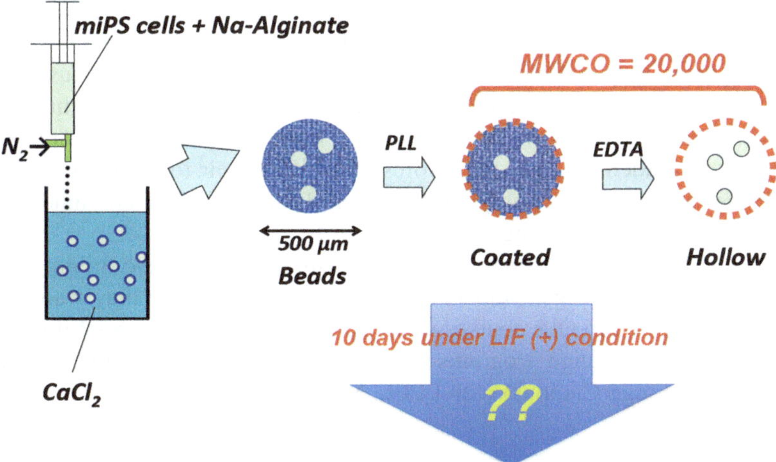

Fig. B.25 Mouse iPS cells encapsulated within different forms of alginate beads. The effects of solid (uncoated) alginate beads, solid alginate beads coated with poly-L-lysine (PLL), or hollow alginate beads coated with PLL on mouse iPS cell pluripotency were examined (Courtesy of Y. Sakai, University of Tokyo)

Sources of Support

The first project presented by Dr. Sakai was supported by funding from "CREST (Core Research for Evolutional Science and Technology) PJ," which is a large JST funding program, and is a collaboration with Prof. Miyajima, from IMCB, Tokyo University, whom we met with earlier in the trip. Additional funding for Dr. Sakai's research also comes from the Japan Science and Technology Agency, Ministry of Education, Culture, Sports, Science & Technology (MEXT), Ministry of Health and Welfare, and the Ministry of Industry and Economy.

Collaboration Possibilities

Several trainees from Dr. Sakai's laboratory met with us as well, including one from Matthias Lutolf's group at École Polytechnique Fédérale de Lausanne (EPFL) who is examining the use of PEG-based hydrogels.

Summary and Conclusions

Although rooted in engineering, the quality and focus of the biology research being performed was very impressive. The University of Tokyo doesn't have a direct relationship with a medical institute, but Dr. Sakai is establishing a collaboration with

Dr. Miyajima. Currently, most collaborations between engineers and biology or clinical faculty are very individual based and motivated.

Twenty years ago, chemical engineering in Japan was heavily focused on the production of biologics. After that, biological research was not continued in the universities, and industry gave up because they were facing difficulties at that time. As a result, in academia, almost all professors shifted their focus to environmental engineering. However, now, only Dr. Sakai and Prof. Kino-Oka continued to conduct research on bioreactor engineering.

Currently there are no formal programs for students to receive training in stem cell engineering or biology. Students have to seek out courses they want/need to take, and are allowed to take courses in life sciences or medical school if they want to do so. Mutual exchange programs have been started by faculty members at the main campus (Hongo Campus) with EPFL, M.D. Anderson, and MIT.

Selected References

Chowdhury, M. M., T. Katsuda, K. Montagne, H. Kimura, N. Kojima, H. Akutsu, T. Ochiya, T. Fujii, and Y. Sakai, 2010. Enhanced effects of secreted soluble factor preserve better pluripotent state of embryonic stem cell culture in a membrane-based compartmentalized micro-bioreactor. *Biomed. Microdev.* 12(6):1097–1105.

Chowdhury, M. M., H. Kimura, T. Fujii, and Y. Sakai, 2012, Induction of alternative fate other than default neuronal fate of embryonic stem cells in a membrane-based two-chambered microbioreactor by cell-secreted BMP4. *Biomicrofluidics* 6:014117.

University of Tokyo, Shirokanedai Campus

Site Address:	The University of Tokyo, The Institute of Medical Science
	Stem Cell Bank/Division of Stem Cell Therapy
	Center for Stem Cell Biology and Regenerative Medicine
	4-6-1 Shirokanedai, Minato-ku, 108-8639 Tokyo, Japan
	http://www.ims.u-tokyo.ac.jp/imswww/About/Map-e.html
Date Visited:	November 14, 2011
WTEC Attendees:	T. McDevitt, J. Loring, D. Schaffer (report author), L. Nagahara, N. Kuhn, H. Ali, M. Imaizumi

(continued)

Host(s):	Professor Makoto Otsu Tel.: +81-3-5449-5129 Fax: +81-3-5449-5451 motsu@ims.u-tokyo.ac.jp http://stemcell-u-tokyo.org/en/sct/ Dr. Motoo Watanabe Tel.: 81-3-5447-7771 Fax: 81-3-5447-7772 mwatanab@ims.u-tokyo.ac.jp

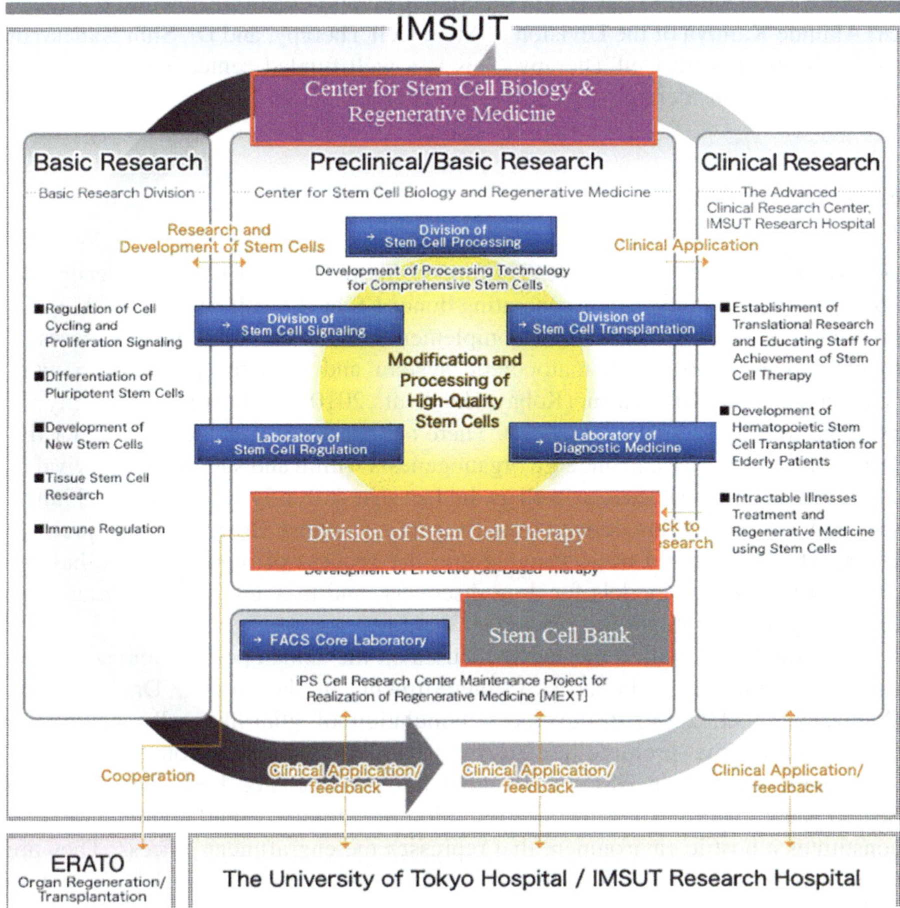

Fig. B.26 Organization of IMSUT (Courtesy of IMSUT, from http://stemcell-u-tokyo.org/en/center/)

Overview

The Center for Stem Cell Biology and Regenerative Medicine lies within the Institute of Medical Science of the University of Tokyo (IMSUT). The center is organized by Professor Hiromitsu Nakauuchi (Director). It houses a number of divisions and core facilities focused on various aspects of basic stem cell biology, translational efforts, and clinical research (Fig. B.26). Our hosts included Prof. Makoto Otsu of the Division of Stem Cell Therapy within the Center for Stem Cell Biology and Regenerative Medicine, Dr. Motoo Watanabe (Manager) of the Stem Cell and Organ Regeneration Project, Dr. Tomoyuki Yamaguchi of the Stem Cell and Organ Regeneration Project, Satoshi Yamazaki of the Stem Cell and Organ Regeneration Project, Dr. Toshihiro Kobayashi of the Stem Cell and Organ Regeneration Project, Dr. Akihide Kamiya of the Division of Stem Cell Therapy, and Dr. Shin Kaneko of the Division of Stem Cell Therapy. This is a well-funded center with impressive resources and considerable organizational synergies.

Research Focus

Dr. Watanabe described the ERATO-funded Stem Cell and Organ Regeneration Project, which has the goal of generating donor ES or iPS cell derived solid organs in xenogenic hosts via blastocyst complementation. This builds upon success in rodent models, with the hematopoietic system and with the pancreas in work described by Dr. Kobayashi (Kobayashi et al. 2010), and will increasingly be explored in large animals (e.g., cow). There are challenges, including whether iPS cells from one species can undergo organogenesis within and support the survival of a host of a different species, as well as the fact that currently vasculature and innervation of solid organs are derived from the xenogenic host. However, the impressive effort, which will yield basic insights into the process of organogenesis, has the potential to generate models for drug discovery, and may ultimately be translated toward organ generation for replacement therapy.

Additional work within the center focuses on the hematopoietic stem cell niche and the promotion of better HSC engraftment. In the former, Dr. Yamazaki (Yamazaki et al. 2009) discovered a population of glial-like cells, potentially Schwann cells, that promote the latency of HSCs via the secretion of TGF-β. In other work focused on HSC engraftment into an irradiated host, Drs. Otsu and Suzuki found that radiation-induced secretion of cytokines—particularly TNF-α—constitutes a hostile environment that represses the engraftment process. They are exploring the use of TNF-α antagonists, including peptides, to block its signaling and promote HSC engraftment.

In addition, Dr. Kamiya is working to address the gap between the need for liver transplants and the number of available donors. This issue is particularly salient for Japan, since there was an incident several decades ago in which a considerable number of blood transfusions were contaminated with hepatitis virus. Investigators are isolating hepatic stem and progenitor cells and conducting genetic screens to identify factors and pathways that may solve the problem that the HSPCs have limited proliferative capacity in culture. In addition, they are working to develop better conditions for hESC and hiPSC differentiation into hepatocytes.

Investigators also work on the development of viral vectors as tools for stem cell research. For example, Dr. Kaneko has developed a lentiviral vector in which herpes simplex virus thymidine kinase (HSVtk, often used as a suicide gene for cancer gene therapy) is placed under the control of a Nanog promoter. Gene delivery to iPS cells then provides the opportunity to ablate pluripotent cells that may otherwise give rise to a teratoma, simply through the administration of the HSVtk substrate gancyclovir. In another application of lentiviral vectors, Dr. Tomoyuki Yamaguchi described the development of a multicistronic lentiviral vector system in which three of the Yamanka factors could be placed under the control of a tetracycline regulation system for inducible control over pluripotency reprogramming. After reprogramming fibroblasts to iPS cells, he generates chimeric mice, then harvests fibroblasts and reprograms these to generate secondary iPS cells. Interestingly, he finds that the efficiency of secondary reprogramming decreases with age, but that ectopic overexpression of c-myc rescues the age-related decline in reprogramming efficiency.

Translation

There are a number of strong translational efforts at various stages of development. The work on HSC transplantation could be translated relatively rapidly, and work with other classes of stem cells and organ development have a strong translational flavor.

Sources of Support

The center is extremely well funded from various governmental sources. These include grants of $21M/6 years from ERATO, $9.4M/5 years from JST/MEXT, $1.2M/2 years from JST/MEXT, and additional grants from JSPS, MHLW, and METI.

Collaboration Possibilities

The center has broad research interests in basic research, translation, and clinical development that are complementary with a number of research efforts within Japan and the United States. While investigators were somewhat cautious of the possibility of collaborating with centers that have closely aligned interests (e.g., Kyoto University), there is substantial complementarity between these programs and the potential for productive interaction.

Summary and Conclusions

This is a well-funded center that effectively integrates basic research with translational capabilities. It is in general a center with strong resources, including sorting, imaging, and animal facilities. Particularly impressive were the efforts directed toward growing xenogenic organs and tissues within animal hosts, which has future implications for human cell and organ replacement therapies. Furthermore, many of their efforts are on a promising translational path.

Selected References

Kobayashi, T., T. Yamaguchi, S. Hamanaka, M. Kato-Itoh, Y. Yamazaki, M. Ibata, H. Sato, Y.S. Lee, J. Usui, A.S. Knisely, M. Hirabayashi, and H. Nakauchi. 2010. Generation of rat pancreas in mouse by interspecific blastocyst injection of pluripotent stem cells. *Cell* 142:787–799.

Yamazaki, S., A. Iwama, S. Takayanagi, K. Eto, H. Ema, and H. Nakauchi. 2009. TGF-beta as a candidate bone marrow niche signal to induce hematopoietic stem cell hibernation. *Blood* 113:1250–1256.

Uppsala University

Site Address:	Meeting conducted at: Radisson Blu Sky City Hotel Arlanda Airport SE-190 45 Stockholm-Arlanda Sweden http://www.uu.se/en
Date Visited:	March 2, 2012
WTEC Attendees:	P. Zandstra (report author), T. McDevitt, D. Schaffer, N. Moore, and H. Sarin
Host(s):	Professor Karin Forsberg Nilsson, Group Leader of Stem Cell Research Department of Immunology, Genetics and Pathology Tel.: +46-18 471 41 58 Karin.nilsson@igp.uu.se Dr. Peetra Magnusson, Assistant Professor, Rudbeck Laboratory Department of Immunology, Genetics and Pathology Peetra.magnusson@igp.uu.se Dr. Oommen Varghese, Assistant Professor, Polymer Chemistry Group of the Angstrom Laboratory Department of Materials Chemistry oommen.varghese@mkem.uu.se

Overview

Uppsala University (UU) is the oldest university in the Nordic countries and consistently ranks highly, internationally, in research productivity. Uppsala alone, and together with the Karolinska Institutet (KI), has undertaken a number of initiatives in the Regenerative Medicine area, with many focusing on translational and bioengineering aspects. These include the Science for Life Laboratory (http://www.scilifelab.se/) and BIODESIGN (http://www.biodesign.eu.com/index.html). Uppsala has fully developed educational programs in both basic/developmental biology and engineering sciences, but appears not to have an international doctoral program in bioengineering or biomedical engineering. New training courses and initiatives in this area are expected to emerge over the next few years. Despite this, significant physical science and translational biology integration, as reviewed below, has occurred within regenerative medicine efforts. These efforts are bolstered by an emerging regenerative medicine-related biotechnology sector and significant participation in the area by multinational corporations such as GE Healthcare and Novo Nordisk. Together with KI, UU has in place many of the pieces needed for leadership in stem cell engineering.

Research Focus

During our visit we met with three representatives from Uppsala, Professors Karin Nilsson, Peetra Magnusson and Oommen P. Varghese. Outlined below is a brief summary of their research programs.

Professor Karin Forsberg Nilsson's research interests are in the area of extracellular matrix (ECM) interactions in neural differentiation and glioma. She outlined the significant challenge in dealing with glioblastoma multiforme (GBM), a very aggressive human neural cell tumor and how she has identified similarities between signaling pathways activated in glioma and in neural progenitor cells. Professor Nilsson is specifically interested in how neural stem cells proliferate and differentiate and what factors govern these processes. She has focused on the role of matrix and matrix-associated signaling proteins in creating environments that control cell fate and migration of normal and malignant neural progenitor cells. Example projects in her group include: examining the role of nuclear receptor binding protein 2 (NRBP2) in neural stem cell self-renewal and differentiation; examining the platelet-derived growth factor (PDGF) overexpression in neural progenitors and its link to neural tumor development; identification of extracellular matrix interactions specific for the early postnatal brain and their role in pediatric tumor development; and studying the role of heparan sulfate proteoglycans in neural stem cell development and tumor invasiveness. As an example of her work in this latter project, Professor Nilsson has examined the properties of neural stem cells in hydrogels (with Jons Hilborn, UU), and is examining a role for hydrogel properties including compliance in neural stem cell differentiation and migration. In building on this and related work Professor Nilsson is involved in a number of initiatives, including the development of more predictive cell models for gliomas and working more closely with hospital and clinical groups to build a well characterized glioblastoma cell bank.

Assistant professor Peetra Magnusson's research is focused on processes to improve islet survival and engraftment after transplantation. It has been recognized that islet cell survival during transplantation is very poor and improvements in these processes could significantly impact the number of donor pancreases/islets needed. An extrahepatic transplantation site is under investigation in the islet transplantation field. As a strategy to improve islet engraftment at alternative transplantation sites, they use supportive components such as matrix proteins or different types of helper cells. One approach is to reformulate donor islet cell composition postharvest with mesenchymal stromal cells (MSC) and endothelial cells (EC). In the engineered pseudo-islet the MSC are hypothesized to play two roles: (1) they could provide support to the EC and (2) they could act in an immunosuppressive capacity to decrease immune activity on the transplanted islets. The EC cells would enhance survival through enhanced neovascularization.

In related research, Magnusson is studying the role of islet vasculature and EC activation on early onset type 1 diabetes. In many infectious and inflammatory diseases the cells of the endothelium are affected, leading to secondary complications such as nephropathy, retinopathy, and coronary artery disease due to endothelial dysfunction. To be able to investigate the interaction between EC and blood cells, she has developed *in vitro* models of blood cell-EC interactions. The Magnusson lab is a key node in the Nordic Network for Clinical Islet Transplantation (http://www.medscinet.com/nordicislets/other_language.aspx?id=1), and is also part of strategic research initiatives such as EXODIAB (http://www.exodiab.se/, an initiative in the diabetes area at Lund University (LU) and Uppsala University (UU) with the aim to create a national leading resource for diabetes research), and part of the National initiative on Stem Cells for Regenerative Therapy led by Jöns Hilborn.

Professor Oommen P. Varghese's research interests are in understanding how cells behave when they contact different extracellular matrix glycosaminoglycans. He is applying this interest to stem cell and regenerative medicine problems, often in collaboration with biological and clinical researches and UU and KI. The goal of Professor Varghese's group is to develop new functional materials (such as hydrogels) that can be used for the delivery of sensitive proteins, drugs, therapeutic genes, or small interfering RNAs (siRNAs) to specific cells or tissues. Examples of well-developed research areas in the Varghese group include the use of injectable extracellular matrix mimetic hydrogels for bone and cartilage tissue regeneration and (working with Dr. Magnusson) to promote pancreatic islet cell survival and function.

Translation

There appears to be a significant interest and effort in clinical and commercial translation at UU. Part of this is enabled through Swedish funding mechanisms (and is true throughout the country) and part is a consequence of the well-structured clinical data and patient database accessible through the integrated healthcare system. Examples of biotechnology companies associated with these efforts are Tikomed (http://www.tikomed.com), Novahep (http://www.novahep.com/), and Neuronova (http://www.neuronova.com/). Large companies such as GE Healthcare, Novo Nordisk, and AstraZeneca also are integrated into some of the research and translation programs.

Sources of Support

A number of investigator-driven interactions have emerged. These appear now to be supported by broader initiatives such as the BIODESIGN and Science for Life Laboratory (www.scilifelab.se) programs.

Collaboration Possibilities

It is very clear that there is extensive collaboration among the Karolinska Institute, Uppsala University, the Royal Institute of Technology, the University of Stockholm, and Lund University that brings together expertise in biology, engineering, and medicine. Additional opportunities for collaborations in engineering areas such as mathematical modeling and materials science (to interact with their existing strength in this area) may exist.

Summary and Conclusions

Uppsala University has strong involvement in clinical and commercial translation of regenerative medicine based technologies. Together with other institutions in the region (KI, Lund, Chalmers) they represent an international center of excellence in regenerative medicine related activities.

Selected References

Brännvall, K., K. Bergman, U. Wallenquist, S. Svahn, T. Bowden, J. Hilborn, and K. Forsberg-Nilsson. 2007. Enhanced neuronal differentiation in a 3D collagen-hyaluronan matrix. *J. Neuroscience Res.* 85:2138–2146.

Cabric, S., J. Sanchez, U. Johansson, R. Larsson, B. Nilsson, O. Korsgren, and P.U. Magnusson. 2010. Anchoring of vascular endothelial growth factor to surface-immobilized heparin on pancreatic islets: implications for stimulating islet angiogenesis. *Tissue Engineering* 16:961–970.

Demoulin, J.B., M. Enarsson. J. Larsson, A. Essaghir, C.H. Heldin, and K. Forsberg-Nilsson. 2006. The gene expression profile of PDGF-treated neural stem cells corresponds to partially differentiated neurons and glia. *Growth Factors* 24(3):184–196.

Forsberg, M., K. Holmborn, S. Kundu, A. Dagalv, L. Kjellen, and K. Forsberg-Nilsson. 2012. Under-sulfation of heparan sulfate restricts the differentiation potential of mouse embryonic stem cells. *J. Biol. Chem.* 287(14):10853–10862.

Johansson, A., J. Lau, M. Sandberg, L.A.H. Borg, P.U. Magnusson, and P.-O. Carlsson. 2009. Endothelial cell signalling supports pancreatic beta-cell function in rat. *Diabetologia* 52:2385–2394.

Johansson, U., I. Rasmusson, S.P. Niclou, N. Forslund, L. Gustavsson, B. Nilsson, O. Korsgren, and P.U. Magnusson. 2008. Composite endothelial-mesenchymal stem- islet cells; a novel approach to promote islet revascularization. *Diabetes* 57:2392–2401.

Larsson J., M. Forsberg, K. Brännvall, X.-Q. Zhang, M. Enarsson, F. Hedborg, and K. Forsberg-Nilsson. 2008. Nuclear receptor binding protein 2 is induced during neural progenitor differentiation and affects cell survival. *Mol. Cell. Neuroscience* 39:32–39.

Martínez-Sanz, E., D.A. Ossipov, J. Hilborn, S. Larsson, K.B. Jonsson, and O.P. Varghese. 2011. Bone reservoir: Injectable hyaluronic acid hydrogel for minimal invasive bone augmentation *J. Controlled Rel.* 152(2):232–240, doi:10.1016/j.jconrel.2011.02.003.

Niklasson, M., T. Bergstrom, X.-Q. Zhang, S. Gustafsdottir, M. Sjögren, P.-H. Edqvist, B. Vennström, M. Forsberg, and K. Forsberg-Nilsson. 2010. Enlarged lateral ventricles and aberrant behavior in mice overexpressing PDGF-B in embryonic neural stem cells. *Exp. Cell Res.* 17:2779–2789.

Ossipov, D.A., S. Piskounova, O.P. Varghese, and J. Hilborn. 2010. Functionalization of hyaluronic acid with chemoselective groups via a disulfide-based protection strategy for *in situ* formation of mechanically stable hydrogels. *Biomacromolecules* 11(9):2247–2254.

Varghese, O.P., W. Sun, J. Hilborn, and D.A. Ossipov. 2009. *In situ* cross-linkable high molecular weight hyaluronan-bisphosphonate conjugate for localized delivery and cell-specific targeting: a hydrogel linked prodrug approach. *J. Am. Chem. Soc.* 131:8781–8783.

Appendix C: "Virtual" Site Visit Reports

The reports in this section were prepared based on interviews with principals or others associated with the site, supplemented by published material. The full panel did not visit these sites.

Bioprocessing Technology Institute

Site Address:	20 Biopolis Way, #06-01 Centros, Singapore 138668
Date Visited:	July 29, 2011
WTEC Attendees:	R.M. Nerem (report author)
Host(s):	Dr. Miranda G.S. Yap Executive Director Tel.: 65-64-78-8888 Fax: 65-6478-9561 Miranda_yap@bti.a-star.edu.sg Dr. Steve Oh Kah Weng Principal Scientist, Stem Cell Group Associate Director for IP and Academic Affairs Tel.: 65-64-78-8888 Fax: 65-6478-9561 steve_oh@bti.a-star.edu.sg

R.M. Nerem et al. (eds.), *Stem Cell Engineering: A WTEC Global Assessment,*
Science Policy Reports, DOI 10.1007/978-3-319-05074-4,
© Springer International Publishing Switzerland 2014

Overview

The visit to the Bioprocessing Technology Institute (BTI) was a follow up to a visit a year earlier by Drs. Todd McDevitt and Robert M. Nerem. Historically BTI came out of the Bioprocessing Technology Unit established in 1990 that subsequently became the Bioprocessing Technology Centre in 1995, and in 2003 became BTI. The stated twin mission of BTI is to develop manpower capabilities and to spearhead research in bioprocess science and engineering that will directly impact and benefit the scientific community and industry. BTI is aligned with the aspirations of Singapore to position itself as the biomanufacturing and biomedical hub in the Asia Pacific region. BTI has a total staff of 150, 30 % of whom have Ph.D.s.

Research and Development Activities

The cornerstone research programs of BTI include novel cell lines and biomolecules, production systems, product purification and analysis, and the profiling of processes. Activities are focused both on biopharmaceutical drugs and cell therapy. For the cell therapy area the platforms at BTI are as follows: (1) cell lines engineering (hESC, hiPSC, hESC-MSC, and hfMSC), (2) culture and media development (microcarrier platform, serum-free media, and defined surfaces), (3) omics technologies (transcriptomics, proteomics, and bioinformatics), (4) downstream purification (cell-cell separation, monoclonal antibody [mAb]-based), and (5) analytics and monitoring (in-process monitoring tools and cell-based assays). There also is an antibody discovery platform that can generate novel, specific highbinding and novel, specific cytotoxic monoclonal antibodies to surface antigens on stem cells and cancer cells. Potential applications include the areas of cell therapy, disease diagnostics, and imaging.

In addition, there is a biologic manufacturing site at Tuas Biomedical Park. This includes a number of companies including the Swiss company Lonza. In 2009 Lonza announced the expansion of its cell therapy business with the construction of a new facility in Singapore. This new facility is located adjacent to its large-scale mammalian cell manufacturing facility at Tuas Biomedical Park.

Manpower development is the other part of the mission of BTI. There currently are 22 Ph.D. students, 14 of whom have a first degree in engineering, either chemical or bioengineering. These are in a variety of degree programs including some outside Singapore. BTI offers the Bioprocess Internship Program (BIP), which has four teaching modules. These are Expression Engineering with a total of 17 h of lectures, Bioprocessing Technology with 27 h of lectures, Analytics for Bioprocessing with 10 h of lectures, and a GMP module of 10 h. In 2011 there were 34 interns.

Sources of Support

Funding comes from the Ministry of Trade and Industry through A*STAR. The FY2010 budget was 30 million Singapore dollars. In FY2011, up to 30 % must be recouped from industry-related activities.

Collaboration Possibilities

BTI represents a research organization that potentially could provide excellent possibilities for collaboration both in research and in education.

Summary and Conclusions

BTI is an impressive institute, one that is front and center in stem cell biomanufacturing and in cell therapies. For 21 years it was led by a chemical engineer, Prof. Miranda Yap. Currently, Prof. Lam Kong Peng is acting director and Dr. Steve Oh is Principal Scientist in the Stem Cell Group.

Selected References

Chen, A.K., X. Chen, A.B. Choo, S. Reuveny, and S.K. Oh. 2011. Critical microcarrier properties affecting the expansion of undifferentiated human embryonic stem cells. *Stem Cell Res.* 7(2):97–111.

Chen, X., A. Chen, T.L. Woo, A.B. Choo, S. Reuveny, and S.K. Oh. 2010. Investigations into the metabolism of two-dimensional colony and suspended microcarrier cultures of human embryonic stem cells in serum-free media. *Stem Cells Dev.* 19(11):1781–1792.

Heng, B.C., J. Li, A.K. Chen, S. Reuveny, S.M. Cool, W.R. Birch, and S.K. Oh. 2011. Translating human embryonic stem cells from 2-dimensional to 3-dimensional cultures in a defined medium on laminin- and vitronectin-coated surfaces. *Stem Cells Dev.* Epub ahead of print Dec. 23, 2011.

Oh, S.K., A.K. Chen, Y. Mok, X. Chen, U.M. Lim, A. Chin, A.B. Choo, and S. Reuveny. 2009. Long-term microcarrier suspension cultures of human embryonic stem cells. *Stem Cell Res.* 2(3):219–30.

Conference for Stem Cell Engineering

Site Address:	International Cooperation Building
	Korean Institute of Science and Technology
	Hwarangno 14-gil 5
	Seongbuk-gu, Seoul, 136-791
	Republic of Korea
Date Visited:	January 17, 2012
WTEC Attendees:	R.M. Nerem (report author)
Host(s):	Professor Soo Hyun Kim
	Center for Biomaterials
	Korea Institute of Science and Technology
	Tel.: 82-2-958-5343
	Fax: 82-2958-5308
	soohkim@kist.re.kr
Organizer(s):	Professor Gilson Khang
	BIN Fusion Technology Department
	Chunbuk National University
	Jeonju, Korea 561-756
	gskhang@jbnu.ac.kr
	Tel.: 82-63-270-2355
	Fax: 82-63-270-2341

Overview

In order to better assess activities in South Korea in the area of stem cell engineering, a 1-day conference was organized. The organizing committee comprised Professors Gilson Khang, Youngsook Son, Jeong Ok Lim, Su Ra Park, Byung-Hyun Min, In Ho Jo, Chun Ho Kim, and Dongwon Lee. The person most responsible for the organization of this conference was Professor Gilson Khang from the BIN Fusion Technology Department in the College of Engineering at Chunbuk National University in Jeonju, Korea. He acted as chair of the organizing committee and worked together with Professor Soon Hyun Kim to ensure the success of this conference.

There were 12 invited talks for this conference (Table C.1). There were approximately 300 attendees.

Table C.1 Speakers, affiliations, and titles of presentations at stem cell engineering conference

Robert M. Nerem (CBNU/ Georgia Tech.)	Stem Cell Engineering
Youngsook Son (Kyung Hee Univ.)	Stem Cell Mobilization for Tissue Repair
Il Hwan Oh (Catholic Univ. of Korea)	Umbilical Cord Blood for Cell and Gene Therapy
Sung-Hun Lee (Hanyang Univ.)	Stem Cell engineering for the Generation of Experimental and Transplantable Mid-Brain Type Dopamine Neurons
Dong-Wook Kim (Yonsei Medical Univ.)	Disease-Specific IPS Cells: A Platform for Human Disease Modeling and Drug Discovery
Sook Ick Chang (Chungbuk National Univ.)	Mass Production of Bone-Tissue Like Structures Constructed from Embryonic Stem Cells by Using the Automatable Perfusion Rotating Wall Vessel Bioreactor
Jong Wook Chang (MEDIPOST Co.)	Therapeutic Potential of Human Umbilical Cord Blood Mesenchymal Stem Cells: Preclinical and Clinical Experience
Byunghyun Min (Ajou Univ.)	Control of Differentiation of Mesenchymal Stem Cells for Cartilage Tissue Engineering
Keun Hong Park (Cha Medical School)	Stem Cell Differentiation by Specific Formulation
Taekyun Kwon (Kyungpook Nat. Univ.)	Stem Cell Therapy for Stress Urinary Incontinence
Seung-Woo Cho (Yonsei Univ.)	Biomaterial Tool Kits for Stem Cell Engineering

Conference Notes

The conference speakers all provided excellent presentations. Following the lead-off talk by Professor Nerem, Professor Youngsook Son talked about Substance P and its role in orchestrating wound healing. As an example, substance P induces IL-10 and M2 macrophages and stimulates tissue repair after spinal cord injury. This was followed by Dr. Il-Hwan Oh addressing the question of how stem cell activity is regulated. Much of his talk focused on the use of human cord blood serum to support stem cell culture. He also raised the question "does functional heterogeneity exist in MSCs?" The morning session concluded with Professor Sung-Hun Lee discussing the generation of experimental and transplantable midbrain-type dopamine neurons. He also discussed autologous cell transplantation, i.e., patient-derived iPS cells.

The afternoon session started with a presentation by Professor Dong-Wook Kim on iPS cells as a platform for human disease modeling and drug discovery. The cell bank in his center has approximately 50 disease-specific iPS cell lines. Of particular interest were the disease-specific neurons and oligodendrocytes which have been carefully characterized. These will be used not only to study disease mechanisms, but to screen new drugs. The next speaker was Professor Soo-Ik Chang, who spoke about the use of an automatable perfusion rotating wall bioreactor to create a bone tissue-like structure. He noted that conventional techniques for the differentiation of stem cells are very much researcher dependent, and noted that the HepG2 cell line secretes signals consistent with endoderm differentiation. He then described a novel perfusion bioreactor developed in the Biological Systems Engineering Laboratory and the application of this in the tissue engineering of a bone tissue-like structure. This bioreactor system demonstrated automatable, reproducible production. This work was in collaboration with Professor Athanasios Mantalaris at Imperial College London. Next was a presentation by Dr. Jong Wook Chang from MEDIPOST who is manager of the NEUROSTEM project. He discussed efforts to develop human umbilical cord blood-derived mesenchymal stem cells (hUBC-MSCs) as a therapy for Alzheimer's disease. He also mentioned the large-scale production of hUBC-MSCs, and discussed a phase 1 clinical trial. The last speaker before the coffee break was Dr. Yong Man Kim from Pharmicell. He discussed his company's thera-peutic approach using bone marrow-derived mesenchymal stem cells. Their strategy is an autologous one, and he discussed the current status of clinical trials using bone marrow-derived MSCs and the progress being made by his company. In the treat-ment of myocardial infarction using the rat as the animal model, echocardiographic data indicated an increase in left ventricular ejection fraction from 35 to 48 %. In these studies the myocardial infarction was caused by coronary ligation, and this may be why this increase is so much greater than what is seen in humans using the same therapy where the increase is less than 10 %.

After the coffee break, the first speaker was Professor Byoung-Hyun Min who talked about cartilage tissue engineering. He discussed the use of different cell types, including MSCs and factors influencing MSC differentiation. This included oxygen tension, mechanical factors, and the biomaterial used. As part of this he discussed the merits of low intensity ultrasound, i.e., as a method for applying mechanical force both to MSCs and to cartilage. The next presentation was by Dr. Keun-Hong Park. He talked about surface modification of a micro-structure. An example is heparinized nanoparticles coated on a PLGA microsphere for cell delivery. Also, nanoparticles polyplexed with plasma DNA may prove effective in the differentiation of stem cells. The next to the last talk was by Dr. Kwon Tae Gyun who discussed stress urinary incontinence and the possibility of an adult stem cell injection therapy for the repair of an impaired sphincter. Fifty percent of patients receiving two injections were continent at 12 months. The final speaker of the day was Dr. Seung-Woo Cho who spoke about biomaterials as genetic engineering vehicles, drug delivery systems, and lineage-specific differentiation tools for the engineering of stem cell nanoparticles and hydrogel microarrays. The latter is to guide differentiation by way of cell-matrix interaction. The microarray is to identify the ideal extracellular matrix components. Finally, he discussed polymer chips for clonal expansion.

Conclusions

From the presentations at this conference it is clear that the quality of the stem cell science in Korea is excellent. However, there was little in the context of stem cell engineering. Of the 11 presentations, only two were made by individuals educated as engineers. These two were Dr. Keun-Hong Park who received his Ph.D. in biomedical engineering from Tokyo Institute of Technology and Dr. Seung-Woo who received his Ph.D. in chemical engineering from Seoul National University. In addition, there were two other presentations where the research could be characterized as using an engineering approach. These two were Professor Soo-Ik Chang who discussed an automatable bioreactor system and Professor Byoung-hyun Min who described the use of ultrasound in his research. Although there were two presentations from companies developing cell therapies, neither appeared to have any engineers involved. It thus appears that the development of stem cell technology in Korea could benefit considerably from a greater involvement of engineers and from the use of engineering approaches.

MEDIPOST, Co., Ltd.

Site Address:	1571-17 Seocho3-dong Seocho-gu Seoul 137-874, Korea http://www.medi-post.com/index.asp
Date Visited:	January 16, 2012
WTEC Attendees:	R.M. Nerem (report author)
Host(s):	Yoon-Sun Yang, M.D., Ph.D. President and CEO MEDIPOST Co. Tel.: 82-2-3456-6677 Fax: 82-2-3465-6688 ysyang@medi-post.co.kr Francis Sung Ho Han, Ph.D. Global Head Business Development and R&D Strategy MEDIPOST Co. Tel.: 82-2-3465-6650 Fax: 82-2-3465-6688 francishan@medi-post.com

(continued)

Antonio S. Lee, Ph.D.
Associate Director
Business Development and R&D Strategy
MEDIPOST Co.
Tel.: 82-2-3465-6657
Fax: 82-2-3465-6688
alee@medi-post.com

Hyuk Jun Nam
Facility Manager/QA Manager
MEDIPOST Co.
Tel.: 82-2-866-7141
Fax: 82-2-866-7144
hnam@medi-post.co.kr

Overview

MEDIPOST is a publicly traded company founded in 2000. It currently has 135 employees. It is based on an allogeneic approach using umbilical cord blood and both hematopoietic stem cells (HSCs) and mesenchymal stem cells (MSCs) from the cord blood. It markets itself as the leading stem cell company in Korea. Its goal is the commercialization of allogeneic, off-the-shelf, adult stem cell products. It currently has four products in the pipeline in the area of stem cell therapeutics— CARTISTEM®, PROMOSTEM®, NEUROSTEM®, and PNEUMOSTEM®. Dr. Yoon-Sun Yang is a clinical pathologist and previously held a clinical position at the Samsung Medical Center in Seoul, Korea. She developed the cord blood bank. Dr. Francis Sung Ho Han received her Ph.D. in neuroscience from St. Jude in Memphis, Tennessee. Dr. Antonio Lee was born in Korea, but moved to New Zealand. He received his Ph.D. in developmental biology from the University of Otago and then did postdoctoral research in Sydney, Australia. The profit making part of MEDIPOST is cord blood banking, a business that they have been in for over 10 years. MEDIPOST has a 40 % share of the private cord blood banking market in Korea.

Research and Development Activities

MEDIPOST pioneered in Korea the use of human umbilical cord blood (hUCB) as a source for mesenchymal stem cells (MSCs). This source is used because of their minimal immunogenicity, their better functionality, and their accessibility. The concept is one of a paracrine action affecting the endogenous cells rather than directly differentiating and becoming replacement cells.

Translation

The lead product for MEDIPOST is CARTISTEM® for degenerative osteoarthritis (OA). This idea came from Professor Chul Woon Ha at the Samsung Medical Center. CARTISTEM® is now approved in Korea and in phase I/IIa clinical trials in the United States.

Sources of Support

MEDIPOST has received a number of publicly funded projects since 2000. Also, its business arm of Private Cord Blood Bank and Nutritional Supplements generates revenue for its R&D and clinical development programs.

Collaboration Possibilities

There are no engineers working in the company, although the CEO Dr. Yang stated that she believes there are many opportunities for cooperation and collaboration with engineers.

Summary and Conclusions

MEDIPOST appears to be a company that will make it in the regenerative medicine area. Its cord blood banking service provides a substantial income.

Selected References

Chang, Y.S., S.J. Choi, and D.K. Sung, et al. 2011. Intratracheal transplantation of human umbilical cord blood derived mesenchymal stem cells dose-dependently attenuates hyperoxia-induced lung injury in neonatal rats. *Cell Transplantation* (2011).

Chang, Y.S., W. Oh, S.J. Choi, D.K. Sung, S.Y. Kim, E.Y. Choi EY, S. Kang, H.J. Jin, Y.S. Yang, and W.S. Park. 2009. Human umbilical cord blood-derived mesenchymal stem cells attenuate hyperoxia-induced lung injury in neonatal rats. *Cell Transplantation* 18(8):869–886.

Jang, Y.K., D.H. Jung, M.H. Jung, D.H. Kim, K.H. Yoo, K.W. Sung, H.H. Koo, W. Oh, Y.S. Yang, and S.-E. Yang. 2006. Mesenchymal stem cells feeder layer from human umbilical cord blood for *ex vivo* expanded growth and proliferation of hematopoietic progenitor cells. *Annals Hematol.* 85(4):212–225.

Jin, H.J., H.Y. Nam, Y.K. Bae, S.Y. Kim, I.R. Im, W. Oh, Y.S. Yang, S.J. Choi, and S.W. Kim. 2010. GD2 expression is closely associated with neuronal differentiation of human umbilical cord blood-derived mesenchymal stem cells. *Cell Mol Life Sci.* 67(11):1845–1858.

Jin, H.J., S.K. Park, W. Oh, Y.S. Yang, S.W. Kim, and S.J. Choi. 2009. Downregulation of CD105 is associated with multi-lineage differentiation in human umbilical cord blood-derived mesenchymal stem cells. *Biochem. Biophys. Res. Commun.* 381(4):676–681.

Kim, D.-S., J.H. Kim, J.K. Lee, S.J. Choi, J.S. Kim, S.S. Jeun, W. Oh, Y.S. Yang, and J.W. Chang. 2009. Overexpression of CXC chemokine receptors is required for the superior glioma-tracking property of umbilical cord blood-derived mesenchymal stem cells. *Stem Cells Dev.* 18(3):511–519.

Kim, E.S., Y.S. Chang, S.J. Choi, J.K. Kim, H.S. Yoo, S.Y. Ahn, D.K. Sung, S.Y. Kim, Y.R. Park, and W.S. Park. 2011. Intratracheal transplantation of human umbilical cord blood-derived mesenchymal stem cells attenuates Escherichia coli-induced acute lung injury in mice. *Respir. Res.* 12(1):108. Published online August 15, 2011, doi:10.1186/1465-9921-12-108.

Kim, J.-Y., D.H. Kim, D.-S. Kim, J.H. Kim, S.Y. Jeong, H.B. Jeon, E.H. Lee, Y.S. Yang, W. Oh, and J.W. Chang. 2010. Galectin-3 secreted by human umbilical cord blood-derived mesenchymal stem cells reduces amyloid-beta42 neurotoxicity *in vitro*. *FEBS Lett.* 584(16):3601–3608.

Kim, S.M., J.J. Oh, S.A. Park, C.H. Ryu, J.Y. Lim, D.-S. Kim, J.W. Chang, W. Oh, and S.-S. Jeun. 2010. Irradiation enhances the tumor tropism and therapeutic potential of tumor necrosis factor-related apoptosis-inducing ligand-secreting human umbilical cord blood-derived mesenchymal stem cells in glioma therapy. *Stem Cells* 28(12):2217–2228.

Kim, S.M., J.Y. Lim, S.I. Park, C.H. Jeong, J.H. Oh, M. Jeong, W. Oh, S.H. Park, Y.C. Sung, and S.S. Jeun. 2008. Gene therapy using TRAIL-secreting human umbilical cord blood-derived mesenchymal stem cells against intracranial glioma. *Cancer Res.* 68(23):9614–9623.

Lee, H.J., J.K. Lee, H. Lee, J.E. Carter, J.W. Chang, W. Oh, Y.S. Yang, J.G. Suh, B.H. Lee, H.K. Jin, and J.S. Bae. 2010. Human umbilical cord blood-derived mesenchymal stem cells improve neuropathology and cognitive impairment in an Alzheimer's disease mouse model through modulation of neuroinflammation. *Neurobiol. Aging* 33(3):588–602.

Oh, W., D.-S. Kim, Y.S. Yang, and J.K. Lee. 2008. Immunological properties of umbilical cord blood-derived mesenchymal stromal cells. *Cellular Immunology* 251(2):116–123.

Yang, S.-E., C.-W. Ha, M. Jung, H.J. Jin, M. Lee, H. Song, S. Choi, W. Oh, and Y.S. Yang. 2004. Mesenchymal stem/progenitor cells developed in cultures from UC blood. *Cytotherapy* 6(5):476–486.

National University of Singapore (NUS)

Site Address:	www.nus.edu.sg http://www.lsi.nus.edu.sg/nustep/
Date Visited:	August 3, 2011
WTEC Attendees:	R.M. Nerem (report author)
Host(s):	Professor Seeram Ramakrishna Director, NUS Center for Nanofibers & Nanotechnology http://serve.me.nus.edu.sg/seeram_ramakrishna/ Former Vice President for Research National University of Singapore seeram@nus.edu.sg

Overview

The National University of Singapore (NUS) is the leading institution of higher education in Singapore. Its stated mission has three components. These are (1) to provide a transformative education, (2) to conduct high impact research, and (3) to add to social, economic, and national development. Its engineering faculty has 11 departments. In addition it has a computer faculty with seven departments including computer engineering and computational biology. NUS also has two faculties of medicine. At the graduate level NUS has the Graduate School for Integrative Sciences and Engineering. This is a university-wide endeavor aimed at providing transdisciplinary graduate education and research in science, engineering, and related aspects of medicine.

Research and Development Activities

Based on a discussion with Professor Seeram Ramakrishna, the following NUS engineering faculty were identified as doing stem cell engineering.

- William Birch (Institute for Materials Research and Engineering): Developing defined matrices or expansion of human ESCs.
- James Goh (Bioengineering): Bone and ligament repair using bone marrow stromal stem cells.
- C.T. Lim (Bioengineering): Developing sorting techniques for MSCs based on inertial microfluidic devices; involved in BioSym
- Michael Raghunath (Bioengineering): Derivation of pericyte progenitor cells; influence of macromolecular crowding; involved in BioSym.

- Seeram Ramakrishna (Medical Engineering): Applying nanotechnology method for cardiac tissue engineering; has major grant.
- Yen Wah Tong (ChBE): *In vitro* culture of adipose-derived stem cells using gelatin microspheres
- Evelyn Yim (Bioengineering): Micro-and nanotypographical regulation of stem cell proliferation and differentiation.
- Zhang Yong (Bioengineering): Fluorescent nanoparticles for bioimaging

There also are some key people in Singapore who collaborate with the engineers. These include the following.

- Ariff Bongso (ObGyn): Pioneering research with human embryonic stem cells
- Jerry Chan (ObGyn): Bone repair through the use of primitive stem cells
- Toan Than Phan (Surgery): Stem cells for skin regeneration
- Shu Wang (Biological Sciences): neural stem cells, stem cell-base delivery vehicles

Translation

A spin-off company named Electrospunra Private Limited (www.electrospunra. com) was established to commercialize 2D and 3D fiber scaffolds for tissue regeneration. This company is housed at CREATE, a new R&D complex next to NUS.

Sources of Support

Funding is provided through a variety of grant mechanisms available through the Singapore government.

Collaboration Possibilities

There are some opportunities for collaborations with the NUS faculty.

Summary and Conclusions

The author of this site visit report first became acquainted with NUS more than 10 years ago when he was a visiting faculty member there. At that time NUS was clearly on the rise as an institution. This has continued over the past decade, and NUS is today one of the top 100 institutions in the world. Several of its faculty members are internationally recognized, and this includes several listed here.

Pharmicell Co., Ltd.

Site Address:	901 SICOX Tower, 513-14 Sangdaewon—1 dong Jungwon-gu, Seongram-si Gyeonggi-do, Korea www.pharmicell.com
Date Visited:	January 16, 2012
WTEC Attendees:	R.M. Nerem (report author)
Host(s):	Dr. Hyun Soo Kim, CEO/M.D. Tel.: 82-2-3496-0115 Fax: 82-2-3496-0110 khsmd@pharmicell.com Hyun Ra Kim Assistant Manager, R&D Division Pharmicell Co. Tel.: 82-2-3496-0142 Fax: 82-2-3496-0149 hyunra@pharmicell.com Dr. Shuning Zhang Tel.: 15921766132 zhang.shuning@zs-hospital.sh.cn

Overview

Pharmicell was founded in 2002. The company has four core businesses: (1) stem cell therapeutics, (2) adult stem cell and a cord blood banking service, (3) By Pharmicell Lab, a cosmetic containing stem cell culture media, and (4) technology consultation and CMO service. There are approximately 103 employees, and the CEO is a medical doctor specializing in hematooncology. Pharmicell markets itself as the No.1 stem cell therapy company. The stem cell therapeutics effort is focused on the heart, acute ischemic stroke, spinal cord injury, and liver failure. Pharmicell has six U.S. and European patents. The overall goal of Pharmicell is life prolongation.

Research and Development Activities

Although results were shown for all of the areas being pursued as part of Pharmicell's stem cell therapeutics program, the focus of the discussion was on their approach to cardiac repair, what is called Hearticellgram. This is an autologous approach to restore damaged heart tissue. The cells are derived from the patient through bone marrow aspiration. This is followed by mononuclear cell separation and then mesenchymal stem cell (MSC) separation. The MSCs are then cultured and expanded to passage 4–5. Once the quantity of cells required per dose is achieved, the MSCs are harvested, remaining medium residue is removed, and 0.9 % saline is added so that the final quantity of cells equals 5×10^6 cells/ml. There are three types of dosage, the choice of which depends on the patient's body weight at the time his or her bone marrow was aspirated. The process is then completed with packaging and sent back to the physician for delivery back into the patient. There are 7×10^7 cells in the package sent back, and this population is 85 % pure based on two markers, CD105 and CD73. There have been extensive clinical trials. The results show that, if the therapy is administered within 6 h of the heart attack, the left ventricular ejection fraction is increased 8.3 %. This is in comparison to only a 5.9 % increase in all patients The increase is due to a paracrine effect. The cost of this procedure is approximately $18,000.

For the therapy for acute ischemic stroke and spinal cord injury, the phase 2 and 3 studies are ongoing. For the liver failure therapy, a phase 1 trial has been completed and phase 2 studies will be started in 2012.

Sources of Support

The parent company, FCB12, is publicly traded. Pharmicell also has investor financial support.

Collaboration Possibilities

There clearly are some possibilities for collaboration. These are, however, long-term.

Summary and Conclusions

In spite of the progress made, Pharmicell is still in a start-up mode. The current capability is one for processing cells for two patients per day. A new facility is being built that will allow scaling up to 10,000 patients per year or 30 patients per day, which is the level required to address the patient need that exists. Also, although no engineers are employed by the company, Dr. Hyun Soo Kim, the CEO, indicated that for stem cell biomanufacturing there is a need for engineers who have expertise in cell manipulation and also information and experience with FDA regulations.

Selected References

Baek, S., C.S. Kim, S.B. Kim, Y.M. Kim, S.W. Kwon, Y. Kim, H. Kim, and H. Lee. 2011. Combination therapy of renal cell carcinoma or breast cancer patients with dendritic cell vaccine and IL-2: results from a phase I/II trial. *J. Transl. Med.* 9:178.

Eom, Y.W., J.E. Lee, M.S. Yang, I.K. Jang, H.E. Kim, D.H. Lee, Y.J. Kim, W.J. Park, J.H. Kong, K.Y. Shim, J.I. Lee, and H.S. Kim. 2011. Rapid isolation of adipose tissue-derived stem cells by the storage of lipoaspirates. *Yonsei Med. J.* 52(6):999–1007, doi:10.3349/ymj.2011.52.6.999.

Kang, K.N., D.Y. Kim, S.M. Yoon, J.Y. Lee, B.N. Lee, J.S. Kwon, H.W. Seo, I.W. Lee, H.C. Shin, Y.M. Kim, H.S. Kim, J.H. Kim, B.H. Min, H.B. Lee, and M.S. Kim. 2012. Tissue engineered regeneration of completely transected spinal cord using human mesenchymal stem cells. *Biomaterials* 33(19):4828–4835, Epub Apr 10, 2012.

Kim, J.H., M. Jung, H.S. Kim, Y.M. Kim, and E.H. Choi. 2011. Adipose-derived stem cells as a new therapeutic modality for ageing skin. *Exp. Dermatol.* 20(5):383–387.

Park, J.H., D.Y. Kim, I.Y. Sung, G.H. Choi, M.H. Jeon, K.K. Kim, and S.R. Jeon. 2012. Long-term results of spinal cord injury therapy using mesenchymal stem cells derived from bone marrow in humans. *Neurosurgery* 70(5):1238–1247, doi:10.1227/NEU.0b013e31824387f9.

Stem Cell Bioengineering Laboratory, Instituto Superior Técnico (IST)

Site Address:	Instituto Superior Técnico
	Technical University of Lisbon
	Taguspark Campus
	Av. Prof. Doutor Aníbal Cavaco Silva—2744-016 Porto Salvo
	Portugal
	http://berg.ist.utl.pt/scbl/
Date Visited:	April 24, 2012

(continued)

WTEC Attendees:	T. McDevitt (report author)

Host(s):   Prof. Joaquim Sampaio Cabral, PI of Stem Cell Bioengineering Laboratory
                        and Director of Institute for Biotechnology and Bioengineering (IBB)
                        Tel.: +351-21-8419063
                        Fax: +351-21-8419062
                        joaquim.cabral@ist.utl.pt

                        Dr. Claudia Lobato da Silva, Assistant Professor
                        claudia_lobato@ist.utl.pt

Overview

The Stem Cell Bioengineering Laboratory (SCBL) recently moved to the Taguspark facility, which is located about 10 km from the main IST campus (www.ist.utl.pt). This brand new laboratory has state-of-the-art facilities and a lot of new equipment to conduct stem cell bioprocessing research. The laboratory contains the complete spectrum of cell culture and bioreactor technologies along with molecular and cellular analysis tools (real-time PCR, microscopy, cytometry, metabolite bioanalyzer) to assay cell phenotype.

Research and Development Activities

The primary research focus of the laboratory is the expansion of stem and progenitor cells for regenerative medicine applications. A major emphasis is placed on the development of scalable approaches for cells in suspension bioreactor systems. Many of the strategies are targeted at using existing bioreactor technologies, such as stirred-tank systems that are commonly used for biochemical engineering purposes, as well as novel bioreactor types (Wave bioreactor).

Fig. C.1 Expansion of human mesenchymal stem cells in stirred-tank bioreactors (Courtesy of Stem Cell Bioengineering Laboratory, IST)

A number of different research projects focus on adult stem cell expansion, primarily working with mesenchymal stem cells (MSCs) on microcarriers and hematopoietic stem/progenitor cell (HSC) for clinical-scale expansion in suspension culture (Fig. C.1). In parallel to MSC and HSC work, there are also projects on pluripotent stem cell expansion and differentiation. Pluripotent work began with mouse ES cells, but has transitioned now almost entirely to human ES as well as iPS cells. Neural progenitor differentiation is the primary phenotype of interest for most of the pluripotent stem cell differentiation studies.

In addition to examining cell culture format (i.e., microcarriers) and bioreactor vessels, the laboratory is also examining the effects of environmental parameters, such as hypoxia, on stem cell phenotype and function. The intent would be to use oxygen tension as an additional parameter in bioprocessing schemes to manipulate the cells in addition to soluble media composition and bioreactor conditions to direct cell growth and fate decisions.

Education and Training

Most of the researchers at the Stem Cell Bioengineering Laboratory are master's or Ph.D. students. Often postdocs take a lead role supervising different elements of the research being conducted by several of the graduate students.

Many students participate in the MIT-Portugal Bioengineering Doctoral program, which provides the opportunity for students to spend a portion of their time conducting thesis research at MIT or international universities. On average, students typically spend 1–2 years at MIT or international universities and the other 2–3 years in Portugal for their Ph.D. theses.

Starting in 2011, IST has created the new Department of Bioengineering (DBE), of which Dr. Cabral is also the head. This department is unique in Portugal with faculty members who have expertise in different areas of life sciences, biological process engineering and biomedical engineering. DBE coordinates the M.Sc. and Ph.D. degrees in Biological Engineering, Biomedical Engineering and Biotechnology (https://fenix.ist.utl.pt/departamentos/dbe/).

Translation

Dr. Cabral and the laboratory are actively collaborating with several industry partners in a variety of different ways. In some instances, the laboratory is trying out new culture reagents such as novel types of microcarriers to transition adherent cells from standard flask culture systems to bioreactors. In addition, some of the laboratory's research is focused on the development of xeno-free culture platforms for MSC, HSC and pluripotent expansion.

Collaboration Possibilities

The laboratory has active international collaborations with investigators from several different U.S. institutions, namely MIT, RPI, UC Berkeley, LSU, and Wake Forest. Collaborations with investigators throughout Europe on stem cell biology and bioprocessing are also currently being supported by funding provided by the 7th Framework program of the European Union.

Summary and Conclusions

This laboratory is one of very few in the world with an exclusive focus on stem cell engineering and bioprocessing. Much of the research activity focuses on the study of existing bioreactor technologies for pluripotent and multipotent stem cell expansion and differentiation, thereby directly examining how to scale-up the production of stem cells for regenerative medicine applications.

Selected References

Fernandes-Platzgummer, A., M.M. Diogo, R.P. Baptista, C.L. da Silva, and J.M. Cabral. 2011. Scale-up of mouse embryonic stem cell expansion in stirred bioreactors. *Biotechnol. Prog.* 27(5):1421–1432.

Rodrigues, C.A.V., M.M. Diogo, C.L. da Silva, and J.M.S. Cabral. 2010. Hypoxia enhances proliferation of mouse embryonic stem cell-derived neural stem cells. *Biotechnol. Bioeng.* 106:260–270.

Rodrigues, C.A.V., T.G. Fernandes, M.M. Diogo, C.L. da Silva, and J.M.S. Cabral. 2011. Stem cell cultivation in bioreactors. *Biotechnology Advances* 29(6):815–829.

Santos, F., P.Z. Andrade, M.M. Abecasis, J.M. Gimble, L.G. Chase, A.M. Campbell, S. Boucher, M.C. Vemuri, C.L. da Silva, and J.M. Cabral. 2011. Toward a clinical-grade expansion of mesenchymal stem cells from human sources: a microcarrier-based culture system under xeno-free conditions. *Tissue Eng. Part C, Methods* 17(12):1201–1210.

Instituto de Engenharia Biomédica (INEB)

Site Address:	Universidade do Porto
	Rua do Campo Alegre, 823
	4150-180 Porto, Portugal
	Tel.: 351 226074900
	Fax: 351 226094567
	http://www.ineb.up.pt
Date Visited:	April 27, 2012
WTEC Attendees:	T. McDevitt (report author)
Host(s):	Prof. Dr. Mário Barbosa, Director of INEB
	and Biomimetic Microenvironments Team Lead
	Tel.: +351 226074981
	Fax: +351 226094567
	mbarbosa@ineb.up.pt
Other Attendees:	Dr. Ana Paula Pêgo, Co-coordinator of NEWTherapies Group
	Tel.: +351 226074981
	Fax: +351 226094567
	apego@ineb.up.pt
	Dr. Pedro L. Granja, Co-coordinator of NEWTherapies Group
	Tel.: +351 226074981
	Fax: +351 226094567
	pgranja@ineb.up.pt

Overview

The Instituto de Engenharia Biomédica (INEB) is a private, nonprofit association that was originally founded in 1989 by six institutions, including the University of Porto (UPorto). The research mission of INEB is to provide advanced training and technology transfer in biomedical engineering in order to apply integrated engineering solutions to improve human health. INEB has adopted the motto "Engineering for Life" to express its philosophy for the development of technologies and devices aimed at improving patients' quality of life. Approximately 140 persons work at INEB, 47 of whom hold a doctoral degree and 55 are postgraduate students. The dynamic and modern nature of the institute is reflected by a large majority of young researchers (79 % of its researchers are less than 40 years old) and women (female/male ratio of 2.6).

INEB has an institutional link with the Institute for Cell and Molecular Biology (IBMC), also a private nonprofit institution of the UPorto. In 2000 the IBMC, INEB was granted the statute of Associate Laboratory by the Ministry for Science and Technology. In 2008 IBMC, INEB, and the Institute of Molecular Pathology and Immunology of the UPorto (IPATIMUP) formed a consortium—the I3S (Instituto de Investigação e Inovação em Saúde [Institute for Research and Innovation in Health]), an integrative research platform that is fostering collaboration among more than 600 researchers (>350 with a doctoral degree). INEB is also a founding member of Health Cluster Portugal (HCP) and a member of the following European Networks: Nanomedicine (European Technology Platform) and the European Institute for Biomedical Imaging Research (EIBIR).

Research and Development Activities

The main aim of the NEWTherapies Group (NEWT) is to develop integrated biomaterials and nanomedicine based approaches for tissue repair and regeneration. The primary interests are on osteoarticular, spine, and neurosciences applications, but research activities also include cell-based therapies for repairing cardiac injuries, as well as novel strategies for prevention, early diagnosis, and treatment of cancer. The interaction between inflammatory cells and biomaterials in the context of tissue regeneration is a major topic of research at INEB (Fig. C.2).

The NEWT group comprises seven complementary research teams, each led by a different principal investigator, that focus on biomimetic microenvironments, bone tissue engineering, biomaterials for neurosciences, bioengineered surfaces, neuro-osteogenesis, stem cell biology, and tumor microecosystems. Many of the current projects are intended to understand and direct cell-matrix and cell-cell interactions by molecularly designing surfaces and matrices capable of promoting stem cell expansion and migration, of directing their differentiation and of promoting their recruitment *in vivo*. The design of hydrogels for cell transplantation in the development of strategies to address spinal cord lesions is one of the topics of research in the field of nerve regeneration (Fig. C.3). By better defining the functional elements of *in vivo* stem cell niches and cancer microenvironments, the hope is to elucidate how stem and progenitor cells participate in the regeneration and repair of adult tissues.

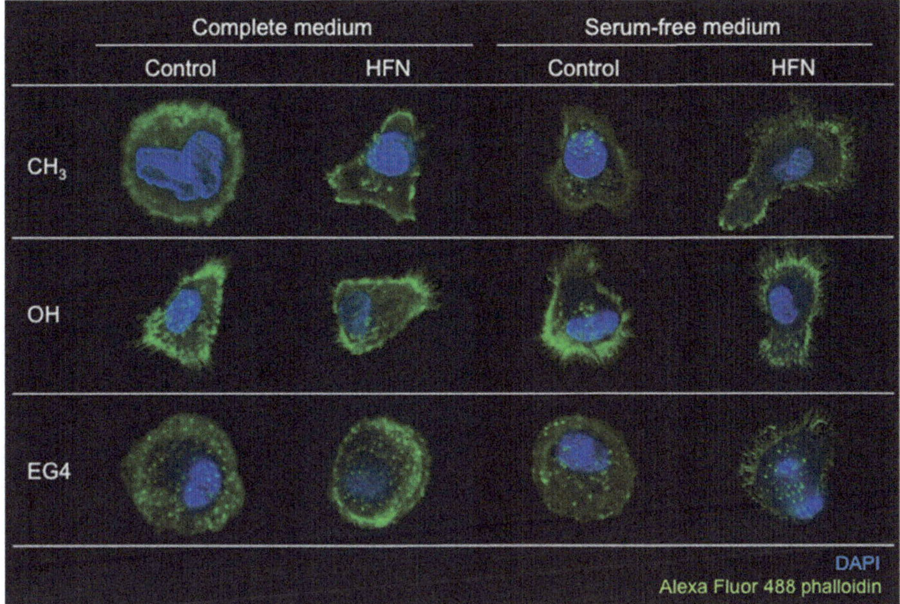

Fig. C.2 Influence of surface chemistry on macrophage polarization (Courtesy of INEB)

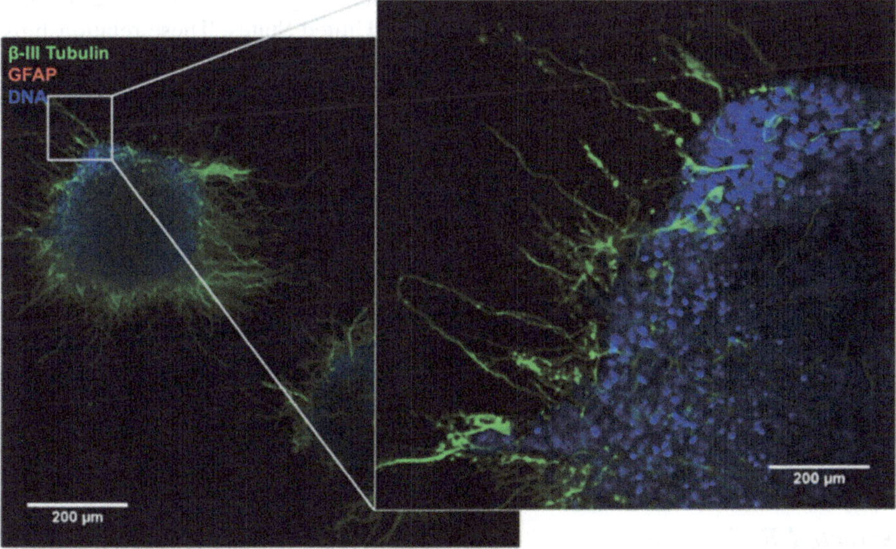

Fig. C.3 Phenotypic characterization of neurospheres cultured in a fibrin hydrogel following non-contact coculture with endothelial cells (Courtesy of INEB)

Education and Training

Since INEB is a research institute it does not have its own graduate program. However, INEB is involved in several graduate student education programs at many levels, and is strongly involved in postgraduate education through advanced training of young researchers. INEB is deeply involved in the Ph.D. and master's programs in Biomedical Engineering at UPorto, both created in 1996, as well as in the Integrated Master in Bioengineering, among others. INEB had an active role in creating the Medical Simulation Center, in partnership with the Faculty of Medicine of UPorto.

Translation

The translational activities are considered vital for the projection of INEB in society, by contributing to the solution of health problems. In the past 10 years, 8 patents, 29 prototypes, 6 products and 3 spin-offs have been generated from the work of INEB researchers.

Collaboration Possibilities

INEB researchers collaborate with ~100 institutions worldwide, including hospitals, companies and many top scientists and research groups in Australia, China, Europe, Japan, South America, Canada and the United States. These relationships result in research exchanges and cosupervision of many postgraduate students and joint publications in international journals. From 2006 to 2010, nearly 25 % of INEB publications were coauthored with researchers from foreign institutions.

Summary and Conclusions

INEB fosters a highly interdisciplinary and collaborative environment that seeks to apply engineering strategies, especially biomaterials-based approaches, to regenerative medicine applications. A very unique aspect of the INEB environment is that regenerative medicine research is being conducted alongside cancer research—two parallel fields that share many common themes.

Selected References

Bidarra, S.J., C.C. Barrias, M.A. Barbosa, R. Soares, and P.L. Granja. 2011. Evaluation of injectable *in situ* crosslinkable alginate matrix for endothelial cells delivery. *Biomaterials* 32:7897–7904.

Fonseca, K.B., S.J. Bidarra, M.J. Oliveira, P.L. Granja, and C.C. Barrias. 2011. Molecularly-designed alginate hydrogels susceptible to local proteolysis as 3D cellular microenvironments. *Acta Biomater.* 7:1674–1682.

Goncalves, R.M., J.C. Antunes, and M.A. Barbosa. 2012. Mesenchymal stem cell recruitment by stromal derived factor-1-delivery systems-based on chitosan/poly(gamma-glutamic acid) polyelectrolyte complexes. *Eur. Cell Mater.* 23:249–261.

Martins, M.C.L., V. Ochoa-Mendes, G. Ferreira, J.N. Barbosa, S.A. Curtin, B.D. Ratner, and M.A. Barbosa. 2011. Interactions of leukocytes and platelets to immobilized poly(lysine/leucine) onto tetraethylene glycol-terminated self-assembled monolayers. *Acta Biomater.* 7:1949–1955.

Oliveira, H., R. Fernandez, L.R. Pires, M.C.L. Martins, S. Simões, M.A. Barbosa, and A.P. Pêgo. 2010. Targeted gene delivery into peripheral sensorial neurons mediated by self-assembled vectors composed of poly(ethylene imine) and tetanus toxin fragment c. *J. Control Release* 143:350–358.

Royan Institute for Stem Cell Biology and Technology (RI-SCBT)

Site Address:	Tehran, Iran www.RoyanInstitute.org
Date Visited:	Report based on email from Dr. Hossein Baharvand, January 15, 2012
WTEC Attendees:	R.M. Nerem (report author)
Host(s):	Hossein Baharvand, Ph.D. Head, Department of Stem Cells and Developmental Biology Tel.: 98-21-22306485 Fax: 98-21-23562507 Baharvand@Royaninstitute.org

Overview

The Royan Institute is a public, nongovernmental, nonprofit organization committed to multidisciplinary, campus-wide integration and collaboration of scientific academic and medical personnel for understanding reproductive biomedicine, stem cells, and biotechnology. Established in 1991 by the late Dr. Saeid Kazemi Ashtiani as a research institute for reproductive biomedicine and infertility treatments, Royan in Persian means "embryo" and "land of continuous growth." The institute focuses on increasing the success rate of infertility treatment and embryo health. In addition to providing a comprehensive and coordinated bench to bedside approach to regenerative medicine, Royan also works in the areas of fundamental biology of stem cells, developmental biology, tissue engineering, stem cell therapeutics, and administration of new cell-therapeutic approaches that can restore tissue function to patients. Today, the Royan Institute is a leader of stem cell research and infertility treatment in Iran and Middle East.

The mission of the Royan Institute covers the following:

- Research and development of science and technology in biology, biotechnology, and medical areas of reproductive and regenerative biomedicine
- Treatment of infertile patients and patients who need to restore tissue function by administration of new cell-therapy approaches
- Commercialization of research findings to be offered as services or biological products
- Education and promotion of scientific findings at national and international levels.

Royan consists of three research institutes, as follows:

- Royan Institute for Reproductive Biomedicine, established in 1991 and including the "Infertility Treatment Center"
- Royan Institute for Stem Cell Biology and Technology (RI-SCBT), established in 2002, and including the "Cell Therapy Center"
- Royan Institute for Animal Biotechnology, established in 2004, and including the "Dairy Assist Center"

RI-SCBT was established in 2002 with the aim of promoting research in Iran on general stem cell biology. It started as the Department of Stem Cells, but was subsequently expanded to 15 main research groups. The vision is to make stem cell research results applicable to the treatment of disease and in a broader way to improve public health. Today, RI-SCBT provides an integrated approach to regenerative medicine that includes basic stem cell research, translational research related to stem cell therapeutics, and the administration to patients of new cell-therapy approaches. Core facilities include:

- Royan Stem Cell Bank
- Molecular Biology
- Electrophysiology
- Flow cytometry and Sorting
- Imaging
- Histology
- Gene Targeting
- Viral Transduction
- Nano- and Bio-materials
- Stem Cells for All, which trains students from primary and high school to university

Research and Development Activities

The active research programs are as follows:

- Biology of Pluripotent Stem Cells
- Epigenetic Reprogramming
- Hepatocytes
- Pancreatic Beta Cells
- Germ Cells
- Tissue- and Nanoengineering
- Neural Cells-Developmental Biology, Neural-Cell Trauma, and Neural Cells-Neurodegenerative Disease
- Bone and Cartilage/Mesenchymal Stem Cells
- Cardiomyocytes and Endothelial Cells
- Skin cells
- Kidney cells
- Regenerative Medicine
- Molecular Systems Biology and Proteome of Y chromosome
- Cancer and Hematopoietic Stem Cells
- Public Cord Blood Bank

These programs together make up the major efforts of the RI-SCBT. There are seven individuals heading these programs. Six of these are Ph.D.s and one is an M.D., Ph.D. All appear to have had their formal education in Iran in the life sciences or medicine, with a few having spent some time abroad. Recently, several engineers joined the Institute in order to bring biomedical or medical engineering approaches to start stem cell and tissue engineering and nanoengineering programs.

Furthermore, the annual report indicates that there is a wide variety of basic research in progress and states that RI-SCBT has a bench-to-bedside integrated approach. Royan also has a Good Manufacturing Practice (GMP) facility, and administration to patients of new cell-therapies is also done through the cooperation with hospitals in Iran.

Translation

RI-SCBT is developing commercial products from its research results to be marketed by pharmaceutical companies.

Sources of Support

There are 60 research assistants at RI-SCBT and more than 50 graduate students. Funding is provided by the government of Iran, commercial companies, and charities.

Collaboration Possibilities

It is noteworthy that RI-SCBT holds national and international workshops and conferences, with one example being the 7th Annual International Congress on Stem Cell Biology and Technology held in Tehran, September 7–9, 2011. There were approximately 800 participants including 16 international scientists from Europe, Japan, China, and the United States. Also, Dr. Hossein Baharvand is the editor of a book entitled *Trends in Stem Cell Biology and Technology* published in 2009 by Humana Press/Springer in the United States. In addition, RI-SCBT has begun international collaborations such as the proteome of Y chromosome and human embryonic stem cell proteome on which they have published papers. They have published more than 180 Institute for Scientific Information (ISI) papers.

Summary and Conclusions

Although it is difficult to assess the state of stem cell biology and technology in Iran, clearly there is major activity in progress in this area. It does appear, however, to be focused on the basic stem cell science with limited involvement to date of engineers and/or an engineering approach.

Selected References

Abbasalizadeh, S., M.R. Larijani, A. Samadian, and H. Baharvand. 2012. Bioprocess development for mass production of size-controlled human pluripotent stem cell aggregates in stirred suspension bioreactor. *Tissue Eng. Part C, Methods*, Epub ahead of print.

Ahmadi, H., M.M. Farahani, A. Kouhkan, K. Moazzami, R. Fazeli, H. Sadeghian, M. Namiri, M. Madani-Civi, H. Baharvand, and N. Aghdami. 2012. Five-year follow-up of the local autologous transplantation of CD133+ enriched bone marrow cells in patients with myocardial infarction. *Arch Iran Med.* 15(1):32–35.

Amirpour, N., F. Karamali, F. Rabiee, L. Rezaei, E. Esfandiari, S. Razavi, A. Dehghani, H. Razmju, M.H. Nasr-Esfahani, and H. Baharvand. 2012. Differentiation of human embryonic stem cell-derived retinal progenitors into retinal cells by Sonic hedgehog and/or retinal pigmented epithelium and transplantation into the subretinal space of sodium iodate-injected rabbits. *Stem Cells Dev.* 21(1):42–53.

Amps, K., P.W. Andrews, G. Anyfantis, L. Armstrong, S. Avery, H. Baharvand, et. al. [119 others]. 2011. Screening ethnically diverse human embryonic stem cells identifies a chromosome 20 minimal amplicon conferring growth advantage. International Stem Cell Initiative. *Nat. Biotechnol.* 29(12):1132–1144, doi:10.1038/nbt.2051.

Asgari, S., M. Moslem, K. Bagheri-Lankarani, B. Pournasr, M. Miryounesi, and H. Baharvand. 2011. Differentiation and transplantation of human induced pluripotent stem cell-derived hepatocyte-like cells. *Stem Cell Rev.*, Epub ahead of print, 11 November 2011.

Faradonbeh, M.Z., J. Gharechahi, S. Mollamohammadi, M. Pakzad, A. Taei, H. Rassouli, H. Baharvand, and G.H. Salekdeh. 2012. An orthogonal comparison of the proteome of human embryonic stem cells with that of human induced pluripotent stem cells of different genetic background. *Mol. Biosyst.* 8(6):1833–1840.

Fathi, A., M. Hatami, V. Hajihosseini, F. Fattahi, S. Kiani, H. Baharvand, and G.H. Salekdeh. 2011. Comprehensive gene expression analysis of human embryonic stem cells during differentiation into neural cells. *PLoS One* 6(7):e22856.

Ghasemi-Mobarakeh, L., M.P. Prabhakaran, M. Morshed, M.H. Nasr-Esfahani, H. Baharvand, S. Kiani, S.S. Al-Deyab, and S. Ramakrishna. 2011. Application of conductive polymers, scaffolds and electrical stimulation for nerve tissue engineering. *J. Tissue Eng. Regen. Med.* 5(4):e17-35.

Gheisari, Y., H. Baharvand, K. Nayernia, and M. Vasei. 2012. Stem cell and tissue engineering research in the Islamic Republic of Iran. *Stem Cell Rev.* Epub ahead of print, 15 February 2012.

Ghoochani, A., K. Shabani, M. Peymani, K. Ghaedi, F. Karamali, K. Karbalaei, S. Tanhaie, A. Salamian, A. Esmaeili, S. Valian-Borujeni, M. Hashemi, M.H. Nasr-Esfahani, and H. Baharvand. 2012. The influence of peroxisome proliferator-activated receptor γ(1) during differentiation of mouse embryonic stem cells to neural cells. *Differentiation* 83(1):60–67.

Hassani, S.N., M. Totonchi, A. Farrokhi, A. Taei, M.R. Larijani, H. Gourabi, and H. Baharvand. 2011. Simultaneous suppression of TGF-β and ERK signaling contributes to the highly efficient and reproducible generation of mouse embryonic stem cells from previously considered refractory and non-permissive strains. *Stem Cell Rev.*, Epub ahead of print, 4 August 2011.

Hosseini, S.M., M. Hajian, M. Forouzanfar, F. Moulavi, P. Abedi, V. Asgari, S. Tanhaei, H. Abbasi, F. Jafarpour, S. Ostadhosseini, F. Karamali, K. Karbaliaie, H. Baharvand, and M.H. Nasr-Esfahani. 2012. Enucleated ovine oocyte supports human somatic cells reprogramming back to the embryonic stage. *Cell Reprogram.* 14(2):155–163.

Karimabad, H.M., M. Shabestari, H. Baharvand, A. Vosough, H. Gourabi, A. Shahverdi, A. Shamsian, S. Abdolhoseini, K. Moazzami, M.M. Marjanimehr, F. Emami, H.R. Bidkhori, A. Hamedanchi, S. Talebi, F. Farrokhi, F. Jabbari-Azad, M. Fadavi, U. Garivani, M. Mahmoodi, and N. Aghdami. 2011. Lack of beneficial effects of granulocyte colony-stimulating factor in patients with subacute myocardial infarction undergoing late revascularization: a double-blind, randomized, placebo-controlled clinical trial. *Acta Cardiol.* 66(2):219–224.

Larijani, M.R., A. Seifinejad, B. Pournasr, V. Hajihoseini, S.N. Hassani, M. Totonchi, M. Yousefi, F. Shamsi, G.H. Salekdeh, and H. Baharvand. 2011. Long-term maintenance of undifferentiated human embryonic and induced pluripotent stem cells in suspension. *Stem Cells Dev.* 20(11):1911–1923.

Masaeli, E, M. Morshed, P. Rasekhian, S. Karbasi, K. Karbalaie, F. Karamali, D. Abedi, S. Razavi, A. Jafarian-Dehkordi, M.H. Nasr-Esfahani, and H. Baharvand. 2012. Does the tissue engineering architecture of poly(3-hydroxybutyrate) scaffold affects cell-material interactions? *J. Biomed. Mater. Res. A*, Epub ahead of print, 12 April 2012, doi:10.1002/jbm.a.34131.

Niapour, A., F. Karamali, S. Nemati, Z. Taghipour, M. Mardani, M.H. Nasr-Esfahani, and H. Baharvand. 2011. Co-transplantation of human embryonic stem cell-derived neural progenitors and Schwann cells in a rat spinal cord contusion injury model elicits a distinct neurogenesis and functional recovery. *Cell Transplant.*, Epub ahead of print, 22 September 2011, doi:10.3727/096368911X593163.

Nikeghbalian, S., B. Pournasr, N. Aghdami, A. Rasekhi, B. Geramizadeh, S.M. Hosseini Asl, M. Ramzi, F. Kakaei, M. Namiri, R. Malekzadeh, A. Vosough Dizaj, S.A. Malek-Hosseini, and H. Baharvand. 2011. Autologous transplantation of bone marrow-derived mononuclear and CD133(+) cells in patients with decompensated cirrhosis. *Arch Iran Med.* 14(1):12–17.

Nourbakhsh, N., M. Soleimani, Z. Taghipour, K. Karbalaie, S.B. Mousavi, A. Talebi, F. Nadali, S. Tanhaei, G.A. Kiyani, M. Nematollahi, F. Rabiei, M. Mardani, H. Bahramiyan, M. Torabinejad, M.H. Nasr-Esfahani, and H. Baharvand. 2011. Induced *in vitro* differentiation of neural-like cells from human exfoliated decidu-ous teeth-derived stem cells. *Int. J. Dev. Biol.* 55(2):189–195.

Ostadsharif, M., K. Ghaedi, M. Hossein Nasr-Esfahani, M. Mojbafan, S. Tanhaie, K. Karbalaie, and H. Baharvand. 2011. The expression of peroxisomal protein transcripts increased by retinoic acid during neural differentiation. *Differentiation* 81(2):127–132.

Pournasr, B., K. Khaloughi, G.H. Salekdeh, M. Totonchi, E. Shahbazi, and H. Baharvand. 2011. Concise review: alchemy of biology: generating desired cell types from abundant and accessible cells. *Stem Cells* 29(12):1933–1041.

Pouya. A., L. Satarian, S. Kiani, M. Javan, and H. Baharvand. 2011. Human induced pluripotent stem cells differentiation into oligodendrocyte progenitors and transplanta-tion in a rat model of optic chiasm demyelination. *PLoS One* 6(11):e27925.

Rahjouei, A., S. Kiani, A. Zahabi, N.Z. Mehrjardi, M. Hashemi, and H. Baharvand. 2011. Interactions of human embryonic stem cell-derived neural progenitors with an electrospun nanofibrillar surface *in vitro*. *Int. J. Artif. Organs* 34(7):559–570.

Ranjbarvaziri, S., S. Kiani, A. Akhlaghi, A. Vosough, H. Baharvand, and N. Aghdami. 2011. Quantum dot labeling using positive charged peptides in human hematopoietic and mesenchymal stem cells. *Biomaterials* 32(22):5195–5205.

Salehi, H., K. Karbalaie, A. Salamian, A. Kiani, S. Razavi, M.H. Nasr-Esfahani, and H. Baharvand. 2012. Differentiation of human ES cell-derived neural progenitors to neuronal cells with regional specific identity by co-culturing of notochord and somite. *Stem Cell Res.* 8(1):120–133.

Salehi, H., K. Karbalaie, S. Razavi, S. Tanhaee, N. Nematollahi, M. Sagha, M.H. Nasr-Esfahani, and H. Baharvand. 2011. Neuronal induction and regional identity by coculture of adherent human embryonic stem cells with chicken noto-chords and somites. *Int. J. Dev. Biol.* 55(3):321–326.

Shahbazi, E., S. Kiani, H. Gourabi, and H. Baharvand. 2011. Electrospun nano-fibrillar surfaces promote neuronal differentiation and function from human embry-onic stem cells. *Tissue Eng Part A* 17(23–24):3021–3031.

Shekari, F., A. Taei, T.L. Pan, P.W. Wang, H. Baharvand, G.H. Salekdeh. 2011. Identification of cytoplasmic and membrane-associated complexes in human embryonic stem cells using blue native PAGE. *Mol. Biosyst.* 7(9):2688–2701.

Taghipour, Z., K. Karbalaie, A. Kiani, A. Niapour, H. Bahramian, M.H. Nasr-Esfahani, and H. Baharvand. 2011. Transplantation of undifferentiated and induced human exfoliated deciduous teeth-derived stem cells promote functional recovery of rat spinal cord contusion injury model. *Stem Cells Dev.*, Epub ahead of print, 5 December 2011.

Vosough, M., M. Moslem, B. Pournasr, and H. Baharvand. 2011. Cell-based therapeutics for liver disorders. *Br. Med. Bull.* 100:157–172.

Zahabi, A., E. Shahbazi, H. Ahmadieh, S.N. Hassani, M. Totonchi, A. Taei, N. Masoudi, M. Ebrahimi, N. Aghdami, A. Seifinejad, F. Mehrnejad, N. Daftarian, G.H. Salekdeh, and H. Baharvand. 2012. A new efficient protocol for directed differentiation of retinal pigmented epithelial cells from normal and retinal disease induced pluripotent stem cells. *Stem Cells Dev.*, Epub ahead of print, 3 February 2012.

Stem Cells Australia

Site Address:	Melbourne Brain Centre
	Cnr Royal Parade and Genetics Lane
	The University of Melbourne
	Victoria 3010, Australia
	http://www.florey.edu.au/about-florey/about-us/melbourne-brain-centre
Date Visited:	December 2011
WTEC Attendees:	R.M. Nerem (report author, with the assistance of Professor Dietmar Hutmacher, Queensland University of Technology, and web site information)
Host(s):	Professor Martin Pera
	Chair in Stem Cell Sciences
	Centre for Neuroscience
	University of Melbourne
	martin.pera@unimelb.edu.an

Overview

On June 30, 2011 the Australia Stem Cell Centre's (ASCC) funding from the Australian Government came to an end at which time ASCC ceased operations. The ASCC thus is now closed after existing since 2002 when it was selected based on a competitive bid process. The ASCC was established to capitalize on Australia's strengths in the stem cell field and to create opportunities for the Australian biotechnology industry and ultimately develop needed solutions for addressing human disease. It was replaced by the establishment of Stem Cells Australia in November 2010, a consortium of Australia's leading universities and research organizations led by Professor Martin Pera.

Research and Development Activities

Stem Cells Australia is a consortium involving the following institutions and research organizations: the University of Melbourne, Monash University, the University of Queensland, the University of New South Wales, the Walter and Eliza Hall Institute for Medical Research, the Victor Chang Cardiac Research Institute, the Florey Neuroscience Institute, and the Commonwealth Scientific and Industrial Research Organization (CSIRO). The University of Melbourne administers funding provided by the Australian government and is the institution where Professor Martin Pera has his appointment.

This initiative brings together Australia's leading experts in stem cell biology, molecular analysis, bioengineering, nanotechnology, and clinical research. The objective is to investigate the mechanisms involved in the regulation of stem cell fate, including differentiation, and then to translate this knowledge into innovative applications including therapeutics. The collaboration among consortium members is not only for advancing research but to lead discussion with the public on important ethical, legal, and societal issues associated with research on stem cells and applications resulting from the advances made.

In addition to Professor Pera, the Stem Cells Australia web site provided profiles on 35 other investigators. Of these four are in the area of bioengineering and nanotechnology. These are as follows:

Professor Peter Gray is the Director and a group leader in the Australian Institute for Bioengineering and Nanotechnology at the University of Queensland. He is applying his extensive experience in bioprocess development for mammalian cell cultures to the development of strategies that will allow for the scalable expansion of pluripotent stem cells under fully defined conditions. As part of the Stem Cells Australia initiative, Professor Gray's group will explore the ability to produce scalable numbers of "spin EB" type human embryonic stem cells and the control of the differentiation of such cells down specific lineage pathways.

Dr. Michael Monteiro's specialty is in the field of nanostructured materials, working on the synthesis, characterization, and the molecular engineering of polymer nanoparticles. He has made major contributions to the understanding of the fundamental mechanisms involved in what is called "living" radical polymerization and new methods to create high order and complex architectures using polymeric building blocks. His role in the Stem Cells Australia initiative is in the synthesis of novel nanoparticles for use by collaborators.

Professor Lars Nielson is the Chair of Biological Engineering and Group Leader for Systems and Synthetic Biology in the Australian Institute for Bioengineering and Nanotechnology. Using his experience in cell culture engineering and in scaling up hematopoietic processes for clinical applications, he will contribute to the development of cell culture processes for hematopoietic stem and progenitor cells derived from pluripotent stem cells. Professor Nielson also has experience in the development of mathematical models of stem cell fate decisions, having developed in the late 1990s the first mathematically consistent model of hematopoietic fate processes. He will contribute to Stem Cells Australia initiative in this area.

Dr. Robert Nordon obtained his Ph.D. in the field of Biomedical Engineering at the University of South Wales in 1994, and following postdoctoral research in Canada, he returned to the department and is now a lecturer. He is considered an authority in the area of mammalian cell bioreactors for clinical applications and therapies. He is the inventor of a hollow fiber bioreactor that was commercialized by Gambro, BCT, now Ceridian BCT. He is currently working on methods for single-cell fate mapping using "lab-on-a-chip" devices. His role in Stem Cells Australia is to collaborate with others in single-cell, real-time analysis of cardiac stem cell growth and differentiation using microfluidics technology.

Sources of Support

The primary source of support is the $21 million received from the Australian Research Council, with this support continuing for up to 7 years.

Collaboration Possibilities

It would appear that there are excellent possibilities for collaboration with the Stem Cells Australia investigators.

Other Research Initiatives

Although Stem Cells Australia is intended to be the main dedicated stem cell research organization in Australia, there are other investigators as well. These include Professor Dietmar W. Hutmacher, who holds the Chair in Regenerative Medicine in the Institute of Health and Biomedical Innovation at Queensland University of Technology, and works on concepts of minimal-invasive injection of adult MSCs into preimplanted custom-made and patient-specific biodegradable scaffolds. He has an active collaboration with Professor Stan Gronthos and Mark

Bartold, University of Adelaide, on the regeneration of the periodontium by using novel scaffolds in combination with human iPS cells.

Professor Richard Boyd is the Director of Immunology and Stem Cell Laboratories at Monash University, and Professor Gregory Dusting is at the Bernard O'Brien Institute in Melbourne. At this institute there is a major tissue engineering laboratory with a focus that includes angiogenesis, matrix biology, and peripheral nerve regeneration. O'Brien Institute investigators have developed a platform technology for vascularizing tissue engineered products and organs. There appears to be some collaboration with faculty in chemical engineering at the University of Melbourne. There is also a Tissue Engineering Research Centre at the University of Western Australia. This organization is in the School of Anatomy and Human Biology, and it does not appear to have any connections/collaborations with engineers.

Professor Jean-Pierre Levesque, from the Mater Medical Research Institute in Brisbane, has main research interests directed toward understanding how the bone marrow regulates the behavior of hematopoietic stem cells that form all blood and immune cells, and how blood forming cells interact with bone forming cells. His research has applications in the field of bone marrow and hematopoietic stem cell transplantation to treat patients with cancer, lymphoma, and leukemia, and provides a better understanding of how normal hematopoietic stem cells can turn into leukemia.

Prof. Nicolas H. Voelcker studies stem cell interactions using novel microarray platforms. His projects include fundamental experimental and theoretical studies of the influence of factors such as surface chemistry, roughness, and topography and substrate elasticity on the behavior of specific stem cells.

Professor Julie Campbell at the University of Queensland, a cell biologist who is a Fellow in the Australian Institute for Bioengineering and Nanotechnology, has been developing an artificial blood vessel grown in the peritoneal cavity of the person into whom it will be implanted, with the tissue derived from the individual's own macrophages, which have undergone transdifferentiation.

Finally, the company Mesoblast, based in Melbourne, focuses on adult stem cell products. Their technology platform relies on the discovery of adult-derived mesenchymal precursor cells (MPCs) and the development of methods to isolate and identify these cells. The company has been granted approval to conduct clinical trials using adult stem cell therapies for a number of conditions. These include congestive heart failure, heart attacks, spinal fusion, and bone marrow regeneration. Some of these are progressing towards Phase 3 clinical trials. In 2010, Mesoblast completed its acquisition of Angioblast Systems, Inc., a U.S. company, and in 2010 it also formed a strategic alliance with Caphalon, a global biopharmaceutical company. Caphalon is focused on late-stage clinical development worldwide for specific products. In August 2011 the author of this site visit report had the opportunity to meet Professor Silviu Itescu, the Chief Executive Officer.

Appendix D: Glossary of Abbreviations and Acronyms

ACP	acid phosphatase
ACTREG	Advanced Center for Translational Regenerative Medicine (Karolinska Institute, Sweden)
AFM	atomic force microscopy
AFM	French Muscular Disease Association
ALP	alkaline phosphatase
AMMS	Academy of Military Medical Sciences (China)
ASSC	Australia Stem Cell Centre
ATMP	advanced therapy medical product
BBS	Foundation Biobank Suisse (Switzerland)
BCRT	Berlin-Brandenburg Center for Regenerative Therapies
BEEH	Biomedical Engineering and Engineering Healthcare [program in CBET division of NSF]
BM	basal membrane
BMBF	Bundesministerium für Bildung und Forschung [Federal Ministry of Education and Research] (Germany)
BMC	Biomedical Center (Lund University, Sweden)
BME	Biomedical Engineering [program in CBET division of NSF]
BMM	BioMedical Materials
BMP	bone morphogenetic protein
BPT	Bioprocessing Technology Institute (Singapore)
BSCN	Basel Stem Cell Network (Switzerland)
BTI	Bioprocessing Technology Institute
CARE	Center for Advanced Regenerative Engineering (Münster, Germany)
CAS	Chinese Academy of Sciences
CAT	Committee for Advanced Therapies [of European Medicines Agency]
CBET	Chemical, Bioengineering, Environmental and Transport Systems [division of NSF]

R.M. Nerem et al. (eds.), *Stem Cell Engineering: A WTEC Global Assessment,*
Science Policy Reports, DOI 10.1007/978-3-319-05074-4,
© Springer International Publishing Switzerland 2014

CECS	Centre pour l'Etude des Cellules Souches [Center for Stem Cell Studies] (France)
CFC	colony forming cell
CH	associated with Roche pharmaceuticals
Chronic MPD	chronic myeloproliferative disorder
CiRA	Center for iPS Cell Research and Application (Kyoto University, Japan)
CLINTEC	Department of Clinical Science, Intervention and Technology (Karolinska Institute, Sweden)
CMB	Department of Cell and Molecular Biology (Karolinska Institute, Sweden)
CML	chronic myelogenous leukemia
CML	chronic myeloid leukemia
CNS	central nervous system
COE	center of excellence
CRBP-1	cellular retinol-binding protein 1
CRC	Clinical Research Center (Malmo, Sweden)
CREST	Core Research for Evolutional Science and Technology (Japan)
CRTD	Center for Regenerative Therapies Dresden (Germany)
CSIRO	Commonwealth Scientific and Industrial Research Organization
CSTEC	Cell Sheet Tissue Engineering Center (Japan)
CSTOF	Cell Sheet-Based Tissue & Organ Factory (Japan)
CSTR	continuous stirred tank reactor
CTC	Clinical Trials Center (unit of SCRM)
CTC	circulating tumor cell
CTI	Commission for Technology and Innovation (Switzerland)
CUHK	Chinese University of Hong Kong
2D, 3D	two-dimensional, three-dimensional
DEP	dielectrophoresis
DFG	Deutsche Forschungsgemeinschaft [German Research Foundation]
DIGS-BB	Dresden International Graduate School for Biomedicine and Bioengineering (Germany)
Dnmts	DNA methyltransferases
DPTE	Dutch Program for Tissue Engineering (Netherlands)
DS	differentiating state
EB	embryoid body
EC	endothelial cell
ECM	extracellular matrix
ECUST	East China University of Science and Technology
EIBIR	European Institute for Biomedical Imaging Research
EMA	European Medicines Agency
EPFL	École Polytechnique Fédérale de Lausanne (Switzerland)
EPO	erythropoietin
ERATO	Exploratory Research for Advanced Technology [of JST] (Japan)
ESC	embryonic stem cell

ETH	Eidgenössische Technische Hochschule [Zurich] (Switzerland)
EU	European Union
FIRST	Funding Program for World-Leading Innovative R&D on Science and Technology (Japan)
FiT	Facility for iPS Cell Therapy (GMP facility at CiRA, Japan)
FMI	Friedrich Miescher Institute [for Biomedical Research] (Basel, Switzerland)
fMRI	functional magnetic resonance imaging
FP7	[European Union] Framework Programme 7
GARDE	General & Age Related Disabilities Engineering [program in CBET division of NSF]
GBM	glioblastoma multiforme
GCP	good clinical practice
GF	growth factor
GIH	Gymnastik- och Idrottshögskolan [Swedish School of Sport and Health Sciences]
GLP	good laboratory practice
GMP	good manufacturing practice
GPU	graphics processing unit [computing]
GSI	gamma secretase inhibitor
GTEC	Georgia Tech/Emory Center
HCP	Health Cluster Portugal
HCS	High Content Screening
HD	Huntington's disease
HDAC3	histone deacetylase 3
hESC	human embryonic stem cell
HFSP	Human Frontier Science Program (granting agency based in Strasbourg, France)
HHT	hereditary hemorrhagic telangiectasia
hiPSC	human induced pluripotent stem cell
hMSC	human mesenchymal stem cell
hMSC	human mesenchymal stromal cells
HSC	hematopoietic stem cell
HSC	hematopoietic stem cell
HSVtk	herpes simplex virus thymidine kinase
HTS	high-throughput screening
hUCB-MSC	human umbilical cord blood-derived mesenchymal stem cell
I3	inhibitor 3
I3S	Instituto de Investigação e Inovação em Saúde [Institute for Research and Innovation in Health] (Portugal)
IBI	Interfaculty Institute of Bioengineering (EPFL, Switzerland)
IF	impact factor [of scientific journal]
I-IMBN	Asia-Pacific International Molecular Biology Network (Korea)
IMB	Institute for Medical Informatics and Biometry (Dresden University of Technology, Germany)

Imec	Interuniversity microelectronics centre (Belgium)
IMI	Innovative Medicines Initiative [of European Union Framework Programme 7]
IMSUT	Institute of Medical Science of the University of Tokyo
INEB	Instituto de Engenharia Biomédica (Portugal)
IOZ	Institute of Zoology (Chinese Academy of Sciences)
iPSC	induced pluripotent stem cell
ISI	Institute for Scientific Information (now Thomson ISI)
IST	Instituto Superior Técnico (Portugal)
I-STEM	Institute for Stem Cell Therapy and Exploration of Monogenic Diseases (France)
JSPS	Japan Society for the Promotion of Science
JST	Japan Science and Technology Agency
KI	Karolinska Institute (Sweden)
KIST	Korean Institute of Science and Technology
KKS	Coordination Centre for Clinical Trials (Dresden, Germany)
KTH	Kungliga Tekniska högskolan [The Royal Institute of Technology] (Sweden)
LIF	leukemia inhibitory factor
LSC	leukemia stem cell
LSCB	Laboratory of Stem Cell Bioengineering (Ecole Polytechnique Fédérale de Lausanne)
lt-NES	long-term neural epithelial stem cells
LU	Lund University (Sweden)
LUMC	Leids Universitair Medisch Centrum [Leiden University Medical Center] (Netherlands)
mAb	monoclonal antibody
ME	mesoendoderm
MEA	multiple-electrode array
MEMS	microelectromechanical systems
mESC	mouse embryonic stem cell
MEXT	Ministry of Education, Culture, Sports, Science & Technology (Japan)
MI	myocardial infarct
MNC	mononuclear cell
MOST	Ministry of Science and Technology (China)
MPC	mesenchymal precursor cell
MPG	Max-Planck-Gesellschaft [Max Planck Society for the Advancement of Science] (Germany)
MPI	Max Planck Institute (Germany)
MRI	magnetic resonance imaging
MSC	mesenchymal stem cell
MSC	mesenchymal stromal cell
NCE	National Center of Excellence (Canada)
NCI	National Cancer Institute (United States)

NE	neuroectoderm
NEWT	NEWTherapies [Group] (Instituto de Engenharia Biomédica, Portugal)
NGF	neural growth factor
NIH	National Institutes of Health (U.S.A.)
NIRM	Netherlands Institute for Regenerative Medicine
NIST	National Institute of Standards and Technology (United States)
NR	neural retina [cells]
NR	neuroretina
NRBP2	nuclear receptor binding protein 2
NSC	neural stem cell
NSF	National Science Foundation (United States)
NSFC	National Natural Science Foundation of China
NT	nuclear transfer
NTEC	National Tissue Engineering Center (Shanghai Jiao Tong University School of Medicine, China)
NUS	National University of Singapore
OA	osteoarthritis
ODE	ordinary differential equation
PA	polyacrylamide
PCR	polymerase chain reaction
PDF	postdoctoral fellow
PDGF	platelet-derived growth factor
PDGFRβ	platelet derived growth factor receptor beta
PDMS	polydimethylsiloxane
PEDOT	poly(3,4-ethylenedioxythiophene)
PEG	poly(ethylene glycol)
PES	polyethersulfone [film]
PI	principal investigator
PLL	poly-L-lysine
PNIPAAm	poly(N-isopropylacrylamide
PSA	prostate-specific antigen
PSC	pluripotent stem cell
PSE	Science Park (EPFL, Switzerland)
RGD [peptides]	synthetic adhesive ligands containing the arginine-glycine-aspartic acid motif
RIKEN	Rikagaku Kenkyūjo [Institute of Physical and Chemical Research] (Japan)
RI-SCBT	Royan Institute for Stem Cell Biology and Technology (Iran)
RM	regenerative medicine
RMB	reminbi (¥, China)
RPE	retinal pigment epithelial [cells] or retinal pigment epithelium
SAMs	self-assembled monolayers
SBML	Systems Biology Markup Language
SC	stem cell

SCBL	Stem Cell Bioengineering Laboratory [of Instituto Superior Técnico] (Portugal)
SCDD	Stem Cells in Development and Disease (Netherlands)
SCE	stem cell engineering
SCI	Science Citation Index
SCN	Stem Cell Network (Canada)
scNT-ESC	somatic cell nuclear transfer-embryonic stem cell
SCRM	Swiss Center for Regenerative Medicine (Zurich)
SDE	stochastic differential equation
SE	surface ectoderm
Shh	sonic hedgehog
SIBCB	Shanghai Institute of Biochemistry and Cell Biology (China)
SIBS	Shanghai Institutes for Biological Sciences (China)
siRNA	small interfering RNA
SKLBE	State Key Laboratory of Bioreactor Engineering (China)
SMC	smooth muscle cell
SNSF	Swiss National Science Foundation
SPR	surface plasmon resonance
T-ALL	T-cell acute lymphoblastic leukemia
TE	tissue engineering
TECS	tilting embryonic culture system
TERM	Tissue Engineering Regenerative Medicine [European FP7-funded collaboration]
TERMIS	Tissue Engineering and Regenerative Medicine International Society
TF	transcription factor
TGF-β	transforming growth factor beta
TjSCRC	Stem Cell Research Center at Tongji University School of Medicine (Japan)
TKI	tyrosine kinase inhibitor
TRM	Translational Center for Regenerative Medicine (Leipzig, Germany)
TUD	Dresden University of Technology (Germany)
TUSM	Tongji University School of Medicine
TWIns	Tokyo Women's Medical University-Waseda University Joint Institution for Advanced Biomedical Sciences (Japan)
TWMU	Tokyo Women's Medical University
UBC	University of British Columbia (Canada)
UHZ	University Hospital of Zurich (Switzerland)
UNIST	Ulsan National Institute of Science and Technology (South Korea)
Uporto	University of Porto (Portugal)
UU	Uppsala University (Sweden)
UZ	University of Zurich (Switzerland)
VE	visceral endoderm
VEGF	vascular endothelial growth factor
WTEC	World Technology Evaluation Center

WTEC Books:

Nanotechnology Research Directions for Societal Needs in 2020: Retrospective and Outlook. Mihail Roco, Chad Mirkin, and Mark Hersam (Ed.) Springer, 2011.

Brain-computer interfaces: An international assessment of research and development trends. Ted Berger (Ed.) Springer, 2008.

Robotics: State of the art and future challenges. George Bekey (Ed.) Imperial College Press, 2008.

Micromanufacturing: International research and development. Kori Ehmann (Ed.) Springer, 2007.

Systems biology: International research and development. Marvin Cassman (Ed.) Springer, 2007.

Nanotechnology: Societal implications. Mihail Roco and William Bainbridge (Eds.) Springer, 2006. Two volumes.

Biosensing: International research and development. J. Shultz (Ed.) Springer, 2006.

Spin electronics. D.D. Awschalom et al. (Eds.) Kluwer Academic Publishers, 2004.

Converging technologies for improving human performance: Nanotechnology, biotechnology, information technology and cognitive science. Mihail Roco and William Brainbridge (Eds.) Kluwer Academic Publishers, 2004.

Tissue engineering research. Larry McIntire (Ed.) Academic Press, 2003.

Applying molecular and materials modeling. Phillip Westmoreland (Ed.) Kluwer Academic Publishers, 2002

Societal implications of nanoscience and nanotechnology. Mihail Roco and William Brainbridge (Eds.) Kluwer Academic Publishers, 2001.

Nanotechnology research directions. M.C. Roco, R.S. Williams, and P. Alivisatos (Eds.) Kluwer Academic Publishers, 1999. Russian version available.

Nanostructure science and technology: R&D status and trends in nanoparticles, nanostructured materials and nanodevices. R.S. Siegel, E. Hu, and M.C. Roco (Eds.) Kluwer Academic Publishers, 2000.

Advanced software applications in Japan. E. Feigenbaum et al. (Eds.) Noyes Data Corporation, 1995.

Flat-panel display technologies. L.E. Tannas, et al. (Eds.) Noyes Publications, 1995.

Satellite communications systems and technology. B.I. Edelson and J.N. Pelton (Eds.) Noyes Publications, 1995.

Selected WTEC Panel Reports:

European Research and Development in Mobility Technology for People with Disabilities (8/2011)

International Assessment of Research and Development in Rapid Vaccine Manufacturing (Part 2, 7/2011)

International Assessment of Nanotechnology Research Directions for Societal Needs in 2020 Retrospective and Outlook (9/2010)

International Assessment of Research and Development in Flexible Hybrid Electronics (7/2010)

The Race for World Leadership of Science and Technology: Status and Forecasts. 12th International Conference on Scientometrics and Informetrics, Rio de Janeiro (7/2009)

Research and development in simulation-based engineering and science (1/2009)

Research and development in catalysis by nanostructured materials (11/2008)

Research and development in rapid vaccine manufacturing (12/2007)

Research and development in carbon nanotube manufacturing and applications (6/2007)

High-end computing research and development in Japan (12/2004)

Additive/subtractive manufacturing research and development in Europe (11/2004)

Microsystems research in Japan (9/2003)

Environmentally benign manufacturing (4/2001)

Wireless technologies and information networks (7/2000)

Japan's key technology center program (9/1999)

Future of data storage technologies (6/1999)

Digital information organization in Japan (2/1999)

Selected Workshop Reports Published by WTEC:

International assessment of R&D in stem cells for regenerative medicine and tissue engineering (4/2008)

Manufacturing at the nanoscale (2007)

(continued)

Building electronic function into nanoscale molecular architectures (6/2007)	Nanoelectronics, nanophotonics, and nanomagnetics (2/2004)
Infrastructure needs of systems biology (5/2007)	Nanotechnology: Societal implications (12/2003)
X-Rays and neutrons: Essential tools for nanoscience research (6/2005)	Nanobiotechnology (10/2003)
Sensors for environmental observatories (12/2004)	Regional, state, and local initiatives in nanotechnology (9/2003)
	Materials by design (6/2003)
Nanotechnology in space exploration (8/2004)	Nanotechnology and the environment: Applications and implications (5/2003)
Nanoscience research for energy needs (3/2004)	Nanotechnology research directions (1999)

All WTEC reports are available on the Web at http://www.wtec.org.